Heritage Tourism and Sustainable City Dynamics

Heritage Tourism and Sustainable City Dynamics

Guest Editors

Fátima Matos Silva
Isabel Vaz de Freitas

Basel • Beijing • Wuhan • Barcelona • Belgrade • Novi Sad • Cluj • Manchester

Guest Editors

Fátima Matos Silva
Tourism, Heritage and Culture Department
Portucalense University
Porto
Portugal

Isabel Vaz de Freitas
Tourism, Heritage and Culture Department
Portucalense University
Porto
Portugal

Editorial Office
MDPI AG
Grosspeteranlage 5
4052 Basel, Switzerland

This is a reprint of the Special Issue, published open access by the journal *Heritage* (ISSN 2571-9408), freely accessible at: https://www.mdpi.com/journal/heritage/special_issues/HM23IPY96A.

For citation purposes, cite each article independently as indicated on the article page online and as indicated below:

Lastname, A.A.; Lastname, B.B. Article Title. *Journal Name* **Year**, *Volume Number*, Page Range.

ISBN 978-3-7258-3365-8 (Hbk)
ISBN 978-3-7258-3366-5 (PDF)
https://doi.org/10.3390/books978-3-7258-3366-5

Cover image courtesy of Sofia Matos Silva

© 2025 by the authors. Articles in this book are Open Access and distributed under the Creative Commons Attribution (CC BY) license. The book as a whole is distributed by MDPI under the terms and conditions of the Creative Commons Attribution-NonCommercial-NoDerivs (CC BY-NC-ND) license (https://creativecommons.org/licenses/by-nc-nd/4.0/).

Contents

About the Editors vii

Preface ix

Pablo Rosser and Seila Soler
From Oblivion to Life: The Recovery of Intangible Cultural Heritage through the Anti-Aircraft Shelters of the Spanish Civil War
Reprinted from: *Heritage* **2024**, 7, 817–828, https://doi.org/10.3390/heritage7020039 1

Alaa M. S. Azzaz and Ibrahim A. Elshaer
Heritage Tourism Resilience and Sustainable Performance Post COVID-19: Evidence from Hotels Sector
Reprinted from: *Heritage* **2024**, 7, 1162–1173, https://doi.org/10.3390/heritage7030055 13

Ibrahim A. Elshaer, Alaa M. S. Azzazz and Sameh Fayyad
Residents' Environmentally Responsible Behavior and Tourists' Sustainable Use of Cultural Heritage: Mediation of Destination Identification and Self-Congruity as a Moderator
Reprinted from: *Heritage* **2024**, 7, 1174–1187, https://doi.org/10.3390/heritage7030056 25

Bintang Noor Prabowo, Alenka Temeljotov Salaj and Jardar Lohne
Urban Heritage Facility Management: A Conceptual Framework for the Provision of Urban-Scale Support Services in Norwegian World Heritage Sites
Reprinted from: *Heritage* **2024**, 7, 1372–1399, https://doi.org/10.3390/heritage7030066 39

Sergio Manzano-Fernández, Camilla Mileto, Fernando Vegas López-Manzanares and Valentina Cristini
Conservation and In Situ Enhancement of Earthen Architecture in Archaeological Sites: Social and Anthropic Risks in the Case Studies of the Iberian Peninsula
Reprinted from: *Heritage* **2024**, 7, 2239–2264, https://doi.org/10.3390/heritage7050106 67

Ibrahim A. Elshaer, Mansour Alyahya, Alaa M. S. Azzazz and Sameh Fayyad
Community Attachment to AlUla Heritage Site and Tourists' Green Consumption: The Role of Support for Green Tourism Development
Reprinted from: *Heritage* **2024**, 7, 2651–2667, https://doi.org/10.3390/heritage7060126 93

Zsuzsanna Bacsi
The Influence of Heritage on the Revealed Comparative Advantage of Tourism—A Worldwide Analysis from 2011 to 2022
Reprinted from: *Heritage* **2024**, 7, 5232–5250, https://doi.org/10.3390/heritage7090246 110

Sonia Mileva and Milena Krachanova
In Search of New Dimensions for Religious Tourism: The Case of the Ancient City of Nessebar
Reprinted from: *Heritage* **2024**, 7, 5373–5389, https://doi.org/10.3390/heritage7100253 129

Lucía Castaño-Prieto, Lucía García-García, Minerva Aguilar-Rivero and José E. Ramos-Ruiz
The Impact of the Sociodemographic Profile on the Tourist Experience of the Fiesta de los Patios of Córdoba: An Analysis of Visitor Satisfaction
Reprinted from: *Heritage* **2024**, 7, 5593–5610, https://doi.org/10.3390/heritage7100264 146

Franklin Omar Zavaleta Chavez Arroyo, Alex Javier Sánchez Pantaleón, Milena Leticia Weepiu Samekash, Jhunniors Puscan Visalot and Rosse Marie Esparza-Huamanchumo
Economic Contribution, Characterization, and Motivations of Tourists: The Raymi Llaqta in Peru
Reprinted from: *Heritage* **2024**, 7, 6243–6256, https://doi.org/10.3390/heritage7110293 164

Makhabbat Ramazanova, Fátima Matos Silva and Isabel Vaz de Freitas
Tourists' Views on Sustainable Heritage Management in Porto, Portugal: Balancing Heritage Preservation and Tourism
Reprinted from: *Heritage* **2025**, *8*, 10, https://doi.org/10.3390/heritage8010010 . . . . **178**

Claudia Patricia Maldonado-Erazo, Susana Monserrat Zurita-Polo, María de la Cruz del Río-Rama and JoséÁlvarez-García
Cultural Dimensions of Territorial Development: A Plan to Safeguard the Intangible Cultural Heritage of Guano's Knotted Carpet Weaving Tradition, Chimborazo, Ecuador
Reprinted from: *Heritage* **2025**, *8*, 60, https://doi.org/10.3390/heritage8020060 . . . . **192**

About the Editors

Fátima Matos Silva

Fátima Matos Silva earned her PhD from the University of Granada in 2008. Since 1988, she has been developing her professional activity at the Portucalense University (Porto, Portugal) as a researcher, professor, and coordinator of cycles of studies na área do Património Cultural, integrating several bodies of the institution. She has directed and collaborated on diversified research projects, both nationally and internationally, and is currently an integrated member of REMIT and CITCEM—FCT R&D units. She has directed multiple archaeological excavations, conducted environmental and heritage impact studies, researched spatial planning and sustainable development, and ensured universal access to cultural heritage, particularly the Santiago Ways. In addition, she has organised bibliographic editions, scientific and cultural events, exhibitions, museological programmes, and other cultural actions. She has collaborated on projects to enhance heritage within a territory, collaborating with several City Councils and local bodies. She is the coordinator of national and international conferences on heritage, culture, education, and tourism. As a member of the Scientific Committee and as chair, she has been part of several International Conferences on various themes. She is a member of the Editorial Board, the Scientific Council, and the Advisory Board of several international journals available on SCOPUS and WOS. The author of several books, she also has published (individually and in partnership) more than ninety articles and book chapters on diverse heritage themes. Her expertise lies in the fields of the humanities and social sciences, with a focus on archaeology, cultural heritage, preventive conservation, heritage management, spatial planning and sustainable development, impacts and threats to cultural heritage, heritage interpretation and enhancement, museology, universal accessibility, the accessibility of heritage, cultural tourism, and religious and accessible tourism.

Isabel Vaz de Freitas

Isabel Vaz de Freitas is Full Professor at Portucalense University. She has a PhD in History, completing research at CIAUD (polo UPT). She was also appointed as Correspondent of the Portuguese Academy of History, Head of the Department of Tourism, Heritage and Culture, and Director of the Conservation and Restoration Center at Portucalense University. She is coordinator of several study cycles in the area of Cultural Heritage and serves as a professor on several undergraduate thesis, internship, and master's degree reports, and master's and doctoral dissertation juries. Editorial Board member, Member of Scientific Council and Advisor Board Member in the international journals available via SCOPUS and the WOS. The coordinator of national and international conferences on heritage, culture, and tourism, she collaborates on other heritage and territory valorisation projects with city councils and local organisms, acting as a researcher in projects financed in Portugal and Spain. Working in the scope of these projects, her published articles and books are centred on themes related to the cultural landscape, territorial development, heritage management, and conservation and heritage risks and values. One of the main areas of research is *water heritage and sustainability*. In this area, she integrated financed projects as a coordinator and researcher, publishing articles and book chapters on this topic. As a researcher and chair, she integrated several International Conferences on thematic *water heritage and sustainability*. She specialises in the humanities and social sciences, focusing on history, cultural heritage, and cultural tourism. Her areas of interest are landscape, heritage safeguard, water heritage, material and immaterial heritage conservation, and heritage and regional development.

Preface

Heritage visits play a pivotal role in shaping the dynamics of historical centres and driving the development of conservation, protection, and sustainability efforts for heritage in these areas. We have witnessed a significant transformation in the landscape of historic cities, accompanied by a growing awareness of the need to preserve both tangible and intangible heritage. This shift is driven by successive trends that promote heritage visits and tourism growth. These trends emphasise heritage and benefit local communities economically, creating new income opportunities and jobs. This, in turn, supports social equity, reduces inequalities, and fosters greater sustainability and responsible behaviour.

However, these dynamics can also pose risks if the balance between tourism and heritage preservation is not adequately managed. It is essential for the various stakeholders involved to engage in open dialogue and cooperation to ensure the sustainable management of historical centres, safeguarding the intrinsic value of cultural heritage. To tackle these challenges, it is essential to identify the primary risks associated with heritage tourism and examine case studies, methodologies, and strategies that encourage the responsible use of cultural heritage along with ongoing monitoring. The risks related to heritage tourism are diverse and can have significant repercussions. The physical deterioration of historic sites due to high visitor traffic is an increasing concern, as is the environmental impact caused by pollution and the destruction of natural habitats. Furthermore, cultural attrition can occur when cultural traditions are commercialised, leading to a loss of authenticity. Overcrowding and excessive economic reliance on tourism are challenges that must be managed prudently to ensure the sustainability of heritage sites.

The impacts of heritage tourism, particularly those of an economic nature, significantly affect local communities' social and cultural dynamics. An increase in tourism may result in gentrification, a phenomenon wherein residents are displaced due to rising real estate prices. Additionally, the pressure on local resources may intensify, thereby affecting the quality of life of residents. Furthermore, the conversion of heritage sites into tourist attractions can alter the manner in which these spaces are experienced and understood by both locals and visitors, thereby compromising the cultural integrity of heritage. A comprehensive understanding of the impacts associated with tourism activities is essential for preserving and protecting heritage sites. This knowledge facilitates the development of practices that enhance economic benefits and ensure the safeguarding of cultural heritage for future generations.

Despite the importance of these issues, there is a lack of scientific literature on these topics. We hope that the various research papers published about diverse types of heritage and diverse countries in a multidisciplinary, interdisciplinary, diachronic, synchronised, and transversal manner in this Special Issue contribute to filling some of these gaps.

The guest editors would like to express their gratitude to the authors for their contributions and their dedication to enhancing their work. We also extend our thanks to the reviewers for their time and effort in analysing the submissions and providing valuable comments and corrections. Finally, we are grateful to the editorial staff for their efficient review and publication process management. We hope the selected publications will significantly impact the scientific community and inspire other researchers to explore these themes and pursue their scientific objectives.

Fátima Matos Silva and Isabel Vaz de Freitas
Guest Editors

Article

From Oblivion to Life: The Recovery of Intangible Cultural Heritage through the Anti-Aircraft Shelters of the Spanish Civil War

Pablo Rosser [1,*] and Seila Soler [2]

1 Faculty of Education, Universidad Internacional de la Rioja, 26006 Logroño, Spain
2 Faculty of Humanities and Social Sciences, Universidad Isabel I, 09003 Burgos, Spain; seilaaixa.soler@ui1.es
* Correspondence: pablo.rosser@unir.net

Abstract: This article examines the rehabilitation of anti-aircraft shelters from the Spanish Civil War in Alicante, Spain. Funded by European resources and managed by local public administration, these shelters have been restored as cultural and tourist attractions. This study aims to analyze their role in preserving and promoting intangible cultural heritage, with a focus on their significance as tangible remnants of a historical period and their reflection on survival practices during the war. This research investigates the impact of public management in rehabilitating these shelters and in disseminating their history and culture. It explores decision-making processes, community engagement, and strategies for promoting cultural tourism. Employing a mixed methodology, this study gathers primary data through interviews with individuals who witnessed the war and secondary data from documentary and bibliographic sources. Findings suggest that the rehabilitation of these shelters has been pivotal in preserving Alicante's historical memory and cultural heritage. Making the shelters accessible to the public facilitates knowledge transmission about the Spanish Civil War, promotes cultural tourism, and engages the local community in the dissemination of history and culture. This study's results and conclusions are relevant for academics, professionals, and cultural heritage managers at both national and international levels, offering insights into the effective preservation and promotion of historical sites.

Keywords: cultural heritage preservation; Spanish civil war education; sustainable cultural tourism

Citation: Rosser, P.; Soler, S. From Oblivion to Life: The Recovery of Intangible Cultural Heritage through the Anti-Aircraft Shelters of the Spanish Civil War. *Heritage* **2024**, *7*, 817–828. https://doi.org/10.3390/heritage7020039

Academic Editors: Fátima Matos Silva and Isabel Vaz de Freitas

Received: 6 January 2024
Revised: 1 February 2024
Accepted: 6 February 2024
Published: 9 February 2024

Copyright: © 2024 by the authors. Licensee MDPI, Basel, Switzerland. This article is an open access article distributed under the terms and conditions of the Creative Commons Attribution (CC BY) license (https:// creativecommons.org/licenses/by/ 4.0/).

1. Introduction

1.1. Preservation of Historical Memory and Cultural Identity

The preservation of historical memory and cultural identity is crucial for community development, cultural tourism, and sustainability. Cities with historical and cultural legacies attract tourism, driving economic and social development while preserving their heritage and cultural identity. The conservation of sites and historical monuments is fundamental for understanding the evolution of the city [1].

The preservation of historical memory includes tangible elements and the collection of personal testimonies through interviews, documents, and photographs, contributing to archives and museums. These narratives enrich historical understanding and strengthen cultural identity, while cultural preservation promotes traditions and education to maintain identity among future generations [1].

1.2. Historical Context: The Air-Raid Shelters of the Spanish Civil War in Alicante and the Second World War

The failed coup d'état of 1936 in Spain against the Government of the Second Spanish Republic initiated a civil war and a dictatorship until 1975 under the orders of Francisco Franco. During the conflict (1936–1939), Spain experienced bombings from Nazi Germany and Fascist Italy, resulting in over 500,000 deaths and devastation, especially in coastal cities like Alicante, a strategic enclave [2,3], being the last province to be conquered.

Starting in November 1936, Alicante faced intense bombings, including the significant '8-h bombing' on November 28. Faced with a shortage of air-raid shelters, they were built in densely populated and strategic areas, equipped with surveillance systems, alarms, and telephone communications. The distribution of these shelters was based on population density and proximity to essential services [2]. By August 1937, Alicante had 41 shelters, with a capacity for 24,020 people [3]. The increase in number, quality, and size of the shelters during the war reflects the escalation of bombings, especially in 1938, marked by the attack on the Central Market on May 25. This event drove an improvement in the city's organizational capacity for defense, resulting in the construction of nearly a hundred shelters [3].

The air-raid shelters were specifically designed to protect the civilian population from bombings. For instance, a study documents the design of private air-raid shelters aimed at safeguarding civilians, analyzing the materials, construction methods, and structural systems of these shelters during the Spanish Civil War [4].

During the Second World War, a study examined the effectiveness of various air-raid shelters, including those used during the Spanish Civil War, with a focus on experiences from Great Britain and Spain. This study highlighted their significance in civil protection, symbolizing resistance and the survival of the population [5,6].

1.3. The Upcycling Process: Impact of Public Management in Shelter Rehabilitation

In our project, we employ the concept of upcycling, which involves the creative and sustainable transformation of existing spaces to serve new purposes. Rather than discarding or demolishing these spaces, upcycling aims to maximize available resources while minimizing environmental impact [7–17].

The air-raid shelters from the Nazi era in Szczecin, Poland, pose a significant planning challenge for the city due to their historical association with World War II [7]. The experience of rehabilitating the air-raid shelters in Szczecin, Poland, provides a valuable parallel for the study of the shelters in Alicante, as it demonstrates how the subsequent history of their existence can extend beyond mere heritage restoration. In the case of Szczecin, the revitalization of the shelters was not only focused on preserving their historical structure but also actively sought to reanimate the neighborhoods and surrounding areas through various activities implemented within them. In this regard, the current rehabilitation of the shelters in Alicante has adopted a similar strategy that not only preserves the heritage but also promotes the reactivation of the surrounding areas through community-involving activities that generate a long-term positive impact on the city. This strategy aligns with one of the objectives of the European funding period from which the project has been financed: the economic, social, and cultural revitalization of its influence zone.

This process of space upcycling involves innovative thinking and the reuse of materials, structures, and existing features within these shelters. Examples include converting old buildings into residences or workspaces, repurposing warehouses as event venues, or rejuvenating abandoned gardens into community parks [12,13].

Furthermore, space upcycling can encompass the integration of sustainable technologies such as solar panels, rainwater harvesting systems, or natural ventilation systems. These additions enhance energy efficiency and environmental friendliness within the space [7,11,15].

Ultimately, space upcycling represents a creative and conscious approach to rejuvenating existing spaces, making the most of available resources while reducing environmental impact [17].

1.4. Study Objectives

This study focuses on analyzing the contribution of the rehabilitation of air-raid refuge to the preservation of intangible cultural heritage in Alicante, Spain. It involves exploring their role as witnesses of the Spanish Civil War, examining interviews with individuals who experienced the war as children to understand how their restoration and opening to the public help maintain historical memory and local cultural identity. Furthermore, it will evaluate the impact of public management on the rehabilitation and promotion of their history and culture, including the analysis of financial decisions, community involvement, and strategies to promote cultural tourism.

2. Materials and Methods

In the design of this research on the rehabilitation of air-raid shelters from the Spanish Civil War in Alicante, a mixed methodological approach was adopted to encompass both the quantitative and qualitative aspects of the study [18]. This dual-front methodology was chosen to capture the complexity and depth of the impact of rehabilitation from a holistic perspective that integrates architectural, historical, and sociocultural data.

2.1. Research Design

The research was structured in two complementary phases: the collection of primary data through interviews with direct witnesses of the Civil War and the compilation of secondary data through the review of historical documents, restoration reports, and the analysis of cultural tourism indicators. The primary data came from structured interviews conducted following a predetermined set of questions to ensure coherence and comparability of the collected information. These interviews focused on gathering personal narratives and perceptions about life experiences during the war and the use of the shelters [19,20].

On the other hand, secondary data were extracted from reliable and reviewed documentary sources, which provided a broad context on the architecture of the shelters, their historical significance, and their role in the collective memory of Alicante [3].

Quantitative information about the extent of the rehabilitation and the increase in cultural tourism, such as figures on rehabilitated surface area and visitor statistics, was included to demonstrate the direct impact of the intervention on the shelters [21].

2.2. Justification of the Methodology

The choice of interviews and hermeneutic analysis as the main methods were guided by the nature of the study's subject. The interviews provided a direct means to access lived experiences and individual perceptions, vital for understanding the intangible significance that the shelters hold for those who lived through the Civil War era [3]. Hermeneutic analysis was chosen for its ability to deeply interpret witness narratives, offering a richer and more detailed understanding of the underlying meanings. This approach allowed for the unraveling of layers of meaning in personal narratives, relating them to a broader cultural context and connecting the past with present interpretations. The combination of these methods provided a solid foundation for a holistic and dynamic interpretation of the shelters as spaces of cultural memory. It allowed for an assessment of how rehabilitation has transformed these spaces into instruments of education and tourism, and how they contribute to the construction of a renewed cultural identity and collective memory [22]. Therefore, the mixed methodological approach and the use of interviews and hermeneutic analysis have been essential for achieving a comprehensive understanding of the role of the air-raid shelters in the cultural fabric of Alicante, ensuring that the study's conclusions are supported by a rigorous and appropriate methodological framework.

2.3. Research Approach

The analysis of the collected data was carried out systematically and rigorously, employing qualitative analysis techniques [23,24]. Patterns, trends, and relationships within the data were examined to address the research objectives. The results obtained are presented clearly and concisely, supported by empirical evidence, and contextualized within the existing theoretical framework with hermeneutic analysis [25–27].

In the realm of interpretative qualitative research, hermeneutics, which focuses on analyzing the cultural context of texts and cultural expressions, plays a crucial role [28]. Within this context, Interpretative Phenomenological Analysis (IPA) stands out as a qualitatively significant method [29–31]. Within the framework of IPA, the selection of interviews has been conducted for meticulous analysis using Atlas.ti software v. 24. Combining textual, contextual, and cultural analysis, it concludes with a comprehensive hermeneutic interpretation.

2.4. Primary Data Collection: Interviews with War Witnesses and Hermeneutic Analysis

This research, therefore, will use a hermeneutic approach to analyze interviews related to the air-raid shelters of the Spanish Civil War in Alicante. It will explore personal perceptions and emotions, such as fear and uncertainty, and daily life during wartime conditions to understand the human impact of the shelters. Changes in routines and urban structures will be examined, along with the interaction with the physical space of the shelters and the long-term emotional and psychological impact. Furthermore, it will study how these narratives contribute to historical memory and the perception of the shelters, seeking interpretations beyond the explicit.

Therefore, to significantly enrich the hermeneutic analysis in the context of our study on the air-raid shelters of the Spanish Civil War, it is necessary to address the following issues:

- Cultural Semiotics: this approach is useful for unraveling the meanings of symbols and metaphors in the interviewee's narratives [32].
- Cultural Comparison: comparing the interview narratives with other perceptions and expectations of that era [33] can provide a more comprehensive insight into how the shelters were perceived and experienced in different cultural and social contexts.
- Hermeneutics Proper: On one hand, Contextual Interpretation is an essential approach to understanding testimonies in their historical and cultural context [34].
- Critical Reflection: on the other hand, the analysis will include questioning biases, as it is crucial to be aware of our prejudices and perspectives when analyzing these interviews [35].

The IPA focuses on exploring how individuals interpret and understand their significant experiences [29,30,36], making it particularly useful for examining how individuals perceive and comprehend events and phenomena, thus allowing for a deep understanding of their subjective experiences [37–39].

2.5. Hermeneutic Analysis of Interviews with Individuals Who Experienced the War

In our research, we have applied hermeneutic analysis to interviews conducted with individuals who spent their childhood in Alicante during the Spanish Civil War. *Due to space limitations, we will briefly introduce the harrowing narrative of a woman over eighty years old, from the San Antón neighborhood, whose experience encapsulates the high scientific potential of these interviews.*

Our findings address various aspects, which we have synthesized into four key categories: Cultural Semiotics, Cultural Comparison, Contextual Interpretation, and Critical Reflection.

Cultural Semiotics reveals that shelters, perceived as sanctuaries of safety, become symbols of tragedy and loss, reflecting the duality and chaos of war. In the community context, shelters stand as monuments to solidarity and collective effort, highlighting the importance of mutual support and resilience. Through Cultural Comparison, we observe that daily routines and family and social dynamics were drastically altered, with shelters playing a central role in everyday life and the perception of safety. Contextual Interpretation shows how the constant threat of war and the need for adaptation transformed the urban infrastructure and social interactions.

In the realm of Critical Reflection, we acknowledge the need to approach these narratives with balance, avoiding projecting current preconceptions onto past experiences, and instead, striving to understand them within their own historical and cultural context.

To facilitate understanding and avoid repetition, we have condensed the findings into Table 1, which summarizes the key values and results identified in our analysis.

Table 1. Table of key values and results identified.

Category	Main Findings
Cultural Semiotics	Shelters as symbols of tragedy and survival.
Cultural Comparison	Alteration of everyday life and perceptions of safety.
Contextual Interpretation	Reconfiguration of urban space and infrastructure.
Critical Reflection	Importance of a balanced and contextual perspective.

We conclude that historical memory and personal narrative are essential for a comprehensive understanding of the human impact of war and that preserving these testimonies enriches both our culture and our understanding of history.

3. Results

3.1. Rehabilitation of Air-Raid Shelters

Under the ERDF (European Regional Development Fund) Operational Program "Sustainable Growth 2014–2020", the Alicante City Council implemented the Integrated Sustainable Urban Development Strategy (EDUSI) "Las Cigarreras Area" for sustainable urban development, co-financed by the European Commission. This project focused on the rehabilitation of air-raid shelters from the Civil War in various neighborhoods, improving their accessibility and contributing to cultural and touristic revitalization. The EDUSI, aligned with the ERDF Thematic Objective 6 (recovery of former industrial, religious, and military spaces), seeks to protect and develop cultural and natural heritage. With a budget of EUR 600,000 and a five-year timeline, the project included the rehabilitation of shelters and activities to reconstruct Historical Memory, aiming to increase sustainable tourism and improve cultural heritage. The first phase involved the planning of the opening and accessibility to six shelters in the Alicante DUSI area. Figure 1.

(a)   (b)

Figure 1. Air-raid shelters in the EDUSI area: (a) photographs of the interior of the R35 Marvá Promenade shelter; (b) photographs of the interior of the R69 Tobacco Factory shelter. Own work.

The Project was to be carried out on the following six shelters:

- ○ R35: Paseo Marvá, next to the steps of IES Jorge Juan. Figure 1a.
- ○ R5: Marqués de Molins (formerly Maestro Bretón Street).
- ○ R4: Padre Mariana Street.
- ○ R38: Músico Tordera Square. Figure 2.
- ○ R3: Central Market.
- ○ R69: Tobacco Factory. Figure 1b.

Figure 2. Air-raid shelters in the EDUSI area: photographs of the interior of the R38 Músico Tordera Square shelter and its layout. Own work.

The justification for calculating the area eligible for rehabilitation was developed in several stages. Initially, the first project estimated an intervention in a total area of 544 m^2. Subsequently, this calculation was adjusted in the modified project, which considered an area of 432.36 m^2, to which an extension of 35 square meters was added due to an opening in Marqués de Molins, resulting in a total of 467.36 m^2. In addition, an extra surface not initially documented in Padre Mariana, of 378.64 m^2, was identified, raising the total to 846 square meters.

Regarding the justification for the calculation of the area to be rehabilitated for the year 2023, an intervention in an interior surface of 531.07 m^2 was planned, while the exterior surface, including ramps for universal accessibility and entrance prisms, covered 416.01 m^2.

Finally, the justification for calculating the rehabilitated surface at the end of the project indicates that a total of 963.43 m^2 was rehabilitated. This total comprises 531.07 m^2 of intervention on the interior surface and 432.36 m^2 on the exterior surface.

The specific details of the surfaces and interventions to be carried out offer a clear perspective of the project's scope.

3.2. Strategies for Promoting Cultural Tourism

Once the mnemonic resources of the project (shelters) were rehabilitated, and following the planned programming, we proceeded to the activation of cultural dissemination and dynamization.

Between 2021 and 2022, several activities took place in refurbished anti-aircraft shelters, including installations such as video mapping in Tordera, exhibitions about the history of water in Palmeretes, and historical photographs in Tabacalera, accompanied by a detailed model. Theatrical visits were also conducted, as shown in Figure 3. These initiatives contribute to the revitalization and promotion of historical heritage through interactive and educational tools.

Figure 3. Theatrical visit in the Port of Alicante area. Self-made.

Specific events have taken place within the rehabilitated shelters, including theatre, music, poetry, dance at Tabacalera, and architectural performances at Marvá. These activities are part of a strategy to enrich the cultural and educational experience, diversifying artistic expressions and enhancing historical heritage (Figure 4).

Figure 4. Micro-theatre in the anteroom of the tobacco factory air-raid shelter, performed by Vicente de Ramón Producciones. Self-made.

Within the EDUSI project, guided tours to air-raid shelters and historical sites were organized, some of which were theatrical, by the Professional Association of Official Guides of the Valencian Community. Short films about significant local events were produced, and educational materials were published, including a comic book for young people. Additionally, a website (https://refugiosalicante360.com, accessed on 6 February 2024) was launched to provide a virtual and immersive experience of the shelters, featuring photographs, floor plans, and 360-degree views, expanding access to a global audience. Figure 5.

At the Marvá shelter, students, particularly from the second GIAT of the IES Miguel Hernández, have developed guided tours and educational exhibitions, including one about the Holocaust.

The assessment of indicators for scheduled activities and their evolution over time is crucial for the management and planning of cultural resources. Initially, we projected 2000 visitors for the years 2022 and 2023, based on an analysis of historical attendance data at various cultural facilities in the city. This reference figure served as a starting point for resource and activity planning.

Figure 5. 360-degree virtual tour of the air-raid shelters in the EDUSI area: (a) poster and 360-degree photographs of some of the air-raid shelters in Alicante that can be visited on the website https://refugiosalicante360.com (accessed on 6 February 2024); (b) 360-degree photograph of the interior of the shelter at Plaza Músico Tordera. Self-made.

In terms of total visits, a total of 5229 people participated in activities during 2022 and the beginning of 2023. This count includes various types of events. Guided tours, organized by the Association of Official Guides of the Valencian Community under a minor contract, constituted a significant portion of the attendance. A breakdown of these guided tours reveals a temporal distribution with 760 visitors between March and August 2022, 644 in November, 336 in December, and an increase to 1732 in the period from April to August 2023, totaling 3472 visitors.

In addition to the guided tours, other cultural dissemination and dynamization activities, as previously mentioned, attracted 999 participants. Similarly, music, theater, and dance events held at the Tabacalera shelter contributed another 758 attendees. The sum of these complementary activities, along with the guided tours, resulted in a total of 5229 attendees.

This substantial increase in attendance, compared to the initial projections, reflects the effectiveness of the programming and promotion strategies implemented by the Alicante City Council in terms of these cultural management initiatives.

3.3. The Cultural Heritage of Memory (CHM) as a Driving Element, Also for the Awareness and Citizen Participation of Adults and Children

Under the DUSI strategy, two objectives were established: to protect and develop cultural and natural heritage in tourist urban areas and to revitalize the area socially, economically, and physically. Faced with the disconnection between residents and their heritage, and the lack of collective identity, a plan for socialization and community involvement was implemented. This plan included environmental, historical, and sustainable workshops, as well as creative activities for children, with a focus on connecting them with Alicante's heritage through imaginary characters based on heritage elements. Figure 6.

Figure 6. Different drawings of the underground air-raid shelters in the city of Alicante, made by students from the public school La Aneja de Alicante: (**a**) drawing of the interior of an air-raid shelter; (**b**) drawing of an air-raid shelter in a public square. Self-made.

4. Conclusions

4.1. Importance of Shelters as Promoters of the Recovery of the Intangible Heritage of the Spanish Civil War

Heritage encompasses current interpretations and symbolism about objects, environments, myths, memories, and customs inherited from the past. It constitutes an essential factor in identity formation, especially in societies characterized by increasing cultural diversity [40–42]. The past is crucial in the creation of identity narratives in the present. The significant influence of monuments and cultural landscapes as spaces of cultural heritage is very important. The view of certain places and constructions with historical value has evolved, as well as their function in the formation and reformulation of specific identities [43]. The air-raid shelters of the Spanish Civil War are an example of this and play a crucial role in the recovery and preservation of the intangible heritage associated with this historical conflict [2,3,44,45]. These underground spaces, used as a refuge by the civilian population during bombings, are tangible witnesses to the experiences of that period, and their rehabilitation and opening to the public allow for reliving and understanding the history and culture of that time [45–47].

Firstly, the air-raid shelters are places that evoke the protective practices and survival strategies that took place during the Spanish Civil War. These underground spaces symbolize the resistance and struggle of the civilian population against bombings and the violence of the war [3]. By rehabilitating and opening these shelters to the public, current and future generations are allowed to learn and understand the strategies used by those who lived during that time, as well as the emotional and psychological impact they had on society [47,48].

The air-raid shelters, as sites of memory from the historical period of the Spanish Civil War, their restoration, and opening promote the preservation of memory and cultural identity in Alicante. They enable a connection with the past and an understanding of its influence on current society and culture [48]. Furthermore, visiting the air-raid shelters can generate a sense of empathy and solidarity with the people who lived through those difficult times, thus fostering appreciation and respect for the intangible heritage of the civil war [49].

Finally, the rehabilitation of the air-raid shelters and their opening to the public contribute to promoting cultural tourism and disseminating the history and culture associated with the Spanish Civil War. These spaces become tourist attractions that draw visitors interested in learning and understanding the history and culture of Alicante during the

Civil War. This, in turn, generates economic and social benefits for the local community, promoting sustainable development and the valorization of its cultural heritage [44,48].

4.2. Depth and Scope of the Project

The rehabilitation of the Spanish Civil War air-raid shelters in Alicante has proven to be a significant cultural and architectural intervention, reflected not only in the physical preservation of historical spaces but also in the revitalization of sustainable tourism and heritage education. The justification for calculating the rehabilitated surface area, which reached 963.43 m^2 in 2023, and the increase in visitor attendance, totaling 5229 people during 2022 and early 2023, is indicative of the project's success.

The expansion of the rehabilitated areas, from an initial estimate of 544 m^2 to 963.43 m^2, not only maximized the educational potential of the shelters but also underscored effective and adaptive management of urban heritage in response to emerging needs for historical memory and cultural education. The integration of diverse activities, such as guided tours, theatrical performances, and cultural events, has broadened the reach and impact of these spaces, attracted a larger number of visitors, and enriched the educational experience.

These shelters, once symbols of conflict and survival, have been transformed into dynamic platforms for learning and reflection. The cultural and educational programming carried out at these sites has significantly contributed to the increase in the number of visits, exceeding initial expectations with a total of 5229 visits, compared to the anticipated 2000. This not only validates the initially set objectives but also signals the effectiveness of the actions undertaken to revalue Alicante's intangible cultural heritage.

Therefore, the conclusions of our study are underpinned by a robust methodology, with a diverse and rich database that includes architectural analysis, historical testimonies, and public participation metrics. The synergy between the physical conservation and cultural revitalization of the air-raid shelters has proven to be a successful strategy, evidenced by productivity indicators and positive community reception. This integrative approach can serve as a model for similar projects that seek to combine heritage preservation with education and cultural tourism, ensuring that living history continues to inform and enrich our contemporary societies.

Author Contributions: Conceptualization, P.R. and S.S.; methodology, P.R. and S.S.; software, S.S.; validation, P.R.; formal analysis, P.R. and S.S.; investigation, P.R. and S.S.; resources, P.R. and S.S.; data curation P.R. and S.S.; writing—original draft preparation, P.R.; writing—review and editing, S.S.; visualization, P.R. and S.S.; supervision, S.S.; project administration, P.R. All authors have read and agreed to the published version of the manuscript.

Funding: This research received no external funding.

Data Availability Statement: Dataset available on request from the authors.

Conflicts of Interest: The authors declare no conflict of interest.

References

1. Smith, L. *Uses of Heritage*; Routledge: Milton Park, UK, 2006; ISBN 9781134368037.
2. Rosser, P.; Soriano, R. *Alicante En Guerra*; Ayuntamiento de Alicante: Alicante, Spain, 2018; Volumes 1 and 2.
3. Rosser, P. *Bombas Sobre Alicante: Los Diarios de la Guerra y las Comisiones Internacionales de No Intervención e Inspección de Bombardeos*; Universidad de Alicante: Alicante, Spain, 2023; ISBN 9788413022215.
4. Guardiola-Villora, A.; Basset-Salom, L.; Pérez-García, A. Private Air-Raid Shelters Designed by the Valencian Architect Joaquín Rieta during the Spanish Civil War. *J. Archit.* **2021**, *26*, 286–315. [CrossRef]
5. Leaflet—"Air Raid Precautions: Shelters", World War II, Jan 1942. Available online: https://collections.museumsvictoria.com.au/items/1987871 (accessed on 4 November 2023).
6. Envelope of Documents—Civilian Air Raid Defence Association, Feb 1942. Available online: https://collections.museumsvictoria.com.au/items/1991076 (accessed on 4 November 2023).
7. Matacz, P.; Świątek, L. The Unwanted Heritage of Prefabricated Wartime Air Raid Shelters—Underground Space Regeneration Feasibility for Urban Agriculture to Enhance Neighbourhood Community Engagement. *Sustain. Sci. Pract. Policy* **2021**, *13*, 12238. [CrossRef]

8. Le Pavec, A.; Zerhouni, S.; Leduc, N.; Kuzmenko, K.; Brocato, M. Friction Magazine: The Upcycling of Manufacture for Structural Design. *Int. J. Space Struct.* **2021**, *36*, 281–293. [CrossRef]
9. Yoo, H.; Lee, J. A Study on the Upcycling Marketing Strategy of the Brand Space Design. *J. Korea Inst. Spat. Des.* **2015**, *10*, 177–188. [CrossRef]
10. Yoo, K.; Kim, J.Y. A Study on the Characteristics of Upcycling Space Design from the Perspective of the Spatial Marketing. *J. Korea Inst. Spat. Des.* **2017**, *12*, 349–362. [CrossRef]
11. Herman, K.; Sbarcea, M.; Panagopoulos, T. Creating Green Space Sustainability through Low-Budget and Upcycling Strategies. *Sustainability* **2018**, *10*, 1857. [CrossRef]
12. Leem, J.Y.; Kim, K.C. A Study on the Characteristics of Upcycling Space Reflecting Newtro Trend Characteristic Focused on Complex Cultural Space. *J. Basic Des. Art* **2020**, *21*, 347–358. [CrossRef]
13. Lee, Y.-R. A Study on the Space Planning of Creative Studio for Upcycling Creators—Focusing on the Case Analysis of the Upcycling Creative Studio. *Korean Inst. Inter. Des. J.* **2022**, *31*, 134–146. [CrossRef]
14. Park, J. A Study on Space Upcycling Design Characteristics in Retrotopia Concept. *Korean Inst. Inter. Des. J.* **2022**, *31*, 10–18. [CrossRef]
15. Lucanto, D.; Nava, C. The Contribution of the Green Responsive Model to the Ecological and Digital Transition in the Built Environment. In *Proceedings of the 2022 International Symposium: New Metropolitan Perspectives (NMP 2022), Reggio Calabria, Italy, 24–26 May 2022*; Lecture Notes in Networks and Systems; Springer International Publishing: Cham, Switzerland, 2023; pp. 357–377. ISBN 9783031342103.
16. Ham, J.; Sunuwar, M. Experiments in Enchantment: Domestic Workers, Upcycling and Social Change. *Emot. Space Soc.* **2020**, *37*, 100715. [CrossRef]
17. Fivet, C.; Baverel, O. Upcycling Space Structures. *Int. J. Space Struct.* **2021**, *36*, 251–252. [CrossRef]
18. Creswell, J.W.; Creswell, J.D. *Research Design: Qualitative, Quantitative, and Mixed Methods Approaches*; Sage Publications: Thousand Oaks, CA, USA, 2017.
19. Crespi, D.R. Fuentes de Contraste y Juego de Espejos. Una Aproximación Metodológica al Estudio de La Experiencia Bélica En La Guerra Civil Española. *Huarte S. Juan Geogr. Hist.* **2023**, *30*, 19–38. [CrossRef]
20. Vázquez, D.G.; Vergés, O.A.; Martín, A.B. Análisis estratégico de un recurso patrimonial territorial: Los refugios antiaéreos de la Guerra Civil española en la provincia de Girona (Cataluña). *Investig. Tur.* **2022**, *23*, 379–401. [CrossRef]
21. Lira, S. *REHAB 2017: Proceedings of the 3rd International Conference on Preservation, Maintenance and Rehabilitation of Historical Buildings and Structures, Braga, Portugal, 15 June 2017*; Amoêda, R., Lira, S., Pinheiro, C., Eds.; Green Lines Institute for Sustainable Development: Barcelos, Portugal, 2017; ISBN 9789898734242.
22. Stone, S. *UnDoing Buildings: Adaptive Reuse and Cultural Memory*; Routledge: Milton Park, UK, 2019; ISBN 9781315397207.
23. Maher, C.; Hadfield, M.; Hutchings, M.; de Eyto, A. Ensuring Rigor in Qualitative Data Analysis: A Design Research Approach to Coding Combining NVivo With Traditional Material Methods. *Int. J. Qual. Methods* **2018**, *17*, 1609406918786362. [CrossRef]
24. Watkins, D.C. Rapid and Rigorous Qualitative Data Analysis: The "RADaR" Technique for Applied Research. *Int. J. Qual. Methods* **2017**, *16*, 1609406917712131. [CrossRef]
25. Guarate, Y.C. Análisis de las entrevistas en la investigación cualitativa: Metodología de Demaziére D. y Dubar C. *Enferm. Investig.* **2019**, *4*, 14–23. [CrossRef]
26. Carabajo, R.A. Formación de investigadores de las ciencias sociales y humanas en el enfoque fenomenológico hermenéutico (de van manen) en el contexto hispanoamericano. *EducXX1* **2016**, *19*, 359–381. [CrossRef]
27. Villa, J.D.; Rodas, S.; Ospina, S.; Restrepo, S.; Avendaño, M. Creencias Sociales y Orientaciones Emocionales Colectivas Sobre La Protesta Social En Ciudadanos de Medellín (Colombia) y Su Área Metropolitana. *Investig. Desarro.* **2023**, *31*, 55–87. [CrossRef]
28. Muganga, L. The Importance of Hermeneutic Theory in Understanding and Appreciating Interpretive Inquiry as a Methodology. *J. Soc. Res. Policy* **2016**, *6*, 65–88.
29. Smith, J.A.; Flowers, P.; Larkin, M. *Interpretative Phenomenological Analysis: Theory, Method and Research*, 2nd ed.; SAGE Publications: London, UK, 2022; ISBN 9781529753806.
30. Smith, J.A.; Nizza, I.E. Essentials of Qualitative Methods. In *Essentials of Interpretative Phenomenological Analysis*; American Psychological Association: Washington, DC, USA, 2021; ISBN 9781433835650.
31. Roberts, T. Understanding the Research Methodology of Interpretative Phenomenological Analysis. *Br. J. Midwifery* **2013**, *21*, 215–218. [CrossRef]
32. De Azevedo, S.F. El recurso metalingüístico y la performatividad en la samba. Una propuesta de análisis desde la semiótica cultural. *REB* **2019**, *6*, 89–102. [CrossRef]
33. De Mariscal, B.L. Viajeros y Transferencia Cultural Gilliam y Flandrau En El México Del Siglo XIX. *Rev. Humanid. Tecnol. Monterrey* **2009**, *26*, 13–29.
34. Romañá, T.; Saura Carulla, M. *Interpretación del Desarrollo de la Arquitectura de Reformatorio en el Contexto Socio-Cultural Brasileño*; Universitat Politècnica de Catalunya: Barcelona, Spain, 2014.
35. Ramírez, J.P.; Becerra, J.I.R. Recursos culturales y objetos contra-patrimoniales: Apuntes exploratorios sobre las posibilidades de una antropología crítica del patrimonio a partir de la reflexión sobre una práctica religiosa transnacional. *Sphera Publica Rev. Cienc. Soc. Comun.* **2010**, 373–394.
36. Smith, J.A.; Breakwell, G.M.; Wright, D.B.; Clark, A.; Weeks, M. *Research Methods in Psychology*; Sage Publications Ltd.: Thousand Oaks, CA, USA, 2012.

37. Zaghi, A.E.; Grey, A.; Hain, A.; Syharat, C.M. "It Seems Like I'm Doing Something More Important"—An Interpretative Phenomenological Analysis of the Transformative Impact of Research Experiences for STEM Students with ADHD. *Educ. Sci.* **2023**, *13*, 776. [CrossRef]
38. Sak, M.; Gurbuz, N. Unpacking the Negative Side-Effects of Directed Motivational Currents in L2: An Interpretative Phenomenological Analysis. *Lang. Teach. Res.* **2022**, 13621688221125995. [CrossRef]
39. Rajasinghe, D. Interpretative Phenomenological Analysis (IPA) as a Coaching Research Methodology. *Coach. Int. J. Theory Res. Pract.* **2020**, *13*, 176–190. [CrossRef]
40. Micic, D.; Ehrlichman, H.; Chen, R.; Ansuini, C.; Begliomini, C.; Ferrari, T.; Castiello, U.; Moustafa, A.A.; Keri, S.; Herzallah, M.M.; et al. Why Do We Move Our Eyes While Trying to Remember? The Relationship between Non-Visual Gaze Patterns and Memory. Available online: https://es.zlib-articles.se/book/16554813/0427b8/why-do-we-move-our-eyes-while-trying-to-remember-the-relationship-between-nonvisual-gaze-patterns.html (accessed on 21 December 2023).
41. Graham, B.; Howard, P. *The Ashgate Research Companion to Heritage and Identity*; Ashgate Publishing, Ltd.: Farnham, UK, 2012; ISBN 9781409487609.
42. Rosser, P. (Ed.) El Concepto de Memoria y La Memoria En Alicante. In *Recuperación de la Memoria Histórica y la Identidad Colectiva del Área de las Cigarreras, Alicante*; Ayuntamiento de Alicante: Alicante, Spain, 2020; pp. 5–54.
43. Moore, N.; Whelan, Y.; Bullard, L.; Beiter, K.D. *Heritage, Memory and the Politics of Identity: New Perspectives on the Cultural Landscape*; Heritage, Culture and Identity; Ashgate Publishing: Farnham, UK, 2007.
44. Avilés, A.B.G.; Millan, M.I.P.; Ruiz, A.L.R. Reuse of Spanish Civil War Air-Raid Shelters in Alicante: The R46 Balmis and R31 Seneca Shelter. In Proceedings of the Defence Sites III: Heritage and Future, Southampton, UK, 4 May 2016; Volume 158, pp. 107–116.
45. Soler, S.; Rosser, P. Empatizar Con Los Conflictos Bélicos Para Trabajar El ODS 16. Creación de Una Situación de Aprendizaje a Partir de La Simulación Urbana. In *Hacia una Educación con Basada en las Evidencias de la Investigación y el Desarrollo Sostenible*; Dykinson: Madrid, Spain, 2023; ISBN 9788411704274.
46. Soler, S.; Rosser, P.; Gavilán, D. La Investigación Del ODS 16 "Paz, Justicia e Instituciones Sólidas": La Necesidad de Su Integración En La Educación. In *Investigación e Innovación Educativa en Contextos Diferenciados*; Dykinson: Madrid, Spain, 2023; pp. 499–509. ISBN 9788411705585.
47. Soler, S.; Rosser, P. Desafiando Los Límites Del Aprendizaje Histórico: Una Propuesta Educativa Innovadora Basada En La Pedagogía Crítica, Ia y Chatgpt Para Comprender La Guerra Civil Española, La Dictadura Franquista y La Transición Democrática. In *Las Ciencias Sociales, las Humanidades y Sus Expresiones Artísticas y Culturales: Una Tríada Indisoluble Desde un Enfoque Educativo*; Dykinson: Madrid, Spain, 2024; ISBN 9788411705844.
48. Rosser, P. (Ed.) *Recuperación de la Memoria Histórica y la Identidad Colectiva del Área de las Cigarreras, Alicante*; Ayuntamiento de Alicante: Alicante, Spain, 2020.
49. Broseta Palanca, M.T. Defence Heritage of the Spanish Civil War: Preservation of Air-Raid Shelters in Valencia. *Int. J. Herit. Archit. Stud. Repairs Maintence* **2017**, *1*, 624–639. [CrossRef]

Disclaimer/Publisher's Note: The statements, opinions and data contained in all publications are solely those of the individual author(s) and contributor(s) and not of MDPI and/or the editor(s). MDPI and/or the editor(s) disclaim responsibility for any injury to people or property resulting from any ideas, methods, instructions or products referred to in the content.

Article

Heritage Tourism Resilience and Sustainable Performance Post COVID-19: Evidence from Hotels Sector

Alaa M. S. Azzaz [1,2,*] and Ibrahim A. Elshaer [3,4]

1. Social Study Department, College of Arts, King Faisal University, Al-Ahsaa 380, Saudi Arabia
2. Tourism Studies Department, Faculty of Tourism and Hotels, Suez Canal University, Ismailia 41522, Egypt
3. Management Department, College of Business Administration, King Faisal University, Al-Ahsaa 380, Saudi Arabia; ielshaer@kfu.edu.sa
4. Hotel Studies Department, Faculty of Tourism and Hotels, Suez Canal University, Ismailia 41522, Egypt
* Correspondence: aazzaz@kfu.edu.sa

Abstract: Heritage tourism in Egypt, differentiated by its distinctive ancient wonders and cultural prosperity, has faced numerous challenges through its history, with political unrest, economic fluctuations, and, most recently, the global COVID-19 pandemic. This research paper investigates the dynamic interplay between planned and adopted resilience within the hotel sector in Egyptian heritage sites and their consequential effects on both social and economic sustainability. A quantitative research method was employed to empirically explore these dynamics. A structured questionnaire was distributed to 550 top and middle managers in hotels located in heritage sites, capturing insights into their perspectives on planned and adopted resilience. The collected data underwent rigorous analysis utilizing "partial least squares structural equation modeling" (PLS-SEM), providing a robust foundation for drawing meaningful conclusions. Findings from the research underscore the necessity of aligning planned and adopted resilience to generate sustainable social and economic performance. The synthesis of planned and adopted resilience was revealed to be pivotal in generating sustainable social and economic performance for hotels. This synthesis catalyzes the hotels' ability to mitigate uncertainties, adjust to changing environment, and ensure long-term viability. This research might contribute to the current literature by suggesting industry-specific awareness for the reciprocal relationship between planned and adopted resilience in the hotel businesses and their combined influence on both sides of sustainability (social and economic). The findings provide actionable recommendations for hotel management, policymakers, and industry stakeholders to enhance resilience, foster social cohesion, and ensure the economic sustainability of heritage tourism in an everchanging environment.

Keywords: heritage tourism; hotel industry; planned resilience; adopted resilience; social sustainability; economic sustainability; tourism resilience

Citation: Azzaz, A.M.S.; Elshaer, I.A. Heritage Tourism Resilience and Sustainable Performance Post COVID-19: Evidence from Hotels Sector. *Heritage* **2024**, *7*, 1162–1173. https://doi.org/10.3390/heritage7030055

Academic Editors: Fátima Matos Silva and Isabel Vaz de Freitas

Received: 26 January 2024
Revised: 17 February 2024
Accepted: 19 February 2024
Published: 22 February 2024

Copyright: © 2024 by the authors. Licensee MDPI, Basel, Switzerland. This article is an open access article distributed under the terms and conditions of the Creative Commons Attribution (CC BY) license (https://creativecommons.org/licenses/by/4.0/).

1. Introduction

The COVID-19 pandemic presented an exceptional challenge to worldwide tourism industry, including the Egyptian heritage tourism business. Egypt's heritage tourism occupies an exceptional and matchless attraction, attracting global visitors to delve into the intricate weave of its cultural and historical architectural riches [1,2]. Egypt stands as an enduring testament to the wonders of ancient civilizations, providing an enthralling voyage spanning thousands of years of history [3]. Beyond its ancient landmarks, Egypt's cultural heritage expands to include a diverse tapestry of traditions, arts, and crafts. Travelers have the opportunity to wander through lively marketplaces, witness traditional performances, and fully immerse themselves in the vibrant living heritage of the Egyptian people [1,2,4–6].

Human history has witnessed one of the most severe crises as the COVID-19 pandemic unfolded, leading to international turmoil and wreaking havoc on the global economy [7,8]. Due to its highly spreadable nature and associated related health risks, the virus has triggered widespread disruptions globally, including lockdowns, airport closures, stringent workforce regulations, limitations on exports and imports, and so forth [8–11]. The economic disruption has resulted in a decline in manufacturing, a surge in layoffs, increased unemployment rates, decreased consumer demand, and diminished business profits [8,9,12]. Numerous businesses, including air travel, tourism, and transportation, are grappling with severe challenges, with many organizations teetering on the brink of closure. Nevertheless, several organizations have demonstrated resilience, experiencing less impact from the pandemic, and some are recovering faster than their counterparts.

This study addresses several crucial research gaps, underscoring its significance. Firstly, there is a distinct necessity for studies that delve into organizational resilience as a multidimensional construct rather than adopting a holistic view. Secondly, the scarcity of research focusing on the two sides of sustainability (social and economic) in developing economies underscores the need to explore these areas. Thirdly, there is a notable absence of research that combines resilience and sustainability, emphasizing preserving sustainable consequences while navigating recovery from uncertainty. Building upon the identified gaps, this research outlines its key objectives as follows: firstly, to investigate how the resilient structure of hotels, located in heritage sites, influences the both sides of sustainability (social and economic) amid a global crisis such as the COVID-19 outbreak. Secondly, to assess the influence of the two factors of resilience (adopted and planned) on sustainability (social and economic factors).

Consequently, the research formulates the following questions: Research Question 1 (RQ1): What measures can be utilized to evaluate different aspects of resilience within the ongoing crisis? Research Question 2 (RQ2): Amidst the crisis, which dimensions of sustainability, both social and economic, witness notable impacts? Research Question 3 (RQ3): How can we operationalize the effect of tourism resilience on both social and/or economic sustainability of hotels located in heritage sites? To contextualize within a developing economy, the study was conducted in Egypt.

2. Egypt as a Context

Following the 2011 revolution, Egypt encountered a substantial surge in terrorism, marked by attacks predominantly directed at security forces and tourists. This had a detrimental effect on the image of Egypt as a secure tourist destination [13,14]. Between 2011 and 2016, the influx of worldwide tourists to Egypt experienced a noteworthy decline, dropping to approximately 65% below the levels reported in 2010, according to the "World Tourism Organization" (WTO) [15]. Faced with this unpredictable and turbulent situation, numerous hotels chose downsizing strategy or even complete closure [14]. However, by late 2017, a resurgence in tourist arrivals took place, prompting Egypt to be recognized by WTO [15] as the world's fastest-growing destination, experiencing a notable percentage increase in international visitors (55.1%). This recovery highlights the resilience and continuous operations of numerous hotels establishments within the country.

Like numerous other countries, the emergence of COVID-19 has significantly influenced Egypt, particularly taking a toll on the crucial tourism sector, a vital component of the country's economy. In 2019, Egyptian tourism generated USD 13 billion in revenue, signifying indicators of recovery after several years of political upheaval following the 2011 revolution [14]. Nevertheless, the current fiscal year (2019/2020) is anticipated to experience a decrease in industry revenues, estimated at approximately USD 11 billion instead of the initially projected USD 16 billion, due to the impact of the COVID-19 spread [14]. Additionally, Egypt's travel and tourism sector stands as one of the nation's foremost economic pillars, injecting approximately USD 32 billion into the GDP in 2022. Within the same timeframe, international tourist expenditure has exceeded domestic spending for the first time since 2020, largely attributable to travel restrictions imposed by the COVID-19 pandemic. Furthermore, the travel and tourism industry serves as a significant employer in the country, boasting a workforce of nearly 2.4 million in 2022 [16].

The Egyptian government has launched a program to combat the infection caused by COVID-19, dedicating USD 6.3 billion [17]. The government has implemented various initiatives to support the tourism sector during these challenging times, including tax decreases for tourism businesses, reduced prices for electricity and gas for tourism establishments, and safeguarding salaries for permanent tourism employees. In the expectancy of the reopening of tourism businesses for domestic Egyptian tourism in June, the "Ministry of Tourism and Antiquities" has circulated safety recommendations and regulations emphasizing a "safety first" approach for hotels. These guidelines encompass a determined rate of 50% occupancy for the reopening stage [18].

3. Theoretical Background and Hypotheses Development

Resilience emerges as a pivotal factor in business survival, defined as the ability to foresee, withstand, and recuperate from a challenging environment, ultimately restoring to an initial or enhanced state [19–22]. Many scholars classify resilience into two primary categories: adaptive resilience and planned resilience [23–26]. Proactive (planned) resilience involves preparedness, while reactive (adopted) resilience pertains to recovering from turbulence [21,27]. Planned proactive resilience initiates before disasters, while adaptive reactive resilience obviously emerges after such events, requiring adept handling, strong networks, collaboration, and learning from failure and past experiences [17,28]. Research indicates that post-crisis recovery approaches significantly impact an organization's performance [23,29]. However, lacking a recovery plan can impede an enterprise's adaptive resilience [30]. Effective crisis planning facilitates the optimal utilization of resources and infrastructure, thereby contributing to resilience post crisis [31]. Lee et al. [24] introduced a tool for assessing business resilience, distinguishing between planned and adaptive resilience in organizations. Their findings highlighted the significance of planned resilience, encompassing recovery precedence and a proactive attitude, as a critical indicator of adaptive resilience. In the current study, we operationalize resilience as a multidimensional construct, consisting of two dimensions: adopted resilience and planned resilience.

The crisis has profoundly impacted the sustainability framework of corporations, particularly in developing nations such as Egypt, Brazil, and India [12,13]. Sustainability, focusing on preserving future resources, encompasses three key dimensions: "environment, society, and economy", often referred to as the "triple bottom line" (TBL) approach. While extensive research has been conducted globally on "environmental sustainability", "social sustainability" (commonly denoted as corporate social responsibility or CSR) and "economic sustainability" (concentrating on cost control through the acceptance of sustainable practices) have not received as much attention from scholars, particularly in developing economies [17,32,33]. Nevertheless, these two components have endured severe crises, with both society and the economy grappling with the aftermath and striving to recuperate. Research addressing these dimensions predominantly reflects the settings of developed countries. Even within these studies, the amalgamation of economic and social aspects typically occurs individually or within the "triple bottom line" (TBL) framework. For example, Barbosa-Póvoa et al. [34] noted a deficiency in holistic economic assessments within sustainability studies, emphasizing the need for comprehensive evaluations of environmental and social aspects. There is a gap in focusing on the operationalization of executing sustainable practices that contribute to either cost control or profit generation. As an illustration, Zhang et al. [35] concentrated on "green supply chain management" (GSCM) as an antecedent of environmental practices and CSR in their empirical model of "sustainable supply chain management" (SSCM). While advocating for the multidimensional nature of SSCM, their framework falls short in adequately explaining economic considerations. In contrast, Esfahbodi et al. [36] incorporated both environmental firm performance and cost control in their empirical framework, primarily focusing on environmental aspects but neglecting the social side. Meanwhile, King and Lenox [37] stressed the importance of environmental performance within the operations environment yet overlooked economic and social practices. In the present study, our emphasis lies on a two-dimensional approach encompassing social and economic sustainability, addressing crucial concerns through the COVID-19 pandemic.

The resilience and sustainable performance of heritage tourism are interconnected and mutually reinforcing. Resilient destinations are better equipped to navigate challenges, thereby contributing to the sustainability of tourism operations. Conversely, sustainable practices enhance a destination's ability to adapt and recover from disturbances, fostering a holistic and enduring tourism experience. The attainment of financial performance, sustainable organizational success, and a competitive advantage can be facilitated through organizational resilience [38–43]. De Carvalho et al. [44] observed that resilient enterprises, particularly those with innovative characteristics, are better positioned to maintain higher performance levels compared to their counterparts. Research, exemplified by studies such as Orchiston et al. [45], affirms the significance of problem solving, planning, and establishing external network in fostering resilience, thereby exerting a positive influence on performance [46]. Prayag et al. [47] discovered that the impact of adaptive resilience on firm performance is particularly evident in the context of small businesses. There is also an argument for a comprehensive understanding of the interrelationships between resilience and overall performance by considering both factors of resilience, namely, planned and adaptive [25,26,48,49].

Organizational resilience serves as a facilitative instrument for sustaining performance by offering fresh perspectives on social and environmental adaptability in the face of a perpetually changing community [50]. Souza et al. [51] asserted the necessity of enduring plans and benchmarking to cultivate corporate resilience for sustainability. Fatoki [43] also identified core and outer factors influencing the connection between corporate resilience and sustainable performance, encompassing both social and economic sustainability. Inner factors encompass planning, managerial competencies, innovation, and creativity, while outer factors incorporate government support and the nation's overall financial performance. Resilience, defined as an the ability to recover from instability, is crucial in ensuring sustainable performance [52]. Consequently, as illustrated in Figure 1, the below hypotheses could be theorized:

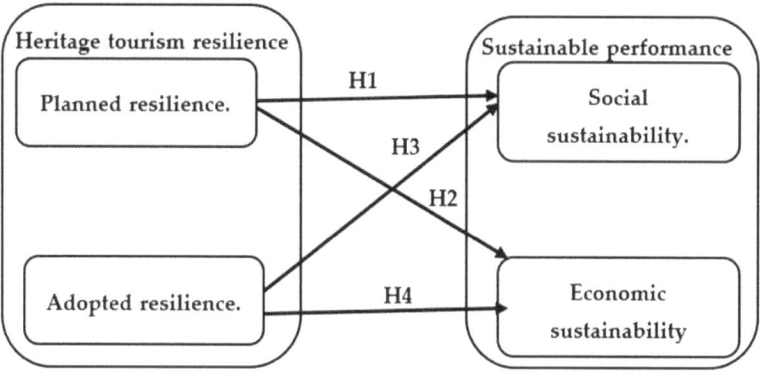

Figure 1. Research hypotheses.

H1. *Planned resilience positively influences social sustainability.*

H2. *Planned resilience positively influences economic sustainability.*

H3. *Adopted resilience positively influences social sustainability.*

H4. *Adopted resilience positively influences economic sustainability.*

The study aims to contribute to practitioners and academia in several ways. Initially, it furnishes an empirical evaluation of organizational resilience as a multidimensional construct, offering valuable insights for firms in enhancing their ability to survive and recover during times of crisis. Secondly, it presents a comprehensive evaluation of social and economic sustainability, two aspects particularly vulnerable during a crisis, often overlooked by organizations focused primarily on survival. Thirdly, the research establishes a cause-and-effect path between hotel resilience and sustainability, aiming to address the question of how organizations can simultaneously survive and act responsibly. Fourthly, it delves into how sustainability and resilience can prove advantageous for companies in the long term.

4. Methods

4.1. The Instrument

A questionnaire-based survey was created to evaluate the resilience of luxury 5-star hotels located in heritage sites, Egypt, in response to the COVID-19 pandemic and its consequences on sustainable social and economic performance. The survey comprised three parts. The first gathered data about the profiles of top and middle managers. Section 2 focused on the companies' status, such as the number and full-time/part-time employees, and years of operation. Section 3 used a five-point scale to examine organizational resilience (both adopted and planned) and sustainable performance (both social and economic).

To obtain suitable measures for this study, standard psychometric procedures were undertaken. All reflective measures were adapted from the existing literature following a thorough review of the current literature, employing five Likert scales. As in previous research [17,23,24,26], two factors of resiliencies (planned and adaptive) were utilized to assess hotel resilience. Each factor consists of five reflective items. The scales for sustainable performance were adapted from Rai et al. [53] and encompass two sub-dimensions. The initial dimension delineates the social facet of sustainable performance, comprising 9 reflective items. Sample items include statements like "We prioritize the health and safety of our employees, even during crises" and "We have maintained our employees' salaries throughout the crisis". Dimension Two concentrates on the economic aspect of sustainable performance, with sample items such as "We curtail resource consumption for sustainability" and "We make investments in quality to enhance the lifecycle of our products".

4.2. Research Population, Sample and Procedures

The study employed a quantitative survey method to gather data from managerial positions, including front office managers, food and beverage managers, rooms division managers, sales managers, and marketing managers, working in 170 five-star hotels located in heritage sites, Egypt. Managerial positions were chosen as the survey targets, given sufficient information and authority to respond to the research questionnaire. Data were collected from a survey involving 700 employees in managerial positions. The participation from each hotel varied from 3 to 5 individuals to prevent the under/overrepresentation of specific hotels. A total of 560 responses were collected, with twenty incomplete questionnaires excluded, resulting in 540 usable responses and a response rate of approximately 77%.

The study involved gathering questions that represent the dependent and independent items from the same set of participants, raising concerns about "common method variance" (CMV). To address potential CMV, various measures were implemented following the recommendations of [54]. Initially, all participants were guaranteed that their replies would be kept confidential. Then, the sequence of the questionnaire placed the dependent questions before the independent questions, following the approach suggested by [55]. Additionally, the questionnaire items were translated from original English to participants' Arabic language by bilingual specialists. Subsequently, it underwent pretesting with 35 experts from the hotel sector and 25 faculty members from institutions specializing in hotel business. Based on the feedback received, appropriate revisions were made to enhance clarity. Thirdly, a comparison was made between early and late responses, utilizing a *t*-test to assess the likelihood of late-reply bias. The analysis exposed no significance differences ($p > 0.05$), indicating that nonresponse bias is not a problem.

5. Data Analysis and Study Results

The current study employed PLS-SEM, which is a variance-based algorithm that can be used for path analysis. This method is an adequate choice that can replace the traditional covariance-based SEM (CB-SEM) [56] Admitted for its appropriateness in exploratory research, PLS-SEM has strong renown [57]. Unrestricted by the normality assumption in the distribution of the study sample, it demonstrates efficacy across both large and small sample sizes [58]. A systematic review of PLS-SEM papers published between 2000 and 2014 in hospitality discipline [59] indicated its underutilization in comparison to the traditional CB-SEM. The selection of this method for the study was driven by its orientation towards exploratory research and its adaptability to accommodate diverse sample sizes. The PLS analysis was executed using SmartPLS 4 [56]. The model estimation employed a bootstrapping process (n = 5000 resamples) utilizing a reflective variable approach [60]. In addition, to investigate and test CMV issue, as suggested by [54], scrutiny was carried out using "Harman's one-factor" test. All 25 items underwent an "exploratory factor analysis" (EFA), disclosing that the initial factor accounted for just 38% of the total variance. This implies that CMV is not a predominant issue in this study. Additionally, all "variance inflation factor" (VIF) values registered below 0.5, signifying the absence of multicollinearity concerns (refer to Table 1).

Table 1. Factors and items' psychometric properties.

Scale	Loadings	VIF
Adoptive Resilience (α = 0.961, CR = 0.962, AVE = 0.864)		
Adpt_Res_1: "People in our organization are committed to working on a problem until it is resolved".	0.957	1.444
Adpt_Res_2: "Our organization maintains sufficient resources to absorb some unexpected change".	0.911	4.319
Adpt_Res_3: "If key people were unavailable, there are always others who could fill their role".	0.947	4.103
Adpt_Res_4: "There would be good leadership from within our organization if we were struck by a crisis".	0.915	3.356
Adpt_Res_5: "We are known for our ability to use knowledge in novel ways".	0.916	3.833
Planned Resilience (α = 0.917, CR = 0.913, AVE = 0.897)		
Plnd_Res_1: "Given how others depend on us, the way we plan for the unexpected is appropriate".	0.958	2.489
Plnd_Res_2: "Our organization is committed to practicing and testing its emergency plans to ensure they are effective".	0.954	3.081
Plnd_Res_3: "We have a focus on being able to respond to the unexpected".	0.937	3.433
Plnd_Res_4: "We have clearly defined priorities for what is important during and after a crisis".	0.941	3.775
Plnd_Res_5: "People in our organization are committed to working on a problem until it is resolved".	0.947	3.361
Economic sustainability (α = 0.929, CR = 0.933, AVE = 0.861)		
Econ_S1: "We invest in CSR without hurting our profits".	0.930	1.460
Econ_S1: "We minimize waste to reduce our material cost".	0.927	3.587
Econ_S3: "We sustainably procure and preserve the materials to increase their lifecycle".	0.926	4.331
Econ_S4: "We reduce resource consumption for sustainability".	0.941	2.110
Econ_S5: "We reuse resources to reduce our costs".	0.915	2.773
Econ_S6: "We invest in quality for the increased life cycle of products".	0.931	3.846

Table 1. Cont.

Scale	Loadings	VIF
Social sustainability (α = 0.934, CR = 0.949, AVE = 0.737)		
Soc_S1: "We pay fair wages to our manpower".	0.890	2.320
Soc_S2: "We have not laid-off workers during the lockdown".	0.901	2.781
Soc_S3: "We invest in our workers' health and safety even during the crisis".	0.876	1.023
Soc_S4: "We have not reduced the salaries of our employees during the crisis".	0.858	4.160
Soc_S5: "We ensure our employees for health issues".	0.776	2.185
Soc_S6: "We focus on protecting our workers' rights".	0.863	4.643
Soc_S7: "We comply with hygiene and social distancing norms".	0.849	2.749
Soc_S8: "We educate and train our employees for new safety requirements".	0.835	4.377
Soc_S9: "We focus on job creation for local and economically, affected society".	0.871	1.058

The evaluation of the measurement model includes the examination of the scale psychometric characteristics, utilizing metrics such as "Cronbach's α, composite reliabilities (CR), and average variance extracted (AVE)". All items demonstrated loadings of 0.7 and above, signifying convergent validity at a satisfactory level. Both CR and Cronbach's α findings outperformed the minimum standard of 0.7, indicating a good internal consistency (Table 1). Additionally, AVE findings for all dimensions exceeded the recommended threshold of 0.5, as proposed by [61]. Consequently, convergent validity was considered acceptable, taken into consideration that all AVEs values were 0.5 and above.

By following the methodology of [61], discriminant validity was validated by ensuring that the root square of the "average variance extracted" (AVE) for each factor surpassed the intercorrelations between that factor and all others in the model (Table 2). Additionally, validity was further calculated using the "heterotrait–monotrait" (HTMT) ratio of correlations, considered a more robust method than Fornell and Larcker's [62]. Concerns about discriminant validity arise when HTMT values exceed 0.9. As indicated in Table 3, all ratios were below the standard value of 0.9, affirming discriminant validity.

Table 2. Factors' discriminating validity employing Fornell and Larcker and HTMT.

	Adaptive Resilience	Economic Sustainability	Planned Resilience	Social Sustainability
Adopted resilience	**0.929**			
Economic sustainability	0.648 [0.670]	**0.928**		
Planned resilience	0.548 [0.565]	0.467 [0.477]	**0.947**	
Social sustainability	0.548 [0.560]	0.424 [0.435]	0.353 [0.364]	**0.858**

Bold figures show the square root of AVE; HTMT ratios are shown in brackets.

Table 3. Path coefficient with t and p values.

Paths	(β)	(T)	(P)
Adopted resilience -> Social sustainability	0.507	12.359	0.000
Adopted resilience -> Economic sustainability	0.561	11.427	0.000
Planned resilience -> Social sustainability	0.075	1.975	0.048
Planned resilience -> Economic sustainability	0.159	3.295	0.001

The R^2 bootstrapped values indicated that adopted and planned resilience collectively has 30.5% of effect size of variance in social sustainability and 43.8% in economic sustainability as shown in Figure 2. As seen in Table 3, it was observed that adopted resilience has a significant and positive impact on social sustainability (β = 0.561, t = 12.359, $p < 0.001$) and economic sustainability (β = 0.507, t = 11.427, $p < 0.001$), supporting H1 and H2. Additionally, adopted resilience has a positive and significant influence on social sustainability (β = 0.075, t = 1.975, $p < 0.05$), corroborating H3. Furthermore, planned resilience has a positive and significant impact on economic sustainability (β = 0.159, t = 3.259, $p < 0.01$), confirming the support for H4.

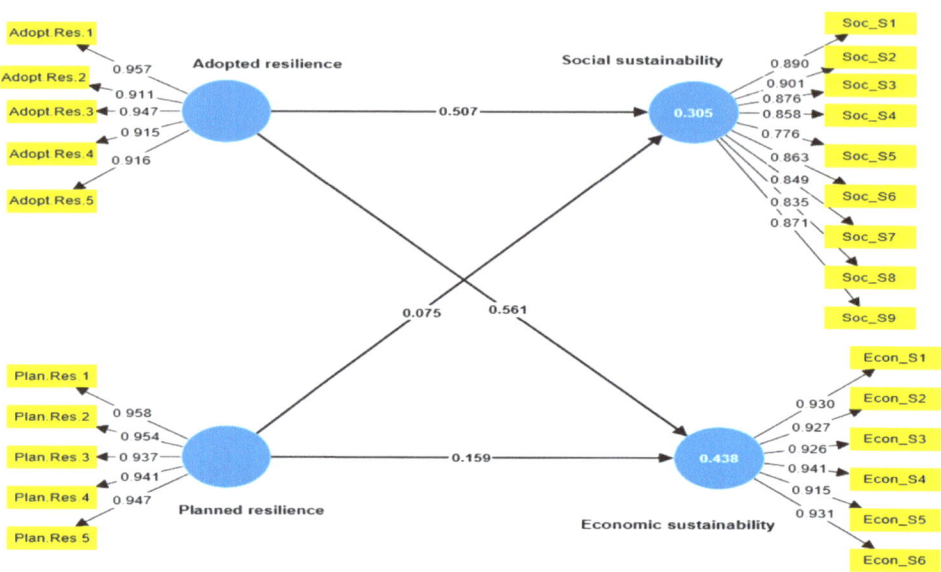

Figure 2. The examined research model.

6. Discussion and Implications

The alignment of the study's findings with the theoretical foundations of resilience, as emphasized by [13,17,23,26,48,52,63,64], underscores resilience's pivotal role in confronting and surmounting organization challenges. Adopted resilience emerges as a crucial factor that allows hotels in heritage sites to weather adversities, and actively contributes to social and economic sustainability [64,65]. In the realm of social sustainability, the positive impact of adopted resilience is manifest in the cultivation of favorable employee relations. Hotels in heritage sites adopting resilient strategies prioritize employee wellbeing, fostering a work environment that promotes job satisfaction, professional growth, and overall employee contentment. This commitment to positive employee relations contributes to the organization's social fabric and creates a ripple effect within the broader community, enhancing the industry's social sustainability [66]. Furthermore, adopted resilience leads to the establishment of robust community partnerships. By prioritizing resilience, Egyptian hotels engage with local communities in meaningful ways, contributing to social development and mutual support. These partnerships extend beyond the business realm, integrating hotels into the social fabric of their surroundings. As hotels actively participate in community initiatives and collaborate with local stakeholders, they become integral contributors to the overall wellbeing of the communities they serve, thereby enhancing social sustainability.

Moving on to economic sustainability, the study results declared that the impact of adopted resilience is equally significant. The proactive nature of resilient strategies enables hotels in heritage sites to manage costs effectively [64]. This involves cost-cutting measures during challenging times and strategic financial planning that anticipates and prepares for potential economic downturns [53]. By adopting resilience, these hotels enhance their financial stability, ensuring their ability to weather economic uncertainties without compromising the quality of their services or the wellbeing of their employees. Moreover, adopted resilience facilitates revenue stability in the Egyptian hotel sector in heritage sites. The ability to adapt to dynamic market conditions, diversify revenue streams, and innovate in response to changing consumer demands positions resilient hotels to maintain a stable and consistent income. This stability is crucial for long-term economic sustainability, allowing hotels to withstand fluctuations in market trends and economic shocks.

Additionally, the alignment of the study's results with existing resilience theories highlights the strategic importance of planned resilience in ensuring economic sustainability. Planned resilience involves proactive measures such as risk assessment, strategic financial planning, and the development of contingency plans to navigate potential challenges [52,67]. One significant impact of planned resilience in the Egyptian hotel industry is its role in revenue management. Strategic planning allows hotels to diversify revenue streams, identify potential market opportunities, and develop pricing strategies that maximize income. This proactive approach positions hotels to optimize revenue even in the face of economic uncertainties. Moreover, planned resilience contributes to cost-effective practices [56,68,69]. The strategic allocation of resources, efficient operational management, and implementing cost-saving measures during stable periods ensure that hotels are well prepared for economic fluctuations [57,58,65]. This foresight and strategic planning contribute to long-term cost stability, crucial for sustained economic viability.

Additionally, the study demonstrates that planned resilience supports strategic financial planning. Hotels that engage in thorough financial forecasting, risk analysis, and scenario planning are better equipped to navigate economic challenges. This approach ensures financial stability, enabling hotels to withstand economic shocks and sustain operations over the long term [64]. Furthermore, the study's output highlights the intricate nature of the interrelationship between planned resilience and social sustainability. The positive impact indicates that strategic planning contributes to fostering social sustainability within the hotel industry in Egypt. However, the weak significance suggests that while planned resilience positively contributes to social sustainability, its impact may be influenced by various contextual factors. For instance, the cultural context of Egypt, the specific community dynamics surrounding each hotel, or the prevailing economic conditions may moderate the relationship between planned resilience and social sustainability.

This research has practical implications for hotel management, policymakers, and industry stakeholders. Hotel managers should prioritize the adoption of resilience strategies to enhance social and economic sustainability. Policymakers can support industry by creating an environment conducive to resilience-building initiatives. Additionally, industry stakeholders should collaborate to share best practices and foster a collective approach to resilience in the Egyptian hotel sector. The positive impact of planned resilience on economic sustainability has practical implications for hotel management and industry stakeholders. Hotel managers should prioritize developing and implementing planned resilience strategies to enhance economic sustainability. Policymakers can support industry by creating an environment conducive to strategic planning and risk management. Collaboration among industry stakeholders is essential to sharing best practices and building a resilient hotel sector in Egypt.

7. Limitations and Future Research Avenues

One limitation of this study is the contextual specificity inherent in the research. The findings are based on a particular industry (e.g., hotel industry) and a specific geographic location (e.g., Egypt). Generalizing the results to other industries or regions requires caution due to potential variations in organizational structures, cultural influences, and economic conditions. Moreover, the research adopted a cross-sectional design, capturing a snapshot of the relationships at a specific point in time. This limits the ability to establish causality and explore the dynamic nature of resilience and sustainability over time. Future research could employ longitudinal designs to understand the temporal aspects of these relationships better. Additionally, the study primarily focuses on economic and social sustainability, leaving out environmental sustainability. Future research should consider a more comprehensive approach by incorporating environmental dimensions to provide a holistic understanding of sustainability in organizations. Conducting longitudinal studies will provide a deeper understanding of how the impact of resilience strategies evolves over time. This could uncover patterns, trends, and long-term sustainability outcomes associated with different resilience planning and adoption approaches. Finally, investigating potential mediating and moderating factors in the relationship between resilience and sustainability could enhance the complexity of the model, for instance, the role of organizational culture, the ongoing war in the nearby Gaza Strip, leadership styles, or external environmental policies in influencing the effectiveness of resilience strategies.

Author Contributions: Conceptualization, I.A.E. and A.M.S.A.; methodology, I.A.E. and A.M.S.A.; software, I.A.E.; validation, I.A.E. and A.M.S.A.; formal analysis, I.A.E. and A.M.S.A.; investigation, I.A.E. and A.M.S.A.; resources, I.A.E.; data curation, I.A.E.; writing—original draft preparation, I.A.E. and A.M.S.A.; writing—review and editing, I.A.E. and A.M.S.A.; visualization, I.A.E.; supervision, I.A.E.; project administration, I.A.E. and A.M.S.A.; funding acquisition, I.A.E. and A.M.S.A. All authors have read and agreed to the published version of the manuscript.

Funding: This work was supported by the Deanship of Scientific Research, Vice Presidency for Graduate Studies and Scientific Research, King Faisal University, Saudi Arabia [Grant No. 5866].

Informed Consent Statement: Informed consent was obtained from all subjects involved in the study.

Data Availability Statement: Data are available upon request from researchers who meet the eligibility criteria. Kindly contact the first author privately through e-mail.

Conflicts of Interest: The authors declare no conflict of interest.

References

1. Mohamed Atef, A. Sustainable Heritage Tourism in Egypt. *Int. J. Multidiscip. Stud. Archit. Cult. Herit.* **2021**, *4*, 89–101. [CrossRef]
2. Mustafa, M.H. Cultural Heritage: A Tourism Product of Egypt under Risk. *J. Environ. Manag. Tour. (JEMT)* **2021**, *12*, 243–257. [CrossRef]
3. Ghanem, M.M.; Saad, S.K. Enhancing Sustainable Heritage Tourism in Egypt: Challenges and Framework of Action. *J. Herit. Tour.* **2015**, *10*, 357–377. [CrossRef]
4. Hang, P.L.K.; Kong, C. Heritage Management and Control: The Case of Egypt. *J. Qual. Assur. Hosp. Tour.* **2001**, *2*, 105–117. [CrossRef]
5. Elgammal, I.; Refaat, H. Heritage Tourism and COVID-19: Turning the Crisis into Opportunity within the Egyptian Context. In *Tourism Destination Management in a Post-Pandemic Context*; Gowreesunkar, V.G., Maingi, S.W., Roy, H., Micera, R., Eds.; Emerald Publishing Limited: Bradford, UK, 2021; pp. 37–48, ISBN 978-1-80071-512-7.
6. Eladway, S.M.; Azzam, Y.A.; Al-Hagla, K.S. Role of Public Participation in Heritage Tourism Development in Egypt: A Case Study of Fuwah City. *WIT Trans. Ecol. Environ.* **2020**, *241*, 27–43.
7. Arora, A.S.; Rajput, H.; Changotra, R. Current Perspective of COVID-19 Spread across South Korea: Exploratory Data Analysis and Containment of the Pandemic. *Environ. Dev. Sustain.* **2021**, *23*, 6553–6563. [CrossRef] [PubMed]
8. Rajput, H.; Changotra, R.; Rajput, P.; Gautam, S.; Gollakota, A.R.K.; Arora, A.S. A Shock like No Other: Coronavirus Rattles Commodity Markets. *Environ. Dev. Sustain.* **2021**, *23*, 6564–6575. [CrossRef]
9. Sharma, A.; Gupta, P.; Jha, R. COVID-19: Impact on Health Supply Chain and Lessons to Be Learnt. *J. Health Manag.* **2020**, *22*, 248–261. [CrossRef]
10. Bherwani, H.; Nair, M.; Musugu, K.; Gautam, S.; Gupta, A.; Kapley, A.; Kumar, R. Valuation of Air Pollution Externalities: Comparative Assessment of Economic Damage and Emission Reduction under COVID-19 Lockdown. *Air Qual. Atmos. Health* **2020**, *13*, 683–694. [CrossRef]
11. Gautam, S. The Influence of COVID-19 on Air Quality in India: A Boon or Inutile. *Bull. Environ. Contam. Toxicol.* **2020**, *104*, 724–726. [CrossRef]
12. Gautam, S.; Hens, L. COVID-19: Impact by and on the Environment, Health and Economy. *Environ. Dev. Sustain.* **2020**, *22*, 4953–4954. [CrossRef]
13. Saad, S.K.; Elshaer, I.A. Justice and Trust's Role in Employees' Resilience and Business' Continuity: Evidence from Egypt. *Tour. Manag. Perspect.* **2020**, *35*, 100712. [CrossRef]
14. Elshaer, I.A.; Saad, S.K. Political Instability and Tourism in Egypt: Exploring Survivors' Attitudes after Downsizing. *J. Policy Res. Tour. Leis. Events* **2017**, *9*, 3–22. [CrossRef]
15. WTO World Tourism Organization (WTO). *UNWTO Tourism Highlights*, 2018th ed.; UNWTO: Madrid, Spain, 2018. Available online: https://www.unwto.org/global/publication/unwto-tourism-highlights-2018 (accessed on 9 December 2023).

16. Statista Topic: Tourism Industry in Egypt. Available online: https://www.statista.com/topics/5767/tourism-industry-of-the-egypt/ (accessed on 17 February 2024).
17. Elshaer, I.A.; Saad, S.K. Entrepreneurial Resilience and Business Continuity in the Tourism and Hospitality Industry: The Role of Adaptive Performance and Institutional Orientation. *Tour. Rev.* 2021, *ahead-of-print*. [CrossRef]
18. Egyptian Ministry of Tourism Home. Available online: http://etf.org.eg (accessed on 11 March 2023).
19. Christopher, M.; Lee, H. Mitigating Supply Chain Risk through Improved Confidence. *Int. J. Phys. Distrib. Logist. Manag.* **2004**, *34*, 388–396. [CrossRef]
20. Brusset, X.; Teller, C. Supply Chain Capabilities, Risks, and Resilience. *Int. J. Prod. Econ.* **2017**, *184*, 59–68. [CrossRef]
21. Chowdhury, M.M.H.; Quaddus, M. Supply Chain Resilience: Conceptualization and Scale Development Using Dynamic Capability Theory. *Int. J. Prod. Econ.* **2017**, *188*, 185–204. [CrossRef]
22. Pettit, T.J.; Croxton, K.L.; Fiksel, J. Ensuring Supply Chain Resilience: Development and Implementation of an Assessment Tool. *J. Bus. Logist.* **2013**, *34*, 46–76. [CrossRef]
23. Prayag, G.; Spector, S.; Orchiston, C.; Chowdhury, M. Psychological Resilience, Organizational Resilience and Life Satisfaction in Tourism Firms: Insights from the Canterbury Earthquakes. *Curr. Issues Tour.* **2020**, *23*, 1216–1233. [CrossRef]
24. Lee, A.V.; Vargo, J.; Seville, E. Developing a Tool to Measure and Compare Organizations' Resilience. *Nat. Hazards Rev.* **2013**, *14*, 29–41. [CrossRef]
25. Elshaer, I.A. Dimensionality Analysis of Entrepreneurial Resilience amid the COVID-19 Pandemic: Comparative Models with Confirmatory Factor Analysis and Structural Equation Modeling. *Mathematics* **2022**, *10*, 2298. [CrossRef]
26. Sobaih, A.E.E.; Elshaer, I.; Hasanein, A.M.; Abdelaziz, A.S. Responses to COVID-19: The Role of Performance in the Relationship between Small Hospitality Enterprises' Resilience and Sustainable Tourism Development. *Int. J. Hosp. Manag.* **2021**, *94*, 102824. [CrossRef] [PubMed]
27. Sheffi, Y. *The Resilient Enterprise: Overcoming Vulnerability for Competitive Advantage*; Pearson Education India: Haldwani, India, 2005.
28. Nilakant, V.; Walker, B.; van Heugen, K.; Baird, R.; De Vries, H. Research Note: Conceptualising Adaptive Resilience Using Grounded Theory. *N. Z. J. Employ. Relat.* **2014**, *39*, 79–86.
29. Corey, C.M.; Deitch, E.A. Factors Affecting Business Recovery Immediately after Hurricane Katrina: Factors Affecting Business Recovery. *J. Contingencies Crisis Manag.* **2011**, *19*, 169–181. [CrossRef]
30. Alexander, D.E. Resilience and Disaster Risk Reduction: An Etymological Journey. *Nat. Hazards Earth Syst. Sci.* **2013**, *13*, 2707–2716. [CrossRef]
31. Faulkner, B.; Vikulov, S. Katherine, Washed out One Day, Back on Track the next: A Post-Mortem of a Tourism Disaster. *Tour. Manag.* **2001**, *22*, 331–344. [CrossRef]
32. Saad, M.H.; Hagelaar, G.; van der Velde, G.; Omta, S.W.F. Conceptualization of SMEs' Business Resilience: A Systematic Literature Review. *Cogent Bus. Manag.* **2021**, *8*, 1938347. [CrossRef]
33. Tang, C.S. Socially Responsible Supply Chains in Emerging Markets: Some Research Opportunities. *J. Oper. Manag.* **2018**, *57*, 1–10. [CrossRef]
34. Barbosa-Póvoa, A.P.; da Silva, C.; Carvalho, A. Opportunities and Challenges in Sustainable Supply Chain: An Operations Research Perspective. *Eur. J. Oper. Res.* **2018**, *268*, 399–431. [CrossRef]
35. Zhang, M.; Tse, Y.K.; Doherty, B.; Li, S.; Akhtar, P. Sustainable Supply Chain Management: Confirmation of a Higher-Order Model. *Resour. Conserv. Recycl.* **2018**, *128*, 206–221. [CrossRef]
36. Esfahbodi, A.; Zhang, Y.; Watson, G. Sustainable Supply Chain Management in Emerging Economies: Trade-Offs between Environmental and Cost Performance. *Int. J. Prod. Econ.* **2016**, *181*, 350–366. [CrossRef]
37. King, A.A.; Lenox, M.J. Lean and green? An empirical examination of the relationship between lean production and environmental performance. *Prod. Oper. Manag.* **2001**, *10*, 244–256. [CrossRef]
38. Akgün, A.E.; Keskin, H. Organisational Resilience Capacity and Firm Product Innovativeness and Performance. *Int. J. Prod. Res.* **2014**, *52*, 6918–6937. [CrossRef]
39. Cavaco, N.M.; Machado, V.C. Sustainable Competitiveness Based on Resilience and Innovation—An Alternative Approach. *Int. J. Manag. Sci. Eng. Manag.* **2015**, *10*, 155–164. [CrossRef]
40. de Oliveira Teixeira, E.; Werther, W.B., Jr. Resilience: Continuous Renewal of Competitive Advantages. *Bus. Horiz.* **2013**, *56*, 333–342. [CrossRef]
41. Gunasekaran, A.; Rai, B.K.; Griffin, M. Resilience and Competitiveness of Small and Medium Size Enterprises: An Empirical Research. *Int. J. Prod. Res.* **2011**, *49*, 5489–5509. [CrossRef]
42. Yi, X.; Lin, V.S.; Jin, W.; Luo, Q. The Authenticity of Heritage Sites, Tourists' Quest for Existential Authenticity, and Destination Loyalty. *J. Travel Res.* **2017**, *56*, 1032–1048. [CrossRef]
43. Bui, H.T.; Jones, T.E.; Weaver, D.B.; Le, A. The Adaptive Resilience of Living Cultural Heritage in a Tourism Destination. *J. Sustain. Tour.* **2020**, *28*, 1022–1040. [CrossRef]
44. De Carvalho, A.O.; Ribeiro, I.; Cirani, C.B.S.; Cintra, R.F. Organizational Resilience: A Comparative Study between Innovative and Non-Innovative Companies Based on the Financial Performance Analysis. *Int. J. Innov. IJI J.* **2016**, *4*, 58–69. [CrossRef]
45. Orchiston, C.; Higham, J.E.S. Knowledge Management and Tourism Recovery (de)Marketing: The Christchurch Earthquakes 2010–2011. *Curr. Issues Tour.* **2016**, *19*, 64–84. [CrossRef]

46. Avery, G.C.; Bergsteiner, H. Sustainable Leadership Practices for Enhancing Business Resilience and Performance. *Strategy Leadersh.* **2011**, *39*, 5–15. [CrossRef]
47. Prayag, G.; Orchiston, C.; Chowdhury, M. From Sustainability to Resilience: Understanding Different Facets of Organizational Resilience. In *BEST EN Think Tank XVII Innovation and Progress in Sustainable*; BEST Education Network: Douglas, Australia, 2017; p. 292.
48. Azazz, A.M.S.; Elshaer, I.A. Amid COVID-19 Pandemic, Entrepreneurial Resilience and Creative Performance with the Mediating Role of Institutional Orientation: A Quantitative Investigation Using Structural Equation Modeling. *Mathematics* **2022**, *10*, 2127. [CrossRef]
49. Jiang, Y.; Ritchie, B.W.; Verreynne, M. Building Tourism Organizational Resilience to Crises and Disasters: A Dynamic Capabilities View. *J. Tour. Res.* **2019**, *21*, 882–900. [CrossRef]
50. Lew, A.A. Scale, Change and Resilience in Community Tourism Planning. *Tour. Geogr.* **2014**, *16*, 14–22. [CrossRef]
51. Souza, A.A.A.; Alves, M.F.R.; Macini, N.; Cezarino, L.O.; Liboni, L.B. Resilience for Sustainability as an Eco-Capability. *Int. J. Clim. Chang. Strateg. Manag.* **2017**, *9*, 581–599. [CrossRef]
52. Annarelli, A.; Nonino, F. Strategic and Operational Management of Organizational Resilience: Current State of Research and Future Directions. *Omega* **2016**, *62*, 1–18. [CrossRef]
53. Rai, S.S.; Rai, S.; Singh, N.K. Organizational Resilience and Social-Economic Sustainability: COVID-19 Perspective. *Environ. Dev. Sustain.* **2021**, *23*, 12006–12023. [CrossRef]
54. Podsakoff, P.M.; MacKenzie, S.B.; Lee, J.-Y.; Podsakoff, N.P. Common Method Biases in Behavioral Research: A Critical Review of the Literature and Recommended Remedies. *J. Appl. Psychol.* **2003**, *88*, 879. [CrossRef]
55. Salancik, G.R.; Pfeffer, J. An Examination of Need-Satisfaction Models of Job Attitudes. *Adm. Sci. Q.* **1977**, *22*, 427–456. [CrossRef]
56. Hair, J.F., Jr.; Matthews, L.M.; Matthews, R.L.; Sarstedt, M. PLS-SEM or CB-SEM: Updated Guidelines on Which Method to Use. *Int. J. Multivar. Data Anal.* **2017**, *1*, 107. [CrossRef]
57. Do Valle, P.O.; Assaker, G. Using Partial Least Squares Structural Equation Modeling in Tourism Research: A Review of Past Research and Recommendations for Future Applications. *J. Travel Res.* **2016**, *55*, 695–708. [CrossRef]
58. Fornell, C.; Larcker, D.F. Structural Equation Models with Unobservable Variables and Measurement Error: Algebra and Statistics. *J. Mark. Res.* **1981**, *18*, 382–388. [CrossRef]
59. Anderson, J.C.; Gerbing, D.W. Structural Equation Modeling in Practice: A Review and Recommended Two-Step Approach. *Psychol. Bull.* **1988**, *103*, 411–423. [CrossRef]
60. Madi Odeh, R.B.; Obeidat, B.Y.; Jaradat, M.O.; Masa'deh, R.; Alshurideh, M.T. The Transformational Leadership Role in Achieving Organizational Resilience through Adaptive Cultures: The Case of Dubai Service Sector. *Int. J. Product. Perform. Manag.* **2023**, *72*, 440–468. [CrossRef]
61. Melián-Alzola, L.; Fernández-Monroy, M.; Hidalgo-Peñate, M. Hotels in Contexts of Uncertainty: Measuring Organisational Resilience. *Tour. Manag. Perspect.* **2020**, *36*, 100747. [CrossRef] [PubMed]
62. Brown, N.A.; Orchiston, C.; Rovins, J.E.; Feldmann-Jensen, S.; Johnston, D. An Integrative Framework for Investigating Disaster Resilience within the Hotel Sector. *J. Hosp. Tour. Manag.* **2018**, *36*, 67–75. [CrossRef]
63. Cheer, J.M.; Lew, A.A. *Tourism, Resilience and Sustainability: Adapting to Social, Political and Economic Change*; Routledge: London, UK, 2017.
64. Mandal, S.; Saravanan, D. Exploring the Influence of Strategic Orientations on Tourism Supply Chain Agility and Resilience: An Empirical Investigation. *Tour. Plan. Dev.* **2019**, *16*, 612–636. [CrossRef]
65. Hao, F.; Xiao, Q.; Chon, K. COVID-19 and China's Hotel Industry: Impacts, a Disaster Management Framework, and Post-Pandemic Agenda. *Int. J. Hosp. Manag.* **2020**, *90*, 102636. [CrossRef] [PubMed]
66. Srijuntrapun, P.; Fisher, D.; Rennie, H.G. Assessing the Sustainability of Tourism-Related Livelihoods in an Urban World Heritage Site. *J. Herit. Tour.* **2018**, *13*, 395–410. [CrossRef]
67. Zhang, G.; Chen, X.; Law, R.; Zhang, M. Sustainability of Heritage Tourism: A Structural Perspective from Cultural Identity and Consumption Intention. *Sustainability* **2020**, *12*, 9199. [CrossRef]
68. Giudici, E.; Melis, C.; Dessì, S.; Francine Pollnow Galvao Ramos, B. Is Intangible Cultural Heritage Able to Promote Sustainability in Tourism? *Int. J. Qual. Serv. Sci.* **2013**, *5*, 101–114.
69. Du Cros, H. A New Model to Assist in Planning for Sustainable Cultural Heritage Tourism. *J. Tour. Res.* **2001**, *3*, 165–170. [CrossRef]

Disclaimer/Publisher's Note: The statements, opinions and data contained in all publications are solely those of the individual author(s) and contributor(s) and not of MDPI and/or the editor(s). MDPI and/or the editor(s) disclaim responsibility for any injury to people or property resulting from any ideas, methods, instructions or products referred to in the content.

Article

Residents' Environmentally Responsible Behavior and Tourists' Sustainable Use of Cultural Heritage: Mediation of Destination Identification and Self-Congruity as a Moderator

Ibrahim A. Elshaer [1,2,*], Alaa M. S. Azazz [3,4,*] and Sameh Fayyad [2,5]

[1] Department of Management, College of Business Administration, King Faisal University, Al-Ahsaa 380, Saudi Arabia
[2] Hotel Studies Department, Faculty of Tourism and Hotels, Suez Canal University, Ismailia 41522, Egypt; sameh.fayyad@tourism.suez.edu.eg
[3] Department of Social Studies, Arts College, King Faisal University, Al-Ahsaa 380, Saudi Arabia
[4] Tourism Studies Department, Faculty of Tourism and Hotels, Suez Canal University, Ismailia 41522, Egypt
[5] Hotel Management Department, Faculty of Tourism and Hotels, October 6 University, Giza 12573, Egypt
* Correspondence: ielshaer@kfu.edu.sa or ibrahim_elshaaer@tourism.suez.edu.eg (I.A.E.); aazazz@kfu.edu.sa (A.M.S.A.)

Citation: Elshaer, I.A.; Azazz, A.M.S.; Fayyad, S. Residents' Environmentally Responsible Behavior and Tourists' Sustainable Use of Cultural Heritage: Mediation of Destination Identification and Self-Congruity as a Moderator. *Heritage* **2024**, *7*, 1174–1187. https://doi.org/10.3390/heritage7030056

Academic Editors: Fátima Matos Silva and Isabel Vaz de Freitas

Received: 6 February 2024
Revised: 19 February 2024
Accepted: 21 February 2024
Published: 23 February 2024

Copyright: © 2024 by the authors. Licensee MDPI, Basel, Switzerland. This article is an open access article distributed under the terms and conditions of the Creative Commons Attribution (CC BY) license (https:// creativecommons.org/licenses/by/ 4.0/).

Abstract: In the face of escalating global concerns surrounding environmental sustainability and the preservation of cultural heritage, this research explores the intricate connection between residents' environmentally responsible conduct (ERB) and tourists' sustainable involvement with cultural heritage sites (SU). Highlighting the pivotal importance of destination identification (DI) as a mediator and self-congruity (SC) as a moderator, our study utilizes a quantitative data approach to investigate the nuanced relationships inherent in the domain of tourism destinations. The data were collected from 324 tourists (visiting Luxor heritage city in Egypt) and analyzed by PLS-SEM, and the results showed a positive correlation between residents who strongly identify with their local environment and an increased dedication to environmentally responsible actions. Moreover, tourists who demonstrate elevated levels of self-congruity with the cultural heritage destination are inclined to embrace more sustainable behaviors, thereby making positive contributions to heritage preservation initiatives. This study enriches the evolving domain of sustainable tourism by providing insights into the intricate interactions between residents and tourists, fostering environmentally responsible behavior, and promoting the sustainable utilization of cultural heritage. Practical applications encompass the formulation of community-based interventions, the design of destination marketing strategies, and the proposal of policy recommendations. These initiatives aim to enhance the engagement of both residents and tourists, fostering the long-term preservation of cultural and environmental assets. Ultimately, the research seeks to guide sustainable tourism practices that strike a balance between the economic advantages of tourism and the essential preservation of cultural heritage and natural environments for future generations.

Keywords: resident; responsible behavior; heritage; destination identification; self-congruity; tourists' sustainability

1. Introduction

In recent years, there has been a noticeable surge in global interest towards sustainable tourism [1–3], driven by escalating concerns surrounding environmental preservation and the safeguarding of cultural heritage [4–7]. At the core of this discourse lies the conduct of residents and tourists within destination locales [6,8]. It is imperative to comprehend how residents engage in environmentally responsible behavior (ERB) and how tourists interact with cultural heritage sites in a sustainable manner to advance sustainable tourism practices. The participation of residents in ERB plays a pivotal role in shaping the overall sustainability of tourist destinations. Often serving as custodians of their local environment and cultural heritage, residents wield significant influence over tourist experiences and perceptions. Their attitudes and behaviors concerning environmental conservation and cultural preservation can have a profound impact on the sustainable development of tourism destinations [9,10].

Concurrently, the sustainable utilization of cultural heritage sites by tourists is imperative for their enduring preservation and enjoyment. Sustainable tourism entails minimizing adverse impacts on cultural and natural resources while offering meaningful experiences for visitors [11]. Tourists who embrace sustainable practices, such as honoring local traditions, reducing waste, and supporting indigenous communities, play a role in conserving cultural heritage and enhancing the welfare of destination residents [2]. Nevertheless, both the ERB of residents and the sustainable conduct of tourists are shaped by diverse psychological factors. One such factor is destination identification, which denotes the degree to which individuals feel connected to and identify with a specific destination [12]. Robust destination identification fosters a sense of affiliation and attachment, prompting individuals to adopt pro-environmental and pro-cultural preservation attitudes and behaviors [10,12]. Another psychological aspect is self-congruity, which concerns the alignment between an individual's self-image and the image of a destination or activity [13]. When tourists perceive a destination as congruent with their self-concept, they are more inclined to engage in behaviors that mirror the destination's values, including the sustainable use of cultural heritage sites [12,13].

Although these psychological factors are crucial, there has been insufficient research exploring their simultaneous impact on residents' ERB and tourists' sustainable behavior within destination settings. Hence, this study endeavors to explore the correlation between destination identification, self-congruity, residents' ERB, and tourists' sustainable utilization of cultural heritage. Through investigating these connections, this research aims to offer valuable insights into how destination-related psychological factors shape the behaviors of both residents and tourists. Ultimately, this endeavor seeks to advance the development of sustainable tourism practices and the conservation of cultural heritage.

2. Literature Review and Developing Hypotheses

2.1. Resident's Environmentally Responsible Behavior (ERB) and Tourists' Sustainable Use of Cultural Heritage (SU)

Residents, as opposed to visitors, are frequent users of the tourist destination setting and may, as a result, have a more consequential influence on the destination's natural environment and resources [14]. It follows that a sufficient emphasis on creating sustainable tourism destinations cannot be reached without considering the residents' responsible attitudes and actions [8]. While prior research has made a significant contribution to our understanding of tourists' sustainable behavior, few of these studies have specifically examined residents' ERBs [15–17]. Thus, studying the residents' ERB and its role in shaping proactively tourists' sustainable use of destination natural and heritage resources has become a subject worth examining. ERB is defined as behaviors that boost individuals' or groups' sustainable usage of natural and heritage resources [18]. In general, the literature has operationalized ERB as individual or group habits and collective activities that involve gaining knowledge and comprehension of environmental attitudes and responsibilities, which are the primary drivers of sustainable tourism development. Specifically, residents' ERBs refer to any activity residents do on a daily basis to limit harmful effects on their local environment or to preserve and safeguard the environment of destinations [19]. Consequently, residents demonstrating ERB will deliberately save resources and limit ecological harm [20].

On the other hand, research has shown that while trying to make tourism more environmentally sustainable, tourists are the primary target [21]. Tourists may contribute to lessening the adverse effects of tourism by choosing eco-friendly travel options and acting sustainably while at heritage sites and destinations [22]. Accordingly, the literature on the destination's sustainability focused on two primary topics: residents and tourists [23]. There is an agreement that residents can affect tourists' behavior [24,25]. Residents' ERBs arise due to their high community attachment and involvement [26] and feeling of psychological ownership [27], so they are more likely to protect the community's interest and go above and beyond the accepted societal norm to ensure the values and heritage of their community are undamaged [8]. Similarly, those residents likely act responsibly as ambassadors of the community in marketing its attractions and features to others by supporting sustainable tourism and motivating tourists to use cultural heritage sites sustainably [28,29]. In this context, tourists tend also to follow social norms and behave in the "usual" way because they think this is what "most people do". Therefore, it is arguable that the sustainability of cultural heritage sites greatly relies on the joint efforts of both residents and tourists. In light of the abovementioned literature on the linkage between residents' ERBs and tourists' SU, we can operate the Stimulus–Organism–Response (S–O–R) framework [30] to formulate the hypotheses of this study. The study suggests that residents' ERBs (stimuli) can affect tourists' SU (response) via destination identification (DI) and self-congruity (SC) of tourists (organism). Consequently, the first hypothesis is set for examination:

Hypothesis 1 (H1). *Resident's ERB positively affects tourists' sustainable use of cultural heritage (SU).*

2.2. Resident's ERB and Destination Identification (DI)

The identity concept, according to the social identity theory (SIT) proposed by Tajfel [31], is a person's feeling of belonging to a particular group. This concept has obtained vast concentration from management researchers with the development of this theory [27]. The identity concept was typically employed in consumer research to investigate the connections between people, brands, and associations [32]. Within the interaction framework between tourists and the tourist destination, destination identification serves as a psychological foundation for building a positive and symbolic link between them [33]. Tourists' DI refers to a physiological state in which visitors perceive likenesses between their self-identity and destination identification [34]. In addition to this definition, SIT states that individuals often formulate a social identity in addition to a personal identity in order to express their sense of self [35,36]. Therefore, studies confirmed that visitors may need to create self-identification with the destinations to achieve the requirement for self-identification in the context of tourism destinations [37]. Residents are often seen as the "spokespeople" of tourism destinations; therefore, their ERB activities may be a reference and base from which a visitor can comprehend a destination's values, duties, and positive identities [38]. In line with that, several studies have shown that travelers might strengthen their feeling of self-identification with tourism destinations that exhibit high levels of ERB which aligns with their own values [33,38]. In other words, tourists have a tendency to identify with those destinations whose principles and shared values are parallel to theirs [39]. Accordingly, the following hypothesis was postulated:

Hypothesis 2 (H2). *Residents' ERB positively affects destination identification (DI).*

2.3. Destination Identification (DI) and Tourists' Sustainable Use of Cultural Heritage (SU)

The tourists' psychological adhesive and attachment to a tourism destination are an identification result [40]. Along these same lines, earlier research has confirmed the link between love and destination identification [41,42]. In this regard, the researchers pointed out that the development and enhancement of positive sentiments towards tourism destinations are crucial for encouraging favorable visitor behavior [5,43] and resistance to a harmful attitude [44]. Thus, tourists who have these feelings resulting from destination identification may be more concerned about the destination environment and have favorable attitudes toward preserving this destination environment because they view themselves as part of it [23]. Some research has explicitly indicated that destination identification positively affects tourists' environmental concerns [45] and drives them to preserve the destination's environment, fight its degradation [46], and even take action and make endeavors to improve its environmental matters [47]. In the context of this framework, we additionally explore this connection between tourism destination identification and tourists' sustainable use of cultural heritage (SU) as follows:

Hypothesis 3 (H3). *Destination identification (DI) positively affects tourists' sustainable use of cultural heritage (SU).*

2.4. The Mediating Role of Destination Identification (DI)

Drawing on prior research and the arguments mentioned earlier that depict the direct linkages between residents' environmentally responsible behavior (ERB) and tourists' sustainable use of cultural heritage (SU) and destination identification (DI), and between destination identification (DI) and tourists' sustainable use of cultural heritage (SU), and in light of the Stimulus–Organism–Response (S–O–R) framework and the social identity theory (SIT), the following hypothesis of mediating effect is proposed:

Hypothesis 4 (H4). *Destination identification (DI) mediates the linkage between residents' ERB and tourists' sustainable use of cultural heritage (TSU).*

2.5. The Moderating Role of Self-Congruity (SC)

Customers purchase products or services for more than just their usefulness or functionality—they purchase objects with attached symbolic connotations [48]. Therefore, the self-concept, defined as "the totality of the individual's thoughts and feelings with reference to the self as an object" [49] played a crucial role in consumer behavior literature [50]. In research linked to self-concept, "self-congruity" was defined as "the combination or degree of alignment between the image of the product/brand and the self-concept of the consumer" [51]. According to the "self-congruity theory", customers tend to prefer those brands—tourism destinations—with personalities identical to theirs [52]. As a result, customers' ties, identification, and loyalty to a destination are now predicated on this congruity [53]. In line with that, self-congruity was presented in the tourism framework as the degree to which the destination brand and the customers' self-concept correspond [52]. Elevated self-congruence can potentially improve customers' perceptions of the brand [54] and foster a stronger emotional attachment to it [55]. Therefore, the self-congruity concept was operated as a moderating variable in several investigations on customer attitudes and behaviors in the tourism field [6,56,57]. In relating a customer's self-congruity to sustainability, the tourist compares his personality and values related to sustainability with the destination's personality and values related to sustainability; if compatible, the tourist's destination identification and sustainable behaviors and attitudes towards the destination are likely to be enhanced [56]. Therefore, this study argues that the self-congruity (SC) can strengthen the positive effects of residents' ERB and destination identification (DI) in our proposed model, and, thus, we suggest the following hypotheses:

Hypothesis 5 (H5). *Self-congruity (SC) moderates the linkage between residents' ERB and destination identification (DI).*

Hypothesis 6 (H6). *Self-congruity (SC) moderates the linkage between destination identification (DI) and tourists' sustainable use of cultural heritage (TSU).*

Figure 1 graphically presents the conceptual model of this investigation in light of the previously listed literature and hypotheses.

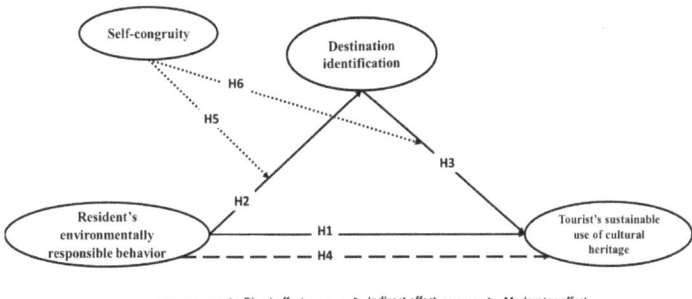

Figure 1. Study hypotheses and conceptual model.

3. Methods

3.1. Measures

All measuring items were adapted from previous literature. Su et al.'s [19] six items were adapted to estimate the residents' ERB variable. The destination identification (DI) was measured by employing a three-item scale developed by Su and Swanson [33]. To assess the construct of tourists' sustainable use of cultural heritage (SU), nine items were adopted from the study of Alazaizeh et al. [4]. Finally, the self-congruity variable was measured by four items from Frias et al. [51] (as shown Appendix A). Twelve academic specialists and ten administrators in cultural heritage tourism evaluated the survey items' validity, and slight changes were made, resulting in paraphrasing some questionnaire statements. Additionally, in order to eliminate the potential for instrument bias, Harman's single-factor test was also employed in this investigation. Given that Harman's single factor value is less than 50%, the result indicates that the single factor retrieved is 24%, suggesting that there are no bias concerns with the current [58].

3.2. Data Collection

The questionnaires were distributed among tourists in Luxor City in Egypt, using convenient samples and drop-off and pick-up approaches. Luxor City offers an exceptional blend of live cultural heritage and archaeological sites, making it a viable choice for our study location. Luxor, in ancient times, called Thebes, was the southern capital of Egypt since the Middle Kingdom. The city is globally renowned for its monuments, which span the Middle and New Kingdoms and include impressive temples, tombs, and towns, as well as archaeological remains from the later Graeco-Roman, Coptic, and Islamic periods [7]. Tourists voluntarily filled out the survey, and their responses were kept private. The surveys were conducted with the assistance of our colleagues registered in our faculty's postgraduate programs and working in Luxor hotels. The data were collected during the winter months (November (2023) to January (2024), during the tourist season in Luxor. The survey was completed by 356 out of 500 targeted tourists, and 324 responses were considered valid after removing 24 unqualified, with a comeback rate of 66.4%. The study sample included 193 males (58.1%) and 139 females (41.9%). The participants' ages ranged between 21 and 56. Also, 269 respondents (81%) had a college degree, followed by 34 respondents (10.4%) with graduate degrees. The questionnaire is structured into five primary sections. The first part focuses on demographic details such as age, gender, and education level. Following this, the second section gathers data on tourists' sustainable practices at heritage sites. The third part addresses inquiries related to residents' environmentally responsible behavior. Subsequently, the fourth section explores aspects of destination identification. Finally, the fifth section comprises items concerning moderating variables, specifically measuring self-congruity. All survey items, except demographic information, were estimated using five-point Likert scales.

3.3. Data Analysis

For a number of reasons, PLS-SEM was carried out using Smart PLS 4 software to evaluate the proposed model. In the first place, this approach facilitates the researchers' assessment of links among factors in the inner model and their associated latent indicators in the outer model. Second, PLS-SEM works well with intricate research models, especially those that include moderation and mediation. Third, PLS features an easier-to-use user graphical interface than other path modelling tools like AMOS. Fourthly, this method is a reliable component-based strategy that has been used widely in earlier research [59]. This method is a two-stage analysis strategy; the measurement (outer) model's validity and reliability are examined in the first stage, and the structural (inner) model is assessed in the second stage to test the proposed hypotheses [60].

4. Results

4.1. The Measurement Model

The measurement (outer) model was evaluated before the hypotheses were tested. Fit indices widely utilized in CB-SEM are unavailable or ill-advised for PLS-SEM because they adopt a different SEM technique [61]. According to Hair et al. [59], the fit of the PLS-SEM model can be evaluated by operating the ensuing standards: loadings of study's items (λ), Cronbach's alpha (a), composite reliability (CR) test, the required cut-offs of all are <0.70, and the threshold of Average Variance Extracted (AVE) must reach 0.50 to achieve Convergent Validity (CV) of the outer model. Regarding the model discriminant validity, the AVE of each variable must be greater than the squared inter-construction correlations [62].

As depicted in Table 1, the outer model satisfies all thresholds of a good Convergent Validity, validating the internal study model's reliability—that is, the consistency of responses to items belonging to the same factor. The AVEs ranged from 0.667 (Residents' ERB to 0.806 (Destination identification (DI)), exhibiting a strong correlation between the items in each factor, and also confirming the model's Convergent Validity. Additionally, Table 2 supports the recommended model's discriminant validity because all AVEs are higher than their related squared inter-construction correlations (Fornell and Larcker, 1981) [62]. This indicated that every factor stood out independently from the rest. In addition, in response to the numerous criticisms of Fornell and Lacker's criterion, some studies examined the Heterotrait–Monotriat ratio of correlation (HTMT) test to confirm the discriminant validity. Table 2 also shows that the discriminant validity is appropriate because all HTMT values are <0.90 [63]. Also, Table 3 exhibits that an item's loading within its factors is higher than any cross-loadings with other factors, guaranteeing the discriminant validity.

Table 1. Psychometric results.

Factors and Items	Loading λ	(a Value)	(C_R)	(AVE)
Residents'ERB		0.898	0.923	0.667
ERB_1	0.857			
ERB_2	0.839			
ERB_3	0.890			
ERB_4	0.776			
ERB_5	0.841			
ERB_6	0.701			
Tourists' sustainable use of cultural heritage (SU)		0.947	0.956	0.730
TSU_1	0.847			
TSU_2	0.826			
TSU_3	0.874			
TSU_4	0.877			
TSU_5	0.874			

Table 1. Cont.

Factors and Items	Loading λ	(a Value)	(C_R)	(AVE)
TSU_6	0.833			
TSU_7	0.857			
TSU_8	0.844			
Destination identification (DI)		0.920	0.943	0.806
DI_1	0.884			
DI_2	0.903			
DI_3	0.892			
DI_4	0.912			
Self-congruity (SC)		0.896	0.928	0.763
SC_1	0.816			
SC_2	0.893			
SC_3	0.881			
SC_4	0.901			

Table 2. "Fornell–Larcker criterion matrix" and HTMT Matrix.

	Fornell–Larcker Criterion Matrix				HTMT Matrix.			
	DI	ERB	(SC)	SU	DI	ERB	(SC)	SU
Destination identification (DI)	0.898							
Residents' ERB	0.553	0.817			0.605			
Self-congruity (SC)	0.578	0.583	0.873		0.632	0.646		
Tourists' SU	0.693	0.587	0.530	0.854	0.736	0.632	0.570	

Note: "Values off the diagonal-line are squared inter-construction-correlations, while values on the diagonal-line are AVEs, and for appropriate DV, all HTMT values need to be <0.90".

Table 3. Factor cross-loadings.

	Residents' ERB	Tourists' SU	Destination Identification (DI)	Self-Congruity (SC)
ERB_1	**0.857**	0.514	0.527	0.538
ERB_2	**0.839**	0.550	0.407	0.460
ERB_3	**0.890**	0.426	0.473	0.469
ERB_4	**0.776**	0.380	0.459	0.437
ERB_5	**0.841**	0.442	0.466	0.448
ERB_6	**0.682**	0.541	0.364	0.491
TSU_1	0.571	**0.847**	0.600	0.499
TSU_2	0.499	**0.826**	0.558	0.305
TSU_3	0.459	**0.874**	0.724	0.458
TSU_4	0.516	**0.877**	0.644	0.523
TSU_5	0.496	**0.874**	0.629	0.453
TSU_6	0.452	**0.833**	0.530	0.363
TSU_7	0.469	**0.857**	0.491	0.467
TSU_8	0.545	**0.844**	0.519	0.535
DI_1	0.565	0.612	**0.884**	0.595
DI_2	0.489	0.624	**0.903**	0.513
DI_3	0.470	0.637	**0.892**	0.485
DI_4	0.453	0.613	**0.912**	0.473
SC_1	0.517	0.426	0.463	**0.816**
SC_2	0.550	0.475	0.536	**0.893**
SC_3	0.425	0.422	0.466	**0.881**
SC_4	0.538	0.520	0.545	**0.901**

4.2. Structural Model and Testing Hypotheses

Given that PLS-SEM lacks the standard fit criteria that CB-SEM does, an inner model must be evaluated using the Variance Inflation Factor (VIF), R2, Q2, and standardized path coefficients using Beta value [59]. For the likelihood of "multi-collinearity" among constructs to be ruled out, VIFs must be less than 5.0 for items, R2 must fulfill norms for the academic area and study situation, standardized path coefficients (p) must be significant, and the Q2 scores also fulfilled the suggested point value of 0.0 [61].

As presented in Table 4, VIFs ranged between 1.532 and 4.111, below the cut-off value. Thus, no multi-collinearity issues exist, which allowed the independent variables' effects on the dependent variables to be separated from one another because there was no substantial correlation between them. As for R2 estimates, tourist's SU displayed a value of 0.562, implying that the remaining constructs in the structural proposed model accounted for 56.0% of the variation in tourist's SU. Similarly, the destination identification (DI)'s R2 was 0.423, satisfying cut-off (0.10 or greater). Q2 exceeded the recommended threshold of 0.0. Additionally, at $p = 0.01$ level, all standardized path coefficients were statistically significant (Table 5). When these criteria were considered collectively, it proved how well the structural model suited the data.

Table 4. VIF, R^2, and Q^2 outcomes.

Name	VIF	Name	VIF	Name	VIF	Name	VIF	Name	VIF
RER_1	2.951	RER_6	1.532	TSU_5	3.825	DI_2	3.181	SC_3	2.973
RER_2	2.658	TSU_1	3.258	TSU_6	3.516	DI_3	3.296	SC_4	3.047
RER_3	3.701	TSU_2	2.792	TSU_7	4.111	DI_4	3.683		
RER_4	2.819	TSU_3	3.793	TSU_8	3.423	SC_1	1.976		
RER_5	3.088	TSU_4	3.854	DI_1	2.816	SC_2	2.760		
Tourist's SU				R2	0.562	Q2	0.376		
Destination identification (DI)				R2	0.423	Q2	0.313		

Table 5. Hypotheses results.

Paths	β Value	t Value	p Value	Result
Direct Paths				
H1—Residents' ERB → Tourists' SU	0.207	2.125	0.034	"Supported"
H2—Residents' ERB → Destination identification (DI)	0.404	5.149	0.000	"Supported"
H3—Destination identification (DI) → Tourists' SU	0.534	8.756	0.000	"Supported"
Indirect Mediating Paths				
H4—Residents' ERB → Destination identification (DI) → Tourists' SU	0.216	3.989	0.000	"Supported"
Moderating Effects				
H5—Residents' ERB × Self-congruity (SC) → Destination identification (DI)	0.146	2.410	0.016	"Supported"
H6—Destination identification (DI) × Self-congruity → Tourists' SU	0.150	2.870	0.004	"Supported"

Additionally, according to Tenenhaus et al. [64], the subsequent equation was presented to assess the Goodness of Fit (GoF) of the PLS-SEM model, where values of 0.1, 0.25, and 0.36 represent low, medium, and high GoF levels, respectively. The GoF index for the proposed model is calculated as 0.604, signifying a high level of Goodness of Fit.

$$GoF = \sqrt{AVE_{avy} \times R^2_{avy}}$$

Using Smart PLS 4, we conducted a bootstrapping procedure with 5000 iterations to test the hypotheses presented in Table 5, following the validation of both the outer and inner models.

The data shown in Table 5 and Figure 2, as extracted from Smart PLS 4.0, indicate that the residents' ERB positively affected tourist's SU and destination identification (DI) at

($\beta = 0.376$, t = 7.918, $p = 0.034$) and ($\beta = 0.404$, t = 5.149, $p = 0.000$), respectively, supporting H1 and H2. The destination identification (DI) positively impacted tourist's SU at $\beta = 0.534$, t = 8.756, and $p = 0.000$, thus proving H3. Additionally, at $\beta = 0.216$, t = 3.989 and $p = 0.000$, the destination identification (DI) mediates the link between the residents' ERB and tourist's SU, thus proving H4.

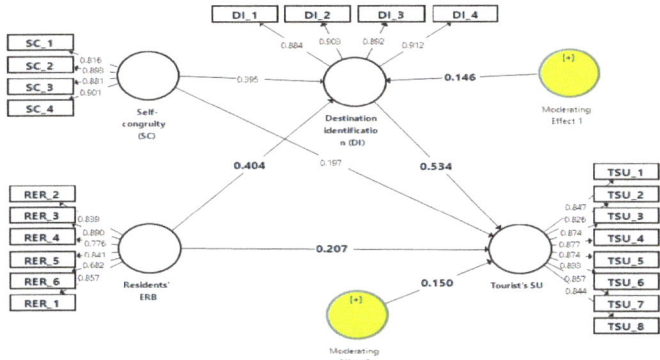

Figure 2. Estimation of structure model.

Regarding moderating effects, Figure 3 illustrates that self-congruity strengthened the impact of the residents' ERB on destination identification, thus proving H5. Similarly, as shown in Figure 4, self-congruity strengthened the impact of the destination identification on Tourist's SU, thus supporting H6.

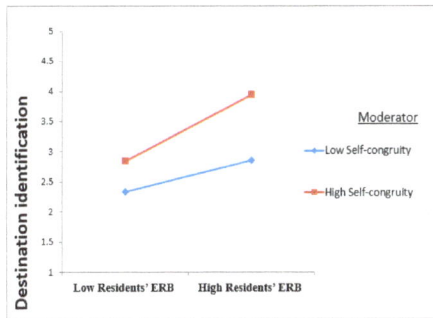

Figure 3. Moderating influence of self-congruity (SC) on residents' ERB toward destination identification (DI).

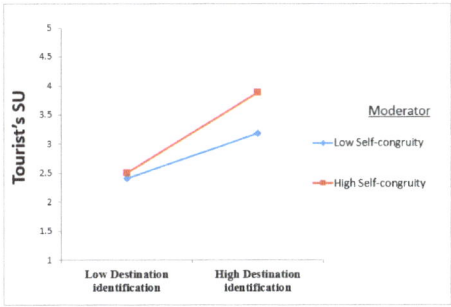

Figure 4. Moderating influence of self-congruity on destination identification (DI) toward Tourists' SU.

5. Discussion and Implication

The discussion section thoroughly explores the implications of the study's discoveries concerning the intricate relationship among residents' ERB, tourists' sustainable utilization of cultural heritage, and the mediating effect of destination identification between residents' ERB and tourists' SU, along with the moderating impact of self-congruity. The study's outcomes unveil noteworthy connections between residents' ERB, tourists' sustainable conduct, destination identification, and self-congruity. The study underscores the importance applications of residents' ERB on tourists' sustainable use of cultural heritage. This result is consistent with [8,22,23,65]. Encouraging residents to participate in decision-making processes and nurturing a feeling of ownership regarding their surroundings enables destination managers to foster a sustainable culture that influences every aspect of the tourism journey. Strategies like community-based tourism endeavors, educational initiatives on environmental matters, and collaborations between local administrations and residents empower communities to actively preserve their cultural heritage while advocating for sustainable tourism. Educational programs and awareness campaigns serve as vital tools in encouraging residents' ERB and promoting tourists' sustainable conduct. By enhancing understanding of the significance of environmental preservation and cultural heritage protection [66–68], destination stakeholders can garner backing from both residents and tourists. These efforts may involve organizing workshops, seminars, and installing interpretive signs at cultural heritage sites to underscore the importance of responsible tourism practices and urge visitors to minimize their environmental footprint. The mediation analysis indicates that residents with a strong identification with their destination are more inclined to practice ERB, thereby influencing tourists to embrace sustainable practices when exploring cultural heritage sites. Furthermore, the moderation analysis suggests that tourists who perceive a strong alignment between themselves and the destination are more likely to demonstrate sustainable behavior at cultural heritage sites.

The results of this study carry significant policy implications for policymakers, and stakeholders in the tourism industry. Implementing policies and regulations that encourage ERB among both residents and tourists can foster the long-term sustainability of tourism destinations. Initiatives such as waste management regulations, carbon offset programs, and eco-certification schemes can serve as incentives for businesses and individuals to embrace sustainable practices. Moreover, sustainable tourism policies should prioritize community involvement, cultural awareness, and environmental stewardship to safeguard cultural heritage for generations to come. The study enhances destination image theory by underscoring the significant role of destination identification in shaping the behaviors of both residents and tourists. It expands our comprehension of self-congruity theory by emphasizing its moderating influence on the relationship between destination identification and tourists' sustainable behavior. These theoretical insights provide valuable understanding of the psychological mechanisms that underpin sustainable tourism and the preservation of cultural heritage. For policymakers, these findings highlight the criticality of nurturing a sense of destination identification among residents through initiatives such as community engagement programs, cultural festivities, and environmental education endeavors. Strengthening residents' attachment to their destination can lead to heightened support for sustainable tourism endeavors and increased participation in ERB. Moreover, destination marketers can capitalize on self-congruity by tailoring promotional messages and experiences to align with tourists' self-perceptions. This approach can bolster tourists' motivation to partake in sustainable practices when visiting cultural heritage sites, thereby fostering a more sustainable tourism environment.

6. Limitations and Future Research

The current study, while providing valuable insights into the correlation between residents' ERB, tourists' sustainable utilization of cultural heritage, and the mediating function of destination identification, along with the moderating impact of self-congruity, is subject to certain limitations. These limitations are crucial to acknowledge as they furnish context for interpreting the findings and suggest avenues for future research in this domain. Initially, the study relied on self-reported data from both residents and tourists, which could introduce common method bias and social desirability bias. Future investigations could enhance reliability by integrating objective measures or observational data to validate self-reported results. Secondly, the study adopted a cross-sectional design, limiting the ability to establish causality between the variables examined. Longitudinal or experimental designs would enable a more comprehensive examination. Moreover, the study focused on a specific geographic area or destination (Egypt/Luxor city), potentially restricting the generalizability of the findings to other settings. Different destinations may possess unique socio-cultural, economic, and environmental characteristics that influence residents' and tourists' behaviors divergently. Future research endeavors could explore these relationships across various destinations to augment the applicability of the findings. Furthermore, while the study explored the mediating role of destination identification and the moderating influence of self-congruity, additional psychological factors could influence residents' and tourists' behaviors in sustainable tourism contexts. Subsequent research could investigate supplementary variables such as environmental attitudes, perceived behavioral control, and social norms to attain a more holistic understanding of the mechanisms underlying sustainable tourism practices. In addition, future studies, in the context of the variables of the current study, could conduct a comparative study between new visitors and returning visitors, as well as between visitors who have knowledge and those who do not have knowledge about the heritage tourist site.

Author Contributions: Conceptualization, I.A.E. and S.F.; methodology, I.A.E., S.F. and A.M.S.A.; software, I.A.E. and S.F.; validation, I.A.E., A.M.S.A. and S.F.; formal analysis, I.A.E. and A.M.S.A.; investigation, I.A.E., S.F. and A.M.S.A.; resources, I.A.E.; data curation, I.A.E.; writing—original draft preparation, S.F., I.A.E. and A.M.S.A.; writing—review and editing, I.A.E., S.F. and A.M.S.A.; visualization, I.A.E.; supervision, I.A.E.; project administration, I.A.E., S.F. and A.M.S.A.; funding acquisition, I.A.E. and A.M.S.A. All authors have read and agreed to the published version of the manuscript.

Funding: This work was supported by the Deanship of Scientific Research, Vice Presidency for Graduate Studies and Scientific Research, King Faisal University, Saudi Arabia [Grant No. 5886].

Data Availability Statement: Data is available upon request from researchers who meet the eligibility criteria. Kindly contact the first author privately through e-mail.

Conflicts of Interest: The authors declare no conflict of interest.

Appendix A

Residents' environmentally responsible behavior

- Residents comply with relevant regulations to not destroy the destination's environment.
- Residents are willing to attend environmental cleaning activities.
- Residents try to convince partners to protect the environment of the place.
- Residents try not to disrupt the heritage features of the destination.
- Residents attach importance to environmental protection.
- Residents try to guide tourists to engage in the protection of the place's heritage.

Destination identification

- I am very interested in what others think about the destination.
- The successes of this destination are my successes.
- When someone praises this destination, it feels like a personal compliment.
- When someone criticizes this destination, I feel embarrassed.

Self-congruity

- When someone criticizes this destination, I feel embarrassed.
- The personality of this destination is consistent with how I see myself.

- The personality of this destination is a mirror image of me.
- The personality of this destination is consistent with how I would like to be.
- The personality of this destination is a mirror image of the person I would like to be.

Tourist's sustainable use of cultural heritage

- I accept the control policy not to enter the sensitive sites.
- I comply with the rules so as to not harm the site environment.
- I help to maintain the local environment quality.
- I report to the site administration any environmental pollution or destruction.
- I spend money in the local area.
- I help other tourists to learn about the site.
- I try to convince others to protect the natural and cultural environment at the site.
- Donating money for protection the site

References

1. Bramwell, B. Theoretical Activity in Sustainable Tourism Research. *Ann. Tour. Res.* **2015**, *54*, 204–218. [CrossRef]
2. Buckley, R. Sustainable Tourism: Research and Reality. *Ann. Tour. Res.* **2012**, *39*, 528–546. [CrossRef]
3. Bramwell, B.; Lane, B. The "Critical Turn" and Its Implications for Sustainable Tourism Research. *J. Sustain. Tour.* **2014**, *22*, 1–8. [CrossRef]
4. Alazaizeh, M.M.; Jamaliah, M.M.; Mgonja, J.T.; Ababneh, A. Tour Guide Performance and Sustainable Visitor Behavior at Cultural Heritage Sites. *J. Sustain. Tour.* **2019**, *27*, 1708–1724. [CrossRef]
5. Elshaer, I.A.; Azazz, A.M.; Fayyad, S. Authenticity, Involvement, and Nostalgia in Heritage Hotels in the Era of Digital Technology: A Moderated Meditation Model. *Int. J. Environ. Res. Public Health* **2022**, *19*, 5784. [CrossRef] [PubMed]
6. Elshaer, I.A.; Fayyad, S.; Ammar, S.; Abdulaziz, T.A.; Mahmoud, S.W. Adaptive Reuse of Heritage Houses and Hotel Conative Loyalty: Digital Technology as a Moderator and Memorable Tourism and Hospitality Experience as a Mediator. *Sustainability* **2022**, *14*, 3580. [CrossRef]
7. Hassan, F. Heritage Tourism Management Plan for Luxor, Egypt. In *Heritage and Cultural Heritage Tourism*; Yu, P.-L., Lertcharnrit, T., Smith, G.S., Eds.; Springer: Cham, Switzerland, 2023; pp. 123–134. ISBN 978-3-031-44799-0.
8. Safshekan, S.; Ozturen, A.; Ghaedi, A. Residents' Environmentally Responsible Behavior: An Insight into Sustainable Destination Development. *Asia Pac. J. Tour. Res.* **2020**, *25*, 409–423. [CrossRef]
9. Jurowski, C.; Uysal, M.; Williams, D.R. A Theoretical Analysis of Host Community Resident Reactions to Tourism. *J. Travel Res.* **1997**, *36*, 3–11. [CrossRef]
10. Ramkissoon, H.; Weiler, B.; Smith, L.D.G. Place Attachment and Pro-Environmental Behaviour in National Parks: The Development of a Conceptual Framework. *J. Sustain. Tour.* **2012**, *20*, 257–276. [CrossRef]
11. UNWTO Sustainable Development | UN Tourism. Available online: https://www.unwto.org/sustainable-development (accessed on 6 February 2024).
12. Kim, S.S.; Lee, C.-K.; Klenosky, D.B. The Influence of Push and Pull Factors at Korean National Parks. *Tour. Manag.* **2003**, *24*, 169–180. [CrossRef]
13. Sirgy, M.J. Self-Concept in Consumer Behavior: A Critical Review. *J. Consum. Res.* **1982**, *9*, 287–300. [CrossRef]
14. Briassoulis, H. Sustainable Tourism and the Question of the Commons. *Ann. Tour. Res.* **2002**, *29*, 1065–1085. [CrossRef]
15. Chen, R.; Zhou, Z.; Zhan, G.; Zhou, N. The Impact of Destination Brand Authenticity and Destination Brand Self-Congruence on Tourist Loyalty: The Mediating Role of Destination Brand Engagement. *J. Destin. Mark. Manag.* **2020**, *15*, 100402. [CrossRef]
16. Lee, J.S.-H.; Oh, C.-O. The Causal Effects of Place Attachment and Tourism Development on Coastal Residents' Environmentally Responsible Behavior. *Coast. Manag.* **2018**, *46*, 176–190. [CrossRef]
17. Su, L.; Huang, S.; Pearce, J. Toward a Model of Destination Resident–Environment Relationship: The Case of Gulangyu, China. *J. Travel Tour. Mark.* **2019**, *36*, 469–483. [CrossRef]
18. Sivek, D.J.; Hungerford, H. Predictors of Responsible Behavior in Members of Three Wisconsin Conservation Organizations. *J. Environ. Educ.* **1990**, *21*, 35–40. [CrossRef]
19. Su, L.; Huang, S.S.; Pearce, J. How Does Destination Social Responsibility Contribute to Environmentally Responsible Behaviour? A Destination Resident Perspective. *J. Bus. Res.* **2018**, *86*, 179–189. [CrossRef]
20. Cheng, T.-M.; Wu, H.C. How Do Environmental Knowledge, Environmental Sensitivity, and Place Attachment Affect Environmentally Responsible Behavior? An Integrated Approach for Sustainable Island Tourism. *J. Sustain. Tour.* **2015**, *23*, 557–576. [CrossRef]
21. Juvan, E.; Dolnicar, S. Measuring Environmentally Sustainable Tourist Behaviour. *Ann. Tour. Res.* **2016**, *59*, 30–44. [CrossRef]
22. Shen, S.; Sotiriadis, M.; Zhou, Q. Could Smart Tourists Be Sustainable and Responsible as Well? The Contribution of Social Networking Sites to Improving Their Sustainable and Responsible Behavior. *Sustainability* **2020**, *12*, 1470. [CrossRef]
23. Hu, J.; Xiong, L.; Lv, X.; Pu, B. Sustainable Rural Tourism: Linking Residents' Environmentally Responsible Behaviour to Tourists' Green Consumption. *Asia Pac. J. Tour. Res.* **2021**, *26*, 879–893. [CrossRef]

24. Wang, C.; Zhang, J.; Cao, J.; Duan, X.; Hu, Q. The Impact of Behavioral Reference on Tourists' Responsible Environmental Behaviors. *Sci. Total Environ.* **2019**, *694*, 133698. [CrossRef] [PubMed]
25. Wang, W.; Wu, J.; Wu, M.-Y.; Pearce, P.L. Shaping Tourists' Green Behavior: The Hosts' Efforts at Rural Chinese B&Bs. *J. Destin. Mark. Manag.* **2018**, *9*, 194–203. [CrossRef]
26. Daryanto, A.; Song, Z. A Meta-Analysis of the Relationship between Place Attachment and pro-Environmental Behaviour. *J. Bus. Res.* **2021**, *123*, 208–219. [CrossRef]
27. Zhang, H.; Cheng, Z.; Chen, X. How Destination Social Responsibility Affects Tourist Citizenship Behavior at Cultural Heritage Sites? Mediating Roles of Destination Reputation and Destination Identification. *Sustainability* **2022**, *14*, 6772. [CrossRef]
28. Stylidis, D. Place Attachment, Perception of Place and Residents' Support for Tourism Development. *Tour. Plan. Dev.* **2018**, *15*, 188–210. [CrossRef]
29. Theodori, G.L. Reexamining the Associations among Community Attachment, Community-Oriented Actions, and Individual-Level Constraints to Involvement. *Community Dev.* **2018**, *49*, 101–115. [CrossRef]
30. Mehrabian, A.; Russell, J.A. *An Approach to Environmental Psychology*; The MIT Press: Cambridge, MA, US, 1974.
31. Tajfel, H. Social categorization, social identity and social comparison. In *Differentiation between Social Groups*; Tajfel, H., Ed.; Academic Press: London, UK, 1978; pp. 61–76.
32. Lichtenstein, D.R.; Netemeyer, R.G.; Maxham, J.G., III. The Relationships among Manager-, Employee-, and Customer-Company Identification: Implications for Retail Store Financial Performance. *J. Retail.* **2010**, *86*, 85–93. [CrossRef]
33. Su, L.; Swanson, S.R. The Effect of Destination Social Responsibility on Tourist Environmentally Responsible Behavior: Compared Analysis of First-Time and Repeat Tourists. *Tour. Manag.* **2017**, *60*, 308–321. [CrossRef]
34. Su, L.; Huang, Y. How Does Perceived Destination Social Responsibility Impact Revisit Intentions: The Mediating Roles of Destination Preference and Relationship Quality. *Sustainability* **2018**, *11*, 133. [CrossRef]
35. Brewer, M.B. The Social Self: On Being the Same and Different at the Same Time. *Personal. Soc. Psychol. Bull.* **1991**, *17*, 475–482. [CrossRef]
36. Ashforth, B.E.; Schinoff, B.S. Identity under construction: How individuals come to define themselves in organizations. *Annu. Rev. Organ. Psychol. Organ. Behav.* **2016**, *3*, 111–137. [CrossRef]
37. Ekinci, Y.; Sirakaya-Turk, E.; Preciado, S. Symbolic Consumption of Tourism Destination Brands. *J. Bus. Res.* **2013**, *66*, 711–718. [CrossRef]
38. Tran, P.K.T.; Nguyen, H.K.T.; Nguyen, L.T.; Nguyen, H.T.; Truong, T.B.; Tran, V.T. Destination Social Responsibility Drives Destination Brand Loyalty: A Case Study of Domestic Tourists in Danang City, Vietnam. *Int. J. Tour. Cities* **2023**, *9*, 302–322. [CrossRef]
39. Scott, S.G.; Lane, V.R. A Stakeholder Approach to Organizational Identity. *Acad. Manag. Rev.* **2000**, *25*, 43. [CrossRef]
40. Hultman, M.; Skarmeas, D.; Oghazi, P.; Beheshti, H.M. Achieving Tourist Loyalty through Destination Personality, Satisfaction, and Identification. *J. Bus. Res.* **2015**, *68*, 2227–2231. [CrossRef]
41. Albert, N.; Merunka, D. The Role of Brand Love in Consumer-Brand Relationships. *J. Consum. Mark.* **2013**, *30*, 258–266. [CrossRef]
42. Alnawas, I.; Altarifi, S. Exploring the Role of Brand Identification and Brand Love in Generating Higher Levels of Brand Loyalty. *J. Vacat. Mark.* **2016**, *22*, 111–128. [CrossRef]
43. Amaro, S.; Barroco, C.; Antunes, J. Exploring the Antecedents and Outcomes of Destination Brand Love. *J. Prod. Brand Manag.* **2020**, *30*, 433–448. [CrossRef]
44. Swanson, K.; Medway, D.; Warnaby, G. 'I Love This Place': Tourists' Destination Brand Love. In *Handbook on Place Branding and Marketing*; Edward Elgar Publishing: Cheltenham, UK, 2017; pp. 88–107. [CrossRef]
45. Podeschi, C.W.; Howington, E.B. Place, Sprawl, and Concern about Development and the Environment. *Sociol. Spectr.* **2011**, *31*, 419–443. [CrossRef]
46. Vorkinn, M.; Riese, H. Environmental Concern in a Local Context: The Significance of Place Attachment. *Environ. Behav.* **2001**, *33*, 249–263. [CrossRef]
47. Scannell, L.; Gifford, R. Personally Relevant Climate Change: The Role of Place Attachment and Local Versus Global Message Framing in Engagement. *Environ. Behav.* **2013**, *45*, 60–85. [CrossRef]
48. Park, C.W.; Macinnis, D.J.; Priester, J.; Eisingerich, A.B.; Iacobucci, D. Brand Attachment and Brand Attitude Strength: Conceptual and Empirical Differentiation of Two Critical Brand Equity Drivers. *J. Mark.* **2010**, *74*, 1–17. [CrossRef]
49. Rosenberg, M. Self-Concept Research: A Historical Overview. *Soc. Forces* **1989**, *68*, 34–44. [CrossRef]
50. Üner, M.M.; Armutlu, C. Understanding the Antecedents of Destination Identification: Linkage Between Perceived Quality-of-Life, Self-Congruity, and Destination Identification. In *Handbook of Tourism and Quality-of-Life Research: Enhancing the Lives of Tourists and Residents of Host Communities*; International Handbooks of Quality-of-Life; Uysal, M., Perdue, R., Sirgy, M.J., Eds.; Springer: Dordrecht, The Netherlands, 2012; pp. 251–261. ISBN 978-94-007-2288-0.
51. Frias, D.M.; Castañeda, J.-A.; Del Barrio-García, S.; López-Moreno, L. The Effect of Self-Congruity and Motivation on Consumer-Based Destination Brand Equity. *J. Vacat. Mark.* **2020**, *26*, 287–304. [CrossRef]
52. Usakli, A.; Baloglu, S. Brand Personality of Tourist Destinations: An Application of Self-Congruity Theory. *Tour. Manag.* **2011**, *32*, 114–127. [CrossRef]
53. Li, Z.; Zhang, J. How to Improve Destination Brand Identification and Loyalty Using Short-Form Videos? The Role of Emotional Experience and Self-Congruity. *J. Destin. Mark. Manag.* **2023**, *30*, 100825. [CrossRef]

54. Aaker, J.L. The Malleable Self: The Role of Self-Expression in Persuasion. *J. Mark. Res.* **1999**, *36*, 45–57. [CrossRef]
55. Malär, L.; Krohmer, H.; Hoyer, W.D.; Nyffenegger, B. Emotional Brand Attachment and Brand Personality: The Relative Importance of the Actual and the Ideal Self. *J. Mark.* **2011**, *75*, 35–52. [CrossRef]
56. Eastman, J.K.; Iyer, R.; Dekhili, S. Can Luxury Attitudes Impact Sustainability? The Role of Desire for Unique Products, Culture, and Brand Self-congruence. *Psychol. Mark.* **2021**, *38*, 1881–1894. [CrossRef]
57. Kumar, V.; Kaushik, A.K. Achieving Destination Advocacy and Destination Loyalty through Destination Brand Identification. *J. Travel Tour. Mark.* **2017**, *34*, 1247–1260. [CrossRef]
58. Podsakoff, P.M.; Organ, D.W. Self-Reports in Organizational Research: Problems and Prospects. *J. Manag.* **1986**, *12*, 531–544. [CrossRef]
59. Hair, J.F.; Risher, J.J.; Sarstedt, M.; Ringle, C.M. When to Use and How to Report the Results of PLS-SEM. *Eur. Bus. Rev.* **2019**, *31*, 2–24. [CrossRef]
60. Leguina, A. A Primer on Partial Least Squares Structural Equation Modeling (PLS-SEM). *Int. J. Res. Method Educ.* **2015**, *38*, 220–221. [CrossRef]
61. Hair, J., Jr.; Hair, J.F., Jr.; Hult, G.T.M.; Ringle, C.M.; Sarstedt, M. *A Primer on Partial Least Squares Structural Equation Modeling (PLS-SEM)*; Sage Publications: New York, NY, USA, 2021.
62. Fornell, C.; Larcker, D.F. Evaluating Structural Equation Models with Unobservable Variables and Measurement Error. *J. Mark. Res.* **1981**, *18*, 39–50. [CrossRef]
63. Gold, A.H.; Malhotra, A.; Segars, A.H. Knowledge Management: An Organizational Capabilities Perspective. *J. Manag. Inf. Syst.* **2001**, *18*, 185–214. [CrossRef]
64. Tenenhaus, M.; Vinzi, V.E.; Chatelin, Y.-M.; Lauro, C. PLS Path Modeling. *Comput. Stat. Data Anal.* **2005**, *48*, 159–205. [CrossRef]
65. Cheng, T.-M.; Wu, H.C.; Wang, J.T.-M.; Wu, M.-R. Community Participation as a Mediating Factor on Residents' Attitudes towards Sustainable Tourism Development and Their Personal Environmentally Responsible Behaviour. *Curr. Issues Tour.* **2019**, *22*, 1764–1782. [CrossRef]
66. Melo, M.P. Cultural Heritage Preservation and Socio-Environmental Sustainability: Sustainable Development, Human Rights and CitizenshipCitizenship. In *Efficiency, Sustainability, and Justice to Future Generations*; Law and Philosophy Library; Mathis, K., Ed.; Springer: Dordrecht, The Netherlands, 2012; pp. 139–161. ISBN 978-94-007-1869-2.
67. Nilson, T.; Thorell, K. *Cultural Heritage Preservation: The Past, the Present and the Future*; Halmstad University Press: Halmstad, Sweden, 2018.
68. Odlyha, M. Introduction to the Preservation of Cultural Heritage. *J. Therm. Anal. Calorim.* **2011**, *104*, 399–403. [CrossRef]

Disclaimer/Publisher's Note: The statements, opinions and data contained in all publications are solely those of the individual author(s) and contributor(s) and not of MDPI and/or the editor(s). MDPI and/or the editor(s) disclaim responsibility for any injury to people or property resulting from any ideas, methods, instructions or products referred to in the content.

Article

Urban Heritage Facility Management: A Conceptual Framework for the Provision of Urban-Scale Support Services in Norwegian World Heritage Sites

Bintang Noor Prabowo [1,2,*], Alenka Temeljotov Salaj [1] and Jardar Lohne [1]

[1] Department of Civil and Environmental Engineering, Faculty of Engineering, Norwegian University of Science and Technology (NTNU), 7034 Trondheim, Norway; alenka.temeljotov-salaj@ntnu.no (A.T.S.); jardar.lohne@ntnu.no (J.L.)

[2] Department of Civil Infrastructure Engineering and Architectural Design, Diponegoro University, Semarang 50275, Indonesia

[*] Correspondence: bintang.n.prabowo@ntnu.no; Tel.: +47-4868-9764

Abstract: This study validated the theoretical keypoints obtained from a previously published scoping literature review within the context of three Norwegian World Heritage sites: Røros, Rjukan, and Notodden. The cross-sectional table of the urban heritage facility management (UHFM) framework, which is based on interviews and correspondence, demonstrates the connection between the tasks of the six clusters of technical departments responsible for the provision of urban-scale support services and the modified critical steps of the Historic Urban Landscape approach, in which an additional step for "monitoring and evaluation" was included. UHFM operates at the intersection of heritage preservation, urban-scale facility management, and stakeholder coordination, which requires a careful balance between urban heritage conservation and sustainable urban management practices, thus enabling the preservation of World Heritage status that, among others, fosters sustainable tourism. The three case studies highlighted the significance of UHFM in preserving heritage value, authenticity, visual quality, and significance. Besides providing comprehensive support services that extend beyond the daily tasks of conservators and World Heritage managers, UHFM also allows feedback mechanisms for continuous improvement. This study highlighted the complex relationship between the provision of urban-scale support services and the preservation of Outstanding Universal Value as the core business of World Heritage sites.

Keywords: urban facility management; support services; urban heritage; urban scale; conservation; World Heritage

1. Introduction

World Heritage (WH) sites are highly valuable assets to humanity because they represent universal value that goes beyond national boundaries [1–3]. To maintain the Outstanding Universal Value (OUV), as the prerequisite of preserving the WH status of protected sites [4,5] and complementary to the daily tasks of conservators, archeologists, academics, and heritage authorities [6], various technical departments in the municipality, county, and national level need to work together in a coordinated manner to achieve the common goals. In accordance with their primary responsibilities, conservators and cultural heritage authorities tend to prioritize the preservation of historic buildings, monuments, and OUV of heritage sites over providing urban-scale support services [7,8]. The delivery of these services is a crucial task that appears not to support conservation efforts directly. However, in order to determine the support services that are required to be provided, it is still crucial to have a comprehensive understanding of the "core business" of the WH site [6].

In the previous study, the scoping literature review of urban heritage facility management (UHFM) highlighted a few discussions and debates amongst academics and

practitioners around urban-scale facility management within urban heritage areas [9]. The previously examined literature mainly discussed facility management (FM) practices of single heritage buildings or a complex of buildings instead of urban-scale facility management (Urban FM). Meanwhile, works of literature in the Urban FM field did not explicitly address historic districts or urban heritage areas nor their relation to urban-scale conservation practices [6,9]. The phenomenon is understandable since Urban FM itself is still a relatively new field in its establishment phase, and it is an expansion of FM discipline within the urban context [10,11]. Most of the heritage-related articles from the examined papers refer to the Historic Urban Landscape (HUL) approach as the latest holistic approach to managing urban heritage [9,12,13]. Although widely recognized as an avant-garde approach, there are many uncertainties in interpreting the HUL approach's operable criteria at the regional and local governance levels [9,13,14]. Many aspects of such an approach could be explained and clarified better using FM and Urban FM as more technical disciplines for the technical departments in charge of providing and delivering urban-scale support services [9].

FM is a branch of management discipline that addresses the tools and services that support the functionality, safety, and sustainability of buildings, grounds, infrastructures, and real estate [9,15,16]. International Facility Management Association (IFMA) also proposed a new definition of FM as a profession, or discipline, that encompasses multiple disciplines to ensure the functionality of the built environment by integrating people, place, process, and technology [15,17,18]. This new definition allowed Urban FM to legitimately become an expansion of the FM discipline since Urban FM is a manifestation of urban-scale facility management. As the definition is applied to a single building, an urban area is also considered a built environment [6,19,20]. The new definition of FM by IFMA also made it possible for the HUL approach, as the latest conservation paradigm, to be incorporated into the Urban FM field since this holistic approach put the people—its main stakeholder—as an important part of the sustainable urban conservation process, especially in reaching consensus on what and how heritage assets should be preserved, within bottom-up heritage policy decision-making [6,9].

UHFM emerged from the expansion of the facility management (FM) discipline into urban-scale facility management (Urban FM) within the context of urban-scale heritage areas [6,9,16]. This development coincided with the emergence of a new paradigm in managing urban heritage areas and historic towns, known as the HUL approach, which was recommended by UNESCO in 2011 [13,21]. This approach advocates for a more holistic and inclusive strategy in managing heritage, aiming to balance the preservation of historical buildings and monuments with the evolving demands of urban development [22–24]. UHFM addresses the complex task of managing urban-scale support services in these unique types of heritage areas. The justification for UHFM establishment is supported by the dual requirement of safeguarding the WH sites' outstanding universal values while ensuring their sustainable development and stakeholders' well-being [6,9]. The HUL approach is a comprehensive framework highlighting the coexistence of heritage preservation and sustainable urban development [22,23]. The HUL approach acknowledged the significance of the historic town as a living environment and dynamic entity. In contrast, the UHFM framework expands on this philosophy by integrating it into the management of urban-scale facilities. WH sites, especially those with urban characteristics, require an advanced approach that goes beyond conventional heritage conservation [25,26], as they preserve exceptional cultural heritage values and attributes. UHFM, as an integration of the HUL approach and Urban FM, provides the opportunity to support the preservation of OUV through the excellent delivery of urban heritage-friendly support services.

UHFM focuses specifically on examining the complex aspects of managing facilities in the context of urban heritage. It acknowledges that the preservation of OUV is not an isolated task but one that requires a coordinated effort in managing various support services crucial for the daily operation of these areas. Thus, UHFM bridges the gap between preserving cultural heritage, ensuring urban functionality, and promoting collaboration among stakeholders. It offers a detailed and practical framework for effectively organizing support services on a large scale in urban areas. Implementing UHFM into the management of historic towns has the potential to complement the conventional conservation measures undertaken by conservators and heritage authorities at various levels, nationally, regionally, and locally. This integration may deliver urban-scale support services that are in compliance with the preservation of OUV as part of the holistic approach recommended by UNESCO through the HUL approach [9,21].

The UNESCO recommendation proposed a paradigm shift in the preservation of historic buildings. Instead of solely focusing on the physical preservation of buildings and monuments, it suggests a broader approach that considers the entire human environment, including both tangible and intangible aspects, such as increased attention to the well-being of the dwellers in urban heritage areas [12,13,26]. This shift in paradigm, together with the emerging concepts of Urban FM as a people-oriented discipline, resulted in an adjustment of the provision of urban-scale support services in establishing a balance between the efficiency and effectiveness of service delivery while simultaneously preserving the heritage integrity and OUV of WH sites. Therefore, there is a necessity for a framework to implement urban heritage facility management that is capable of adapting to the dynamic characteristics of urban environments. This framework is essential for achieving a balance between preserving heritage values and meeting the demands and standards of modern society. By taking into account the roles and responsibilities of various stakeholders, technical departments, and governance structures, the UHFM framework serves as a tool that allows the involvement of urban-scale support services to contribute and align with the protection of the WH status of the areas under study.

Urban heritage facility managers' tasks extend beyond the routine tasks of conservators and heritage authorities. Support services that may not appear directly connected to historical aspects, in practical terms, might have significant impacts on the visual esthetics, cultural value, and the OUV of protected heritage sites. Tasks such as placing waste containers, choosing between cobblestone or asphalt for road construction, conducting excavation work for underground infrastructure, and installing street furniture in the protected core area of WH sites can present significant complexities. These challenges necessitate both heritage and technical skilled and knowledgeable human resources, which can be managed within the proposed UHFM framework in this study. The UHFM provides clear guidance for support service providers and technical departments, overcoming the difficulty of interpreting the HUL approach, which often showed itself to be confusing at the tactical and operational levels. UHFM operates at the intersection of heritage conservation, urban-scale facility management, and collaboration among stakeholders.

This study examines the complexities of UHFM by analyzing information gathered from three Norwegian World Heritage sites: Røros, Rjukan, and Notodden. The study takes a comprehensive approach, integrating insights obtained from interviews and correspondence with key individuals responsible for managing certain aspects of the studied World Heritage sites, including officials from technical departments, heritage authorities, and governmental bodies at the local, regional, and national levels. Document studies were conducted as an additional source to supplement the interviews and correspondences. The information collected provides valuable qualitative data, insights into challenges, achievements, and collaborative efforts related to managing urban-scale support services in urban heritage areas.

The primary objective of this study is to propose a conceptual framework for UHFM that effectively addresses the complexities of organizing urban-scale support services in World Heritage sites. In order to achieve this, this study aimed to address two research questions: (RQ1) "How can urban-scale support services be efficiently organized in an urban heritage area or World Heritage site by technical departments and other stakeholders, without compromising the Outstanding Universal Value (OUV), visual quality, authenticity, and significance of the protected heritage site?" and (RQ2) "How do the processes and coordination functions of urban-scale facility management support services contribute to preserving the World Heritage status of a protected urban heritage area, considering the roles of multiple layers of governance, technical departments, stakeholders, and feedback mechanisms for continuous improvement?".

This study investigated the urban heritage facility management practices in the three Norwegian world heritage sites as the case study to validate the theoretical keypoints on how to conduct urban-scale facility management within urban heritage areas.

2. Methods
2.1. Research Design

This research undertakes three case studies in the Norwegian World Heritage sites: Røros Bergstaden, Rjukan Company Town, and Notodden Industrial Heritage area. The selection of case studies has gone through a long process by taking into account many factors, including representing urban heritage areas or historic towns and aspects of comparability, which makes them relevant to be studied to validate the theoretical keypoints obtained from the urban heritage facility management's scoping review process [9]. Urban heritage areas with World Heritage status were selected due to their compliance with international standards in conservation management and the implementation of a comprehensive periodic reporting system at the local, national, and international levels, thus ensuring the availability of standardized and structured data and documented information. Norway was selected as a nation to be studied based on its unique architectural characteristics, extensive experience in managing World Heritage sites, close proximity to the home base of this study research laboratory, well-established network, ease of access, and budget limitations. The main approach chosen was based on (1) semi-structured interviewing, (2) detailed correspondence with technical departments, and (3) document studies of the investigated cases. The results were organized according to (1) a clustering of technical departments and (2) the validation of the 33 UHFM theoretical keypoints.

The urban-scale support services that form the UHFM foundation in the World Heritage context [6] have been incorporated into corresponding technical departments at the municipality *(kommune)* level. Furthermore, interviews were conducted, and correspondences were exchanged with technical departments at the county *(fylkeskommune)* level regarding urban-scale service delivery at WH sites. As an illustration, the WH coordinator *(verdensarvkoordinator)* for Røros Bergstaden and its surrounding areas operates under the jurisdiction of the local municipality *(Røros kommune)* with some coordination function between counties *(verdensarvrådet)* where the circumference of Røros is situated, whereas the WH coordinators for Rjukan and Notodden operate under the organizational structure of the county level *(Vestfold og Telemark fylkeskommune)*. This study is aware that in 2020, Telemark County underwent a merger with Vestfold County to establish the new Vestfold og Telemark Fylkeskommune (VTFK). Nevertheless, in 2024, Telemark was again restored as a county. This study will use VTFK in conjunction with both Vestfold County and Telemark County, considering the specific timeframe of its data collection. In this study, it is noteworthy that all coordinators of WH sites in the Norwegian context collaborate closely with *Riksantikvaren*, the Directorate for Cultural Heritage of Norway. The support services were categorized into six clusters: planning and zoning, public works and infrastructure, tourism, conservation and cultural heritage, environment and sustainability, and urban safety and security. The data for this research were collected and analyzed employing the three selected Norwegian World Heritage sites as case studies and the six categories mentioned earlier. The 33 theoretical keypoints of UHFM, obtained from the UHFM scoping literature review [9], were utilized in this study to provide guidance for the development of interview protocols, correspondences, coding for qualitative analysis, and cross-sectional tables.

Røros Mining Town, located in Trøndelag County (Figure 1), was designated as a UNESCO World Heritage site in 1980 and extended to its circumference in 2010 due to its exceptional universal value under criteria (iii) for bearing unique witness to the adaptation of technology to the requirements of the natural environment and the remoteness of the situation, (iv) for illustrating in an outstanding manner how people adapted to the extreme circumstances in which they had to live and how they used the available indigenous resources to provide shelter, produce food for their sustenance, and contribute to the national wealth of the country, and (v) for constituting a totality that is an outstanding example of traditional settlement and land use [27,28]. Røros is a remarkable reminder of a lost cultural tradition and an important period in Norwegian history. This picturesque mountainous mining town has been recognized for its well-preserved architectural ensemble, which reflects the socio-economic systems and mining practices of the 17th and 18th centuries, earning it a place on the World Heritage List. Røros, which is distinguished by wooden houses painted in traditional colors, is a remarkable example of how people have adapted to a harsh environment. It plays a crucial role in the Røros Municipality because the town is a thriving hub for community life, cultural traditions, and heritage preservation [28]. Røros is important to Trøndelag County, even outside of its immediate vicinity. It adds to the area's cultural diversity and draws tourists eager to experience the distinctive mining history and charming architecture that characterize this remarkable World Heritage site.

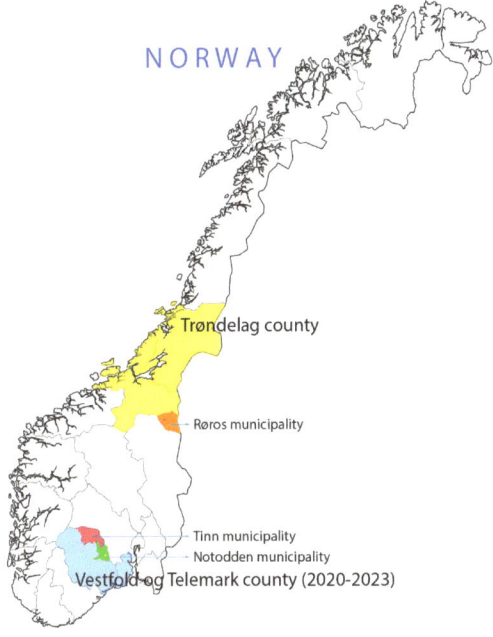

Figure 1. Location of Røros municipality (Trøndelag County) and Tinn and Notodden Municipality (Vestfold og Telemark County).

Meanwhile, an important period in Norway's industrial history is represented by the Rjukan and Notodden Industrial Heritage area, which was inscribed as a UNESCO World Heritage site in 2015. This cultural landscape in Telemark County was essential to the early 20th-century production of fertilizers through the use of hydroelectric power and nitrogen extraction [29,30]. The two towns, Rjukan and Notodden (Figure 1), show how human activity shaped the landscape and are prime examples of inventive industrial urban planning and architecture. This site is inscribed under UNESCO criteria (ii) for demonstrating an exceptional combination of industrial themes and assets tied to the landscape, which exhibit an important exchange on technological development in the early 20th century, and (iv) for its outstanding industrial ensemble comprising dams, tunnels, pipes, power plants, power lines, factory areas and equipment, the company towns, railway lines, and ferry service, located in a landscape where the natural topography enabled hydroelectricity to be generated in the necessary large amounts, stands out as an example of new global industry in the early 20th century [29,30]. This site serves as a testament to the economic and social changes brought about by the development of hydroelectric power and industrialization. The Rjukan and Notodden Industrial Heritage area in Telemark is a living heritage site today, contributing to the identity of the area and drawing tourists eager to learn more about the industrial and architectural legacy of this distinctive cultural landscape.

2.2. Data Collection

The data needed for this study were collected from semi-structured interviews, exchanging correspondences, and document studies. The interviews and correspondences were conducted from 21 January 2022 to 30 December 2023 and were registered to and approved by the Norwegian Center for Research Data (NSD), which later merged with two other Norwegian organizations to establish the new Norwegian Agency for Shared Services in Education and Research (SIKT).

2.2.1. Semi-Structured Interviews

This study used in-depth semi-structured interviews to address the research questions adequately [31]. A predetermined interview protocol was created to ensure alignment with the research questions, and it has undergone pre-testing and peer review by an academic who also works as a researcher and has a particular interest in one of the World Heritage sites in Norway. The feedback was then integrated into the final interview protocol.

The interviewees were chosen based on their roles and/or administration function in the protected urban heritage sites. The main interviewees comprised eight individuals who have specialized knowledge in conservation and World Heritage site management in the Norwegian context, such as city antiquarians *(byantikvar)*, WH coordinators *(verdensarvkoordinator)*, academics, and staff members of the Directorate for Cultural Heritage *(Riksantikvar)* of Norway (Table 1). The *byantikvar* and *verdensarvkoordinator*, part of the technical department cluster responsible for cultural heritage and conservation in the municipality and county, were given special interviews as they agreed to do so. There are several challenges during the data collection, such as conflicted schedules, language barriers, and impracticalities due to the COVID-19 pandemic. It was then decided to conduct some of the interviews via online platforms (i.e., Zoom meetings, Google Meet, and MS Teams) to overcome most of the challenges. Two interviews were conducted in person, while the remaining six interviews were conducted through one-on-one meetings through live video conferences. Minutes of the meetings were taken, and voice notes and/or video conferences were recorded with the interviewees' consent. Automatic transcription was generated and used to transcribe the interviews roughly, but further careful audio rechecks were conducted manually to guarantee the accuracy of the transcription. All interviews were recorded in both video and audio formats, except for the two physical interviews, which were recorded solely in audio format.

Table 1. Distribution of interviewees and correspondence.

Institution/Background	n	Knowledge		
		General	Heritage	Technical
Municipality (Kommune)	18	Yes	Some	Yes
County (Fylkeskommune)	7	Yes	Some	Yes
Academic/University	3	Yes	Yes	Some
National Authority (Riksantikvaren)	1	Yes	Yes	Some

2.2.2. Correspondence with Technical Departments

Nevertheless, a written correspondence method [32,33] was adopted to increase participation and data collection from the technical departments, especially regarding specific tasks and support services. The correspondence technique was employed in this study due to the disinclination of the technical departments' resources to accept interview requests, resulting in low response rates during the initial data collection stage. One possible explanation for the low response rate is that the semi-structured interview material included with the interview request application was too broad for certain specific technical departments. This assumption can be drawn based on the frequent comments made during email correspondence, later, where they expressed their reluctance to address questions that belong to the responsibilities and expertise of other technical departments. However, questions related to the responsibilities, authorities, and duties of the respective departments and sections were addressed comprehensively by the contact persons during the follow-up email correspondence. Another possible cause is that language barriers, cultural differences, and the hectic work schedules of the interviewees in various technical departments at the municipality and county levels posed challenges, making conducting lengthy or repeated interviews impractical. As a result, the electronic correspondence method via email was adopted as a more effective and efficient substitute for the interviews. Questions that remained unresolved or those that generated intellectual curiosity needed by this study were investigated further through a series of exchanged emails. The follow-up inquiries were typically answered in written form with explanations or by providing URL links to relevant documents, reports, or official websites.

A more focused set of questions, specifically tailored to each technical department, was developed from the initial semi-structured interview questions. These inquiries were subsequently sent to the relevant technical department responsible for addressing the specific inquiry. Out of the 72 emails in total sent to the academics, *Riksantikvaren*, and various levels of technical staff in the municipality and county of the studied area, 28 emails were responded to and utilized for further communication and data collection for this study. Among those 28 replies, only 21 of them should be considered as correspondence since 7 of the other email responses agreed to participate in the interviews. Another interviewee was being contacted by phone (Tables 1 and 2). The correspondence data and archives were saved in PDF format and categorized based on the different labels and locations of the study case.

Table 2. Interviewees and correspondence coding.

		PLZ	PWI	TOU	CCH	ESU	USS
Røros (RO)	Røros kommune	RO-PLZ	RO-PWI	RO-TOU	RO-CCH	RO-ESU	RO-USS
	Trøndelag fylkeskommune	TR-PLZ	TR-PWI	-	TR-CCH	TR-ESU	-
	Academics	AC1, AC2	AC1, AC2	AC1, AC2	AC1, AC2	AC1, AC2	AC1, AC2
	Riksantikvaren	RI	RI	RI	RI	RI	RI
Rjukan (RJ)	Tinn kommune	RJ-PLZ	RJ-PWI	RJ-TOU	RJ-CCH	RJ-ESU	RJ-USS
	Vestfold og Telemark fylkeskommune	VT-PLZ	VT-PWI	-	VT-CCH	-	-
	Academics	AC3	AC3	AC3	AC3	AC3	AC3
	Riksantikvaren	RI	RI	RI	RI	RI	RI
Notodden (NO)	Notodden kommune	NO-PLZ	NO-PWI	NO-TOU	NO-CCH	NO-ESU	NO-USS
	Vestfold og Telemark fylkeskommune	VT-PLZ	VT-PWI	-	VT-CCH	-	-
	Academics	AC3	AC3	AC3	AC3	AC3	AC3
	Riksantikvaren	RI	RI	RI	RI	RI	RI

RO = Røros, RJ = Rjukan, NO = Notodden, AC = Academics, RI = *Riksantikvaren*/Directorate for Cultural Heritage, PLZ = planning and zoning, PWI = public works and infrastructure, TOU = tourism, CCH = conservation and cultural heritage, ESU = environment and sustainability, USS = urban safety and security.

The complete responses of the interviewees and correspondences were transcribed and utilized for analysis and coding in NVivo 12 Pro.

2.2.3. Document Studies

During the process of conducting interviews, some interviewees and correspondents occasionally supplied tools, data, information, files, and URL links to provide supplementary information pertinent to this study. Publicly available data were acquired from official websites through the Internet, online databases, and libraries (see Appendix B). The documents consist of nomination dossiers, periodic reporting, Planning and Building Acts, Cultural Heritage Acts, evaluation by advisory bodies, etc. The documents were examined for their capacity to provide a comprehensive analysis of existing records, plans, and reports related to World Heritage sites. Through careful examination of nomination dossiers, periodic reports, management plans, and other documents, researchers can discover valuable insights regarding the historical development, conservation strategies, and difficulties encountered by these sites. These documents serve as a basis for understanding the context, objectives, and recommended management practices for protecting the WH properties. Furthermore, conducting document studies allows for the detection of challenges, inconsistencies, or successes in implemented strategies, providing insights for future improvements [34]. The document studies also enabled this study to understand institutional knowledge, policy frameworks, and the interactions between stakeholders.

2.3. Data Analysis

The empirical analysis primarily relies on an iterative and inductive process [31,35] that involves reading, coding, interpreting, and re-evaluating the transcribed interview notes from the three case studies and their six technical departments. Additionally, it includes input from the national authority *(riksantikvaren)* and academics who have previously been involved or are currently working on the studied and specified World Heritage sites in Norway. The analysis of each case study involved the utilization of open and axial coding techniques in the NVivo 12 Pro environment. The author manually allocated codes, categories, or clusters to each interview during this stage. The coding process utilized the six crucial steps established by the HUL approach, including its additional last UHFM step, and the 33 theoretical keypoints of UHFM as guidance indicators. Furthermore, certain categories were employed in accordance with the research framework. The author and co-authors of this study internally reviewed each case study's coding and transcript. Last, the data were employed for cross-case analysis, pattern matching, grouping, and frequency analysis. In general, there was a strong confidence level in the accuracy of the spoken words during the interviews and the written responses in electronic correspondence.

In order to ensure a high degree of reliability, this study distinguished between construct, internal, and external validity [31,36]. Multiple sources are used for cross-case analysis to ensure construct validity, and a chain of evidence is established through transcripts, as well as visual data and documents presented during the interviews. In addition, the interview and correspondence protocol includes both open-ended and closed questions to ensure the accuracy and reliability of the answers. Internal validity is established by employing pattern matching and constructing explanations based on each individual case. In order to ensure external validity, this study employed a multi-case approach across three Norwegian WH sites, incorporating replication logic within each case. To ensure reliability, this study utilized a comprehensive database containing all interviews, correspondences, interview protocols, and audio and video recordings.

3. Results

3.1. UHFM Cross-Sectional Matrix

The process leading to developing the conceptual framework for urban heritage facility management exposed the complex interconnections and relationships essential for providing urban-scale support services within WH sites (see Appendix A). The cross-sectional table visualized the seven steps of UHFM with the six clusters of technical departments that are responsible for managing the strategic, tactical, and operational levels of urban-scale support services. The table contains a narrative representing the simplified and summarized results of interviews and correspondence with the key stakeholders involved. This comprehensive matrix acts as the primary framework of the study, facilitating the broad spectrum of insights gathered during interviews and correspondence from the stakeholders involved in managing three Norwegian World Heritage Sites: Røros, Rjukan, and Notodden. The table simplifies complex interactions, tasks, and responsibilities into a visually understandable format through data and narratives, with each element symbolizing an important role in providing urban-scale support services.

The UHFM conceptual framework also revealed several missing theoretical keypoints, indicating the unavailability of actions, tasks, or information during the data collection process. The lack of UHFM keypoints revealed considerable facts and information regarding the complexity and challenges involved in providing support services. This framework made it possible to see the big picture and comprehend the narrative of complexities, gaps, and strategic alignments that characterize the UHFM framework in the context of urban-scale Norwegian WH sites. The empirical outcomes of interviews and correspondence were translated and brought concretely to allow for a comprehensive interpretation and discussion in the subsequent sections.

3.2. UHFM Organizational Framework

The organizational framework for UHFM illustrates the complexities involved in managing urban heritage facilities. Due to the complex nature of these organizations, especially in the context of WH sites, it is important to simplify the illustrated interaction to prevent overwhelming the general audience in understanding the framework (Figure 2).

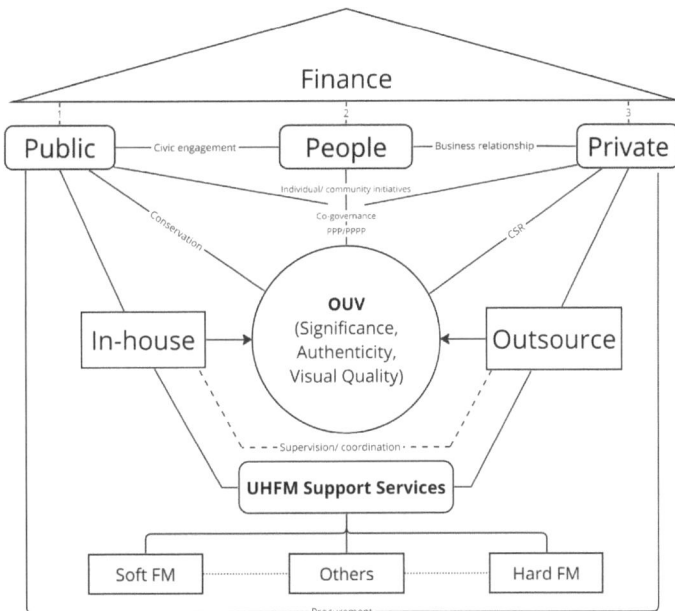

Figure 2. UHFM organizational framework. (1) International, national, regional, and local government funding; private to public funding; sovereign bonds/government paper, etc. (2) Government grant; incentive funds; special taxation; private loan/banking; community funding; self-funding. (3) Private loan/banking; international, national, regional, and local government funding; public to private funding; crowdfunding (people to private funding); public–private partnership (PPP); public–private–people partnership (PPPP).

The UHFM organizational framework prioritizes heritage values as the central focus of urban heritage area conservation. Within the context of WH sites, the OUV serves as the foundation for inscribing cultural heritage on the WH list, making its preservation and care of utmost importance. The OUV, as the "core business" of the WH site, should not be compromised for the sake of efficiency, budget, or effectiveness as traditionally understood in facility management, including Urban FM. Urban-scale support services must be dedicated to ensuring that urban heritage areas, as a component of the built environment in FM defined by ISO41001 [17], continue to uphold their heritage significance, authenticity, and esthetic quality. The delivery of support services, both in terms of soft FM and hard FM (see Appendix A), by in-house teams and outsourced service providers should be rooted in heritage values and attributes that carry those values.

The key stakeholders in UHFM are categorized into three clusters: the public, people, and private sectors. Generally, technical departments under the municipality *(kommune)* and, to a lesser extent, the county *(fylkeskommune)* administration are responsible for providing urban-scale support services. In the UHFM framework, the public sector includes local, regional, national, and international governing authorities, particularly those with direct responsibilities for cultural heritage preservation. The community plays a role in heritage preservation through various initiatives, both at the individual and collective levels [37,38]. Individuals can support cultural heritage preservation efforts in general or take direct action in caring for cultural heritage, particularly if they own or occupy heritage buildings. Individuals' involvement in support services often entails providing feedback or participating in public hearings on support services related to heritage assets and properties [39]. The private sector is also a significant stakeholder, actively utilizing cultural heritage properties and engaging in corporate social responsibility (CSR) within the cultural heritage context [40].

Civic engagement plays a central role in the interaction of public sector interactions with individuals [38]. The level of community involvement in the conservation of urban heritage areas often determines the success of cultural heritage preservation. While the relationship between the private sector and individuals is usually centered around customer–business interactions, there are instances where the private sector directly supports heritage communities. The partnership between the public and private sectors, known as public–private partnership (PPP), can be expanded to include elements of people through the public–private–people partnership (PPPP) model [41], which involves crowdfunding and co-governance mechanisms for funding and managing urban heritage areas.

Funding is crucial for both general conservation efforts and the provision of urban-scale support services [42]. National, regional, and local policies strictly regulate funding sources for managing urban heritage. Government budgets can be allocated to fund private sector service providers and technical departments. Government grants and subsidies may also be provided to individuals and communities to support the preservation of tangible and intangible cultural assets. However, funding for individuals and communities typically does not directly address urban-scale support services. On the other hand, the private sector is directly involved in providing various types of urban heritage support services through outsourcing mechanisms supervised and/or coordinated by the relevant technical department. Establishing a UHFM organization responsible for coordinating and orchestrating all urban-scale support services in the urban heritage district is one of the recommendations proposed in this study. UHFM professionals hold positions similar to facility managers in the context of large-scale building complexes.

3.3. UHFM Process Flowchart

A process flowchart serves as a simplified representation of a specific process within the realm of urban heritage facility management. It provides a model that depicts the sequential steps and decision points involved in delivering support services on an urban scale within an urban heritage area. Such areas are characterized by specific heritage regulations that differentiate them from other types of urban environments. The flowchart offers a graphical representation of the workflow, interactions among stakeholders, and the sequence of activities (Figure 3). By illustrating and facilitating the comprehension of stages and procedures in urban heritage facility management, the process flowchart becomes a valuable tool for analysis, communication, and process improvement.

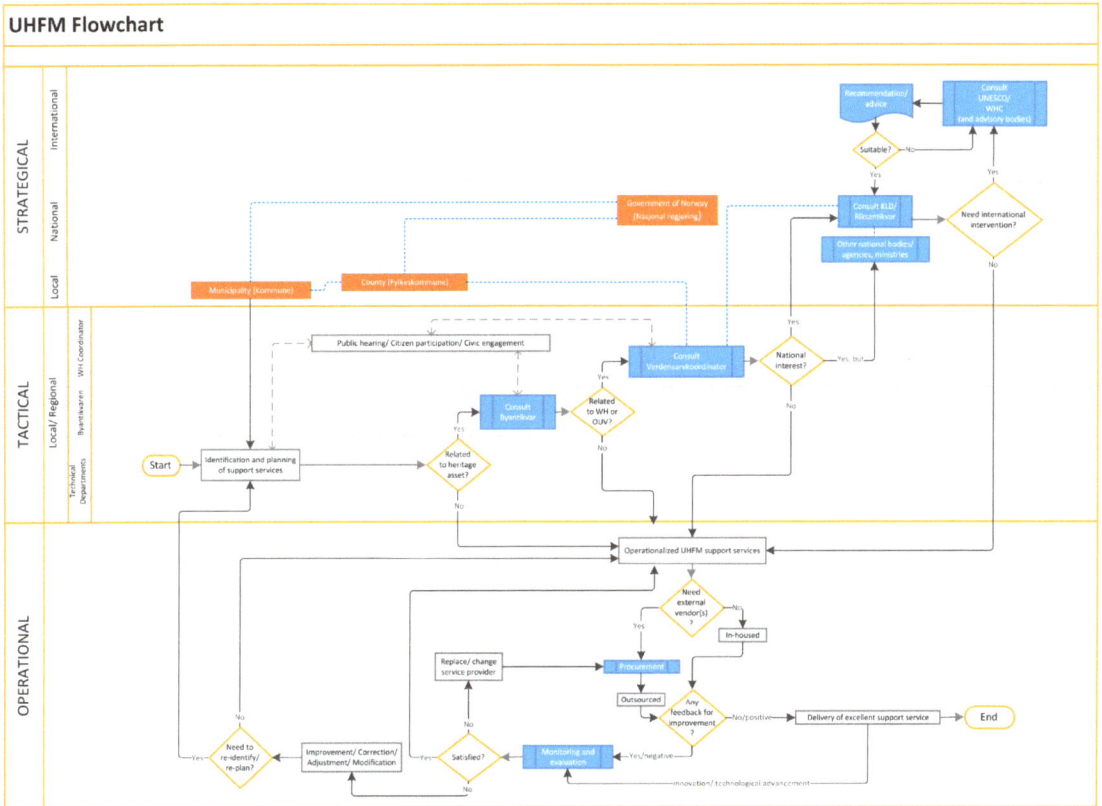

Figure 3. UHFM process flowchart.

The provision of urban-scale support services for urban heritage areas, particularly World Heritage (WH) sites in urban contexts, typically commences with identifying and planning potential support services at the strategic and tactical levels (Figure 3). The responsibility for this initial identification generally lies with governing authorities, such as municipalities and counties, adhering to principles of effective urban governance. Engaging multiple stakeholders, especially through participatory planning processes and public hearings, plays a crucial role in this procedure. Public participation can occur early in the process or be reintroduced through hierarchical consultation involving the cultural heritage department and the WH coordinator, particularly when planned support services may impact the heritage values and characteristics of a World Heritage Site. The identification and planning of support services may undergo a continuous loop based on monitoring and evaluation results, indicating the need for improvement, correction, adjustment, or modification, thereby requiring re-identification or re-planning of these support services. For instance, in the case of Røros, Rjukan, and Notodden, the provision of cobblestone as a substitute for asphalt to enhance visual quality led to complaints from wheelchair and bicycle users, necessitating the re-identification and re-planning of road infrastructure provision to meet the needs of residents through a combination of flat surfaces and cobblestone.

WH coordinators maintain communication forums with their colleagues at other sites and have extensive interactions with *Riksantikvar*, an agency under the Ministry of Climate and Environment (KLD). If the identification and planning of support services have national significance, the WH coordinator will engage in national-level consultations. KLD serves as a communication and coordination channel with UNESCO, the World Heritage Committee (WHC), and their advisory bodies, such as the International Council on Monuments and Sites (ICOMOS), the International Union for the Conservation of Nature (IUCN), and the International Center for the Study of the Preservation and Restoration of Cultural Property (ICCROM), should intervention and consultation from international institutions be required.

While the identification and planning of urban-scale support services originate at the municipal level, the strategic level in Norwegian WH practice also involves coordination functions with the county level *(fylkeskommune)* and the national level through KLD and *Riksantikvar*. Additionally, several national bodies, agencies, and ministries outside of KLD, including those responsible for railways, education, energy, health, and more, may participate in the coordination hierarchy. Once agreements on the provision of urban-scale support services are reached at the strategic and tactical levels, UHFM support services operationalize at the operational level, considering available resources and potential obstacles. Some support services are performed in-house, while others are outsourced to businesses, professionals, contractors, vendors, and private service providers through a procurement process. During the operationalization of support services, feedback for improvement is typically received from the operational level task forces as the avant-garde team and citizens as end users. This feedback mechanism involves various formal and informal procedures. The absence of feedback may indicate inadequacies in the delivery of support services. Enhancing the process of delivering urban-scale support services in an urban heritage area, particularly within the context of World Heritage Sites, requires continuous stakeholder engagement.

4. Discussion

The ambition of the discussion section was to elaborate the findings from the results section by addressing the research questions regarding the efficient organization of urban-scale support services in an urban heritage area, as well as the processes and coordination functions of the six clusters of UHFM technical departments in preserving the World Heritage status of the studied sites following the proposed UHFM steps as the structure (Table 3 and Figure 4).

Table 3. UHFM cross-sectional matrix.

UHFM Steps	Department Planning, Zoning, and Land Use	Public Works and Infrastructure	Tourism	Conservation and Cultural Heritage	Environment and Sustainability	Urban Safety and Security
	Accurate mapping of the topographical features and heritage assets as base maps for all departments					
Mapping Resources	Mapping of land use, values, development zones, building types/patterns, population density	Mapping of infrastructure (roads, bridges, utility networks, urban facilities, etc.)	Mapping of visitor facilities, public space, tourism flow management, interpretation points	Detailed mapping of core and buffer zone of WH sites, archeological sites, cultural routes	Mapping of green spaces, energy consumption patterns, waste management facilities	Mapping of vital infrastructure, emergency services locations, potential natural disasters, surveillance
Missing keypoint(s)	Mapping of the existing partnership and mapping resources using information modeling/BIM-based tools					
	Citizen awareness and engagement, participatory planning, and consensus building for effective decision-making					
Reaching Consensus	Facilitate public input; work with developers for zoning decisions in privately owned development and property	Facilitate public input; collaborate with community groups, academics, and planners to align infrastructure needs	Engage stakeholders in tourism planning, involving local communities and businesses	Collaborate with heritage experts, academics, and communities in heritage management planning; education/developing heritage knowledge; heritage interpretation	Collaborate with environmental advocates and the public for sustainable practices in WH management; education/developing knowledge	Collaborate with law enforcement and communities to identify potential hazards; enhance safety and security measures

Table 3. Cont.

UHFM Steps \ Department	Planning, Zoning, and Land Use	Public Works and Infrastructure	Tourism	Conservation and Cultural Heritage	Environment and Sustainability	Urban Safety and Security
Missing keypoint(s)	N/A					
	Assess the vulnerabilities specific to the technical department's interaction with heritage assets					
Assessing vulnerabilities	Assessing vulnerabilities in zoning decisions; social economic assessment	Assess infrastructure vulnerabilities, utility, and maintenance assessment	Identify vulnerabilities in tourist areas; tourism impact assessment	Assess vulnerability of heritage sites; Heritage Impact Assessment (HIA); heritage policy assessment	Assess vulnerability to climate change; Environment Impact Assessment (EIA)	Assess safety and security vulnerabilities; Risk assessment
Missing keypoint(s)	Citizen satisfaction assessment and digital assessment utilizing BIMs (HBIM, UIM/CIM)					
	Balancing preservation with development and modern needs					
Integrating values and vulnerabilities	Ensure zoning regulations align with urban character and heritage preservation	Integrate infrastructure development into urban esthetics and heritage context	Balance heritage preservation with modern urban development needs; improving public participation	Integrate cultural heritage into development plans; adaptive reuse strategies; improving human resources and public participation; improve heritage regulation	Integrate sustainable practices and green infrastructure into urban planning; improving health and well-being	Integrate safety and security measures into urban design; historic preservation guidelines; improving health, safety, and well-being
Missing keypoint(s)	Enhancing efficiency using information modeling (BIM, HBIM, UIM/CIM), IoT, AI, and sensors					
	Preserving the OUV of the WH sites through the implementation of sustainable cultural heritage management through the efficient delivery of support service(s)					
Prioritizing actions	Zoning regulations enforcement; provide development guidance	Infrastructure maintenance and development; preventive maintenance	Sustainable tourism; visitor experience enhancement; cultural heritage interpretation; preserving cultural identity; increasing citizen participation	Heritage conservation; adaptive reuse; preventive maintenance; cultural value preservation; increasing citizen participation	Environmental protection; sustainable heritage practices; enhance physical and social well-being; increasing citizen participation	Public safety and security; emergency response; preventive maintenance; heritage protection from threats
Missing keypoint(s)	Enabling information modeling (BIM, HBIM, UIM/CIM) integration approach					
	Forming partnerships with stakeholders, experts, local businesses, and community groups aligned with the specific goals of each department (collaborative governance and decision-making)					
Establishing Partnerships	Partners with urban planners, community stakeholders, and developers	Work with contractors, utility providers, and community groups for infrastructure and maintenance	Collaborate with heritage organizations, local businesses, tourism boards; public–private partnership in tourism	Collaborate with cultural experts, historians, and conservationists for preservation, adaptive reuse approach; public–private partnership in heritage preservation	Partners with environmental organizations and sustainable businesses for initiatives; public–private partnership in sustainability	Collaborate with law enforcement, emergency services, and community groups for safety
Missing keypoint(s)	Digital information and information modeling optimization and automation					
	Monitoring and evaluation of support services provided by each technical department					
Monitoring and Evaluation of support service provision	Monitoring and evaluation of urban development impact and zoning/land use compliance	Monitoring and evaluation of urban infrastructure performance, maintenance, and effectiveness	Monitoring and evaluation of tourism flows, visitor satisfaction, tourism support services, and impact of tourism on heritage preservation	Monitoring and evaluation of conservation and WH status, and cultural heritage preservation (reconstruction, restoration, and adaptive reuse)	Monitoring and evaluation of energy consumption, carbon footprint, air quality, environment, and waste management practices	Monitoring and evaluation of emergency preparedness and surveillance effectiveness
Missing keypoint(s)	N/A					

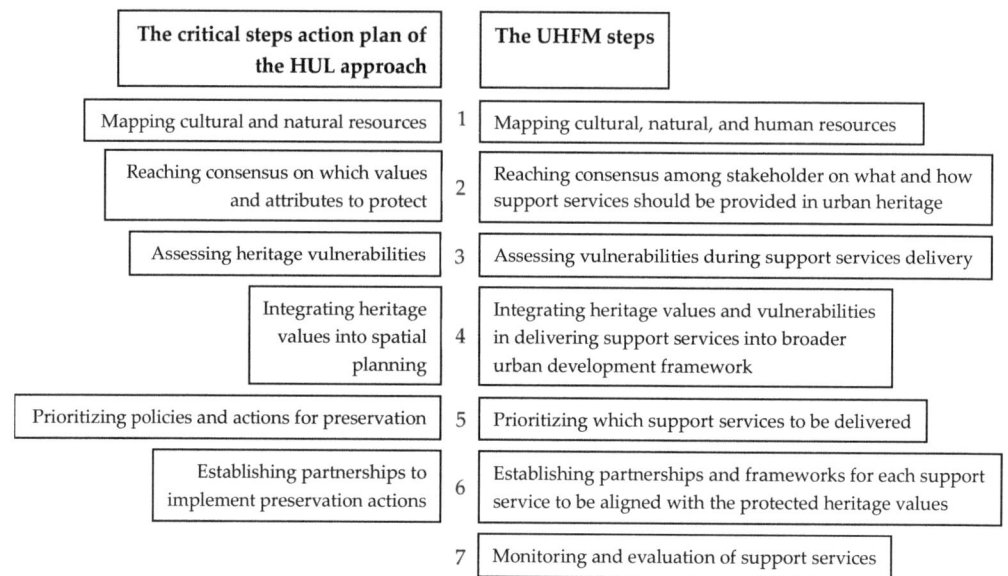

Figure 4. The six critical steps in the action plan of the HUL approach to the UHFM steps.

This section explores various aspects and components of urban heritage facility management (UHFM) using the HUL approach's six critical steps, as reviewed and theoretically studied previously [9], which resulted in 33 UHFM keypoints. Adapting these steps allows for the recognition, identification, and formulation of urban-scale support services in the urban heritage area, which is the focus of this research study. The section is divided into seven main sections to ensure a systematic discussion according to the UHFM steps (Figure 3). Based on the research interviews and the model developed for potential urban-scale support services [6], a comparison is made among three Norwegian World Heritage (WH) sites with urban characteristics, which are Røros Bergstaden—the core city in Røros mining town and its surroundings—The Company Town in Rjukan, and the Notodden Industrial Heritage area in Notodden (see Appendix A). This comparison provides an overall illustration of the UHFM process and its management within the context of good governance in Norway in terms of providing people-oriented urban-scale support services within urban-scale heritage areas without compromising the protected sites' OUV.

As discussed through interviews and correspondence, the conditions shed light on the daily practice of providing urban-scale support services at the three Norwegian World Heritage (WH) sites. Criticisms and potential improvements regarding the provision and delivery of services, as well as coordination between agencies and technical departments, were also explored. Notably, the dynamics and mechanisms of the relationship between public authorities (public), dwellers, citizens, inhabitants, visitors (people), and the private sector (private) emerged as significant aspects in the realm of UHFM.

4.1. Mapping Resources for UHFM

Mapping resources, as the first step in the UHFM steps, serves as a critical foundation for informed decision-making and coordinated efforts across various technical departments. This step involves the accurate mapping of topographical features and heritage assets to create comprehensive base maps for all departments involved in urban management. The cluster of planning and zoning departments ensures precision in mapping land use, development zones, population density, and building types, laying the groundwork for comprehensive urban development. The public works and infrastructure department cluster focuses on mapping vital infrastructure elements such as roads, bridges, utility networks, and other urban facilities. This type of mapping is crucial for the daily practice of infrastructure development and maintenance. The Tourism department's cluster mainly mapped the visitor facilities, public spaces, and the tourism movement to ensure sustainable tourism planning and to avoid overtourism, thus safeguarding a balance between visitor experience and heritage preservation. The conservation and cultural heritage department's cluster provides detailed maps of the WH sites' core and buffer zones, which is essential for heritage conservation, future adaptive reuse strategies, and general conservation initiatives. The environment and sustainability department cluster contributed to mapping green spaces, energy consumption patterns, waste management facilities, and other environment-related tasks. This mapping integrated sustainable practices into urban planning, promoting environmental health and the dweller's well-being. Based on the raw maps provided by the planning and zoning departments, the cluster of urban safety and security departments mapped the vital infrastructure, emergency services locations, and potential natural disaster zones such as flooding, landslides, and fire hazards. This type of mapping is crucial for enhancing public safety measures, emergency response planning, and safeguarding heritage assets from potential threats. The interconnection between these technical departments ensures a holistic approach to managing the studied WH sites.

The unavailability of utilization of the BIM-based tools to map existing resources and mapping partnerships in the urban-scale support services of the three studied Norwegian World Heritage sites—Røros, Rjukan, and Notodden—during the data collection process can be attributed to various factors, such as the limited technological adoption within the technical departments. Moreover, an inadequate level of awareness regarding the potential advantages of utilizing BIM-based tools to map current resources and partnerships could be a contributing factor. The studied WH sites were also a part of national regulatory and policy frameworks that do not explicitly require or incentivize integrating BIM technologies in managing historic towns in Norway.

4.2. Reaching Consensus on What and How Urban-Scale Support Services Should Be Provided

Throughout the reaching-consensus step, each cluster of technical departments adjusted their specific tasks in providing urban-scale support services to be aligned with the WH mission in maintaining OUV as the prerequisite of the WH status. Collaborative decision-making in the cluster of planning and zoning departments relies on the incorporation of citizen awareness, participatory planning, and consensus-building, which highlighted the significance of integrating the citizens' opinions into the city planning and master plan to guarantee their compatibility with the preference of the WH site's inhabitants.

The cluster of planning and zoning departments, together with public works and infrastructure departments, actively sought public input and collaborated with private developers to establish the land use, planning, and zoning decisions that should be aligned with community goals and preservation of OUV. Meanwhile, the tourism departments' cluster involves stakeholders in the tourism planning process by acknowledging the importance of including local communities and businesses during the reaching-consensus step. By adopting such a collaborative approach, tourism initiatives can be aligned with local interests and positively contribute to the community, thus increasing the sustainability of the WH sites economically, socially, and environmentally. The conservation and cultural heritage department cluster engaged in collaborative efforts with heritage experts, academics, and local communities to develop a strategic heritage management plan, focusing on historical education and the advancement of heritage knowledge, which showed a long-term strategy towards conserving heritage. The environment and sustainability department cluster works with environmental advocates and citizens who are interested in promoting sustainable practices in the WH sites. The urban safety and security department cluster prioritizes cooperation with law enforcement and the dwellers to identify potential risks and improve safety and security protocols to protect the integrity of WH assets as a collective duty to guarantee a safe and protected urban heritage setting.

The presence of all necessary theoretical keypoints obtained from the scoping literature review process in the reaching consensus step within the three studied cases of Røros, Rjukan, and Notodden indicated that these sites have effectively implemented comprehensive strategies for engaging the community and building consensus in the delivery of urban-scale support services. As mandated by the Nordic model, the three sites' authorities have placed citizen awareness as their primary concern, actively engaging in efforts to proactively inform the public about current and future development and urban-scale support services. Consensus-building is a commonly accepted practice in Nordic countries, including Norway, that involves collaborative efforts in planning and decision-making processes. The municipalities in charge of managing these studied WH sites have adopted a participatory planning approach, enabling local communities, developers, and other relevant stakeholders to be involved. Furthermore, the emphasis on developing heritage technical knowledge and heritage interpretation indicates a commitment to open and transparent communication among the stakeholders.

The absence of missing theoretical keypoints in the reaching-consensus step suggests successfully integrated community-centric approaches in managing urban-scale support services within the studied Norwegian WH sites in Røros, Rjukan, and Notodden. The Nordic model, characterized by a trusting community and a commitment to equality, serves a significant role in this step. However, a further study of community involvement approaches and decision-making processes would be required to validate these interpretations.

4.3. Assessing the Vulnerabilities of the WH Sites and Their Relationships with UHFM

An assessment step is necessary to address the potential risks and challenges of delivering urban-scale support services within the context of the studied WH sites in Norway. The assessment of vulnerabilities of the WH sites necessitates a comprehensive assessment of various vulnerabilities tailored to the specific functions of each technical department in providing the required urban-scale support services. This is particularly important for addressing the socio-economic pressures and impacts of climate change, besides the strict compliance to the conservation regulations.

Vulnerability assessment in the cluster of planning and zoning focuses on land use, zoning decisions, and socio-economic factors, which suggests acknowledging the commitment to mitigating potential vulnerabilities that may arise from these decisions. The municipal and county authorities must work together to harmonize zoning regulations in broader urban development initiatives. In the meantime, the assessment of infrastructure vulnerabilities has become an important task performed by the cluster of public works and infrastructure departments. Urban-scale utility and maintenance assessments are conducted to identify vulnerabilities and potential hazards in the urban infrastructure, necessitating the cooperation of various technical departments in the local government to work together within more extensive urban development strategies and ensure the infrastructure's long-term functionality. The cluster of tourism departments assessed the impact of tourism to identify particular vulnerabilities in tourist destinations. This approach acknowledges the importance of tourism in World Heritage sites while aiming to minimize any possible adverse effects on the WH assets. Heritage Impact Assessments (HIAs) are essential in assessing the vulnerabilities of heritage sites for the conservation and cultural heritage department cluster. This action shows a commitment to protecting WH sites' cultural and historical significance. Collaboration with heritage experts, academics, and national heritage authorities is important to ensure the precision and efficacy of these assessments. The environment and sustainability department cluster assessed the vulnerabilities related to climate change in the studied WH sites by carrying out Environmental Impact Assessments (EIAs). Effective vulnerability assessment requires collaboration with environmental advocacy groups and national environmental authorities. Last, the urban safety and security department cluster emphasized the importance of conducting comprehensive risk assessments to identify any vulnerabilities related to the safety and security of residents and visitors, which includes cooperating with law enforcement agencies, emergency services, and community groups. Working with local, regional, and national authorities helps ensure that urban safety and security measures align with broader urban development and heritage preservation objectives.

The missing theoretical keypoint found in this step during the data collection is the lack of a mechanism to assess citizen satisfaction and stakeholder feedback. Including citizen feedback in vulnerability assessments could provide valuable insights regarding the effectiveness of urban-scale support services from the end-user's perspective. The operational level of the UHFM team may also provide useful inputs for improving support service delivery in this step. Implementing digital assessment tools and information modeling tools has the potential to bridge this gap, thus improving the overall vulnerability assessment step.

4.4. Integrating Values and Vulnerabilities

Heritage authorities and technical departments employ various measurements to incorporate heritage sites' significance and susceptibilities. One approach involves employing a SWOT analysis, which examines strengths, weaknesses, opportunities, and threats. This analysis allows for the development of strategies by simulating different potential scenarios and determining appropriate solutions. The *Verdensarvkoordinator* and *Riksantikvar*, who are responsible for heritage preservation, can effectively collaborate with the technical departments overseeing road and bridge construction at the local, regional, and national levels. The UHFM organizational framework, obtained from the interview and exchanging correspondence, includes a complex strategy that integrates heritage preservation and urban development. Each technical department serves a distinctive function in this integration, showcasing an awareness of the complex inter-relationship between outstanding universal values and vulnerabilities in WH site management.

The primary responsibility of the cluster of planning and zoning departments is to align land use and zoning regulations with preserving the protected heritage area. This integration acknowledges the importance of land use and zoning decisions in shaping the physical and cultural environment within the core area, buffer zone, and broader urban development. Therefore, the governing stakeholders must work together to ensure that zoning regulations align with the heritage conservation objectives. The cluster of public works and infrastructure departments contributes to urban heritage areas' functional, visual, and historical aspects by integrating infrastructure and physical development vulnerabilities to align with the WH sites' cultural and historical value. The cluster of tourism departments acknowledges that involving the community in tourism planning improves the relationship between tourism initiatives and broader heritage conservation goals to ensure that heritage tourism policies have beneficial impacts on the stakeholders' and citizens' well-being. The cluster of conservation and cultural heritage departments has the role of integrating cultural heritage into development plans and implementing adaptive reuse strategies, thus requiring certain degrees of flexibility in the decision-making process. The flexible approach emphasizes the dynamic nature of conserving cultural heritage, with adaptive reuse being an important strategy. These strategies may ensure alignment with national and international conservation objectives by working closely with heritage experts, academics, and national heritage authorities. Incorporating sustainable practices and green infrastructure into urban planning by the cluster of environment and sustainability departments is essential for promoting the dwellers' health and well-being. This step illustrates an acknowledgment of the mutual reliance between preserving the environment and safeguarding cultural heritage. Coordination with environmental advocacy groups and relevant authorities guarantees the successful incorporation of sustainable practices. The cluster of urban safety and security departments integrates safety and security measures with heritage conservation to protect cultural and historical resources while simultaneously ensuring the well-being, safety, and security of inhabitants and tourists. Coordination with national law enforcement and emergency services is essential to ensure that the safety and security measures align with urban development and heritage preservation strategies.

The keypoint lacking in this step is the systematic integration of information modeling tools or other digital asset management tools to improve efficiency in the integration process. Utilizing digital tools may improve the process of integrating values and identifying vulnerabilities, leading to a more organized and data-driven approach. Incorporating information modeling tools at this step can optimize the overall integration process.

4.5. Prioritizing UHFM Actions

Through the data collection, the respondents were asked about the important factors that need to be taken into account when providing urban-scale support services. Furthermore, they were requested to determine the urban-scale support services that should be prioritized to maintain the WH sites' OUV, heritage significance, authenticity, and visual quality. The respondents from various clusters, in general, emphasized prioritizing maintaining the urban infrastructure, physical urban fabric, accessibility and mobility, and environmental sustainability when planning and implementing urban-scale support services within the realm of UHFM. Several other respondents raised other issues to be prioritized, including matters related to interpretation and education, cleanliness, and waste management.

During the prioritizing actions step, each technical department cluster strategically targets specific aspects that align with their domain as the cluster's priority. The planning and zoning department cluster prioritizes ensuring adherence to zoning regulations and providing guidance for development. This necessitates a robust focus on guaranteeing that development complies with the established regulations and contributes to preserving the urban heritage areas. Effective implementation of zoning regulations requires intensive coordination with other municipal and county sections and bodies.

The public works and infrastructure department cluster prioritizes routine maintenance, development, and preventive infrastructure maintenance. Collaborating with other relevant departments guarantees that infrastructure developments align with the overarching goals of urban-scale heritage preservation. The cluster of tourism departments' priorities are establishing sustainable tourism, enhancing visitor experiences, interpreting cultural heritage, preserving cultural identity, and promoting citizen participation. This comprehensive strategy acknowledges the impact of tourism in shaping the perception and experience of visitors and dwellers of WH sites. The conservation and cultural heritage department cluster prioritizes heritage conservation, adaptive reuse, preventive maintenance, preservation of cultural value, and promoting citizen participation. This comprehensive approach acknowledges the dynamic nature of conserving cultural heritage, integrating preventative measures and strategies for adaptive reuse. Working in collaboration with heritage experts and actively involving the local community in the decision-making related to WH sites ensures a comprehensive approach to preserving urban heritage areas. The priority of the environment and sustainability department cluster is to protect the urban environment within the vicinity of WH sites, improve physical and social well-being, and promote citizen engagement in participating in sustainable heritage practices. The cluster of urban safety and security departments responded with the statement that their priorities are to ensure public safety, security, emergency response, preventive maintenance, and the protection of heritage sites from potential threats. This approach also highlights the commitment to ensuring residents' and visitors' safety and security while protecting valuable heritage assets. Collaboration with national law enforcement and emergency services is necessary for integrating safety measures with broader urban development and heritage preservation strategies.

The keypoint lacking in this step is the intentional incorporation of information modeling tools (such as BIM/HBIM/CIM) into the integration approach to improve efficiency and prioritize actions. Utilizing digital tools could optimize the decision-making and prioritization process, ensuring a more systematic and data-driven approach. Integrating information modeling at this step has the potential to enhance the overall efficiency of prioritizing actions by improving coordination and communication among technical departments and other stakeholders.

4.6. Establishing Partnerships and Frameworks for Each Support Service and Technical Department's Cluster

Throughout the establishing partnerships step, the majority of respondents from each technical department cluster acknowledges the significance of collaborative governance and establishes strategic partnerships to improve the provision of urban-scale support services in urban heritage areas.

The planning and zoning departments cluster plays a crucial role in establishing partnerships with stakeholders, specialists, local businesses, and community groups. This collaborative approach ensures that zoning decisions and urban planning are in accordance with the diverse needs and viewpoints of the community and other stakeholders. The public works and infrastructure departments cluster establishes partnerships with urban planners, community stakeholders, and private developers. This collaborative effort ensures that the construction of infrastructure is aligned with the visual quality of urban heritage areas, historical context, and the preservation of OUV as the core business of WH sites. The cluster of tourism departments establishes partnerships with contractors, utility providers, and community groups through implementing the PPP scheme. The necessary framework for each partnership was developed accordingly to promote sustainable tourism. Effective communication with a wide range of stakeholders, including local communities and businesses, is crucial for successfully implementing tourism initiatives. The conservation and cultural heritage department cluster establishes PPP specifically focused on preserving heritage through collaboration with heritage organizations, local businesses, and tourism boards. However, the respondents did not mention any form of public–private–people partnership (PPPP) practices in the studied WH sites Røros, Rjukan, and Notodden. This collaborative activity ensures that conservation strategies, adaptive reuse programs, and preventive maintenance are in harmony with the objectives of safeguarding cultural heritage. Coordination with heritage organizations enhances the specialized knowledge contributed to conservation initiatives. The environment and sustainability department cluster forms partnerships with environmental organizations and sustainable businesses, participating in PPP to advocate for sustainable practices. The collaborative approach integrates ecological infrastructure into urban heritage development. The urban safety and security departments cluster establish partnerships and coordination with law enforcement, emergency services, and community groups to improve safety measures. The collective endeavor guarantees incorporating safety and security factors into urban design and historic preservation guidelines.

The crucial aspect not found throughout the interviews and correspondence process in this step is the intentional incorporation of digital information modeling optimization and automation to improve the effectiveness of forming partnerships. Incorporating information modeling tools at this step could improve the overall efficiency of collaborative governance, ensuring a more systematic approach to establishing partnerships and developing a framework with a broader city management plan.

4.7. Monitoring and Evaluation

Within the monitoring and evaluation step, as the proposed additional step differs from the HUL approach, each cluster of technical departments has a crucial role in monitoring and evaluating the efficiency of their specific tasks in providing urban-scale support services to ensure continuous improvement and compliance with heritage preservation goals.

The responsibility of the planning and zoning department cluster is to monitor and evaluate the impact of urban development surrounding WH sites and ensure compliance with zoning and land use regulations, especially in the protected sites' core area and buffer zone, which includes evaluating the impacts of zoning decisions on the broader urban development, including their impact on the urban heritage area. The public works and infrastructure department cluster primarily monitors and evaluates urban infrastructure's performance, maintenance, and functionality, including roads, streets, bridges, and other infrastructures. Through real-time monitoring, these departments might identify specific areas and objects requiring maintenance or improvement, ensuring that the infrastructure works comply with the WH sites' heritage conservation regulations and guidances. The cluster of tourism departments monitors and evaluates tourism patterns, providing visitor satisfaction and preventing overtourism that might compromise the preservation of WH sites. The cluster of conservation and cultural heritage departments primarily conducts the monitoring and evaluation of the maintenance of WH status and the preservation, reconstruction, restoration, and adaptive reuse of cultural heritage. The environment and sustainability departments monitor and evaluate energy consumption, air and water quality, environmental conditions, and waste management strategies. The urban safety and security departments monitor and evaluate the efficacy of emergency preparedness and surveillance measures. However, none of the respondents mentioned using an urban command center to conduct surveillance and real-time monitoring to improve the safety of the dwellers and visitors, not to mention the security of the protected assets from vandalism and irresponsible tourist activity. The urban safety and security department cluster monitors and evaluates the effectiveness of emergency preparedness and surveillance measures. This comprehensive approach ensures continuous improvement in managing urban heritage areas and WH sites.

The absence of theoretical keypoints in the UHFM scoping literature review process, specifically regarding the "monitoring and evaluation" step in the management practices of Norwegian World Heritage sites, although being mentioned repeatedly by the respondents during data collection, suggests three possible circumstances during the conception of UHFM keypoints. Firstly, it is possible that academic discussions on the "monitoring and evaluation" step were not identified during the scoping literature review process. Secondly, the absence of this important step in the discussion may be attributed to its unintentional oversight during the scoping literature review, which follows a rigorous protocol incorporating the HUL approach as one of the search criteria for filtering relevant literature. Lastly, the process of conducting a scoping literature review might include adding and classifying "monitoring and evaluation" in academic discussions within the category of "assessment", the third critical step of the HUL approach. Subsequently, during the data collection phase, the respondents, through interviews and correspondences, placed particular emphasis on "monitoring and evaluation" in providing urban-scale support services to ensure continuous improvement in service delivery. Assessments are typically conducted at the beginning to determine the type and manner in which support services will be provided. Meanwhile, "monitoring and evaluation" is usually carried out during the operational phase, where inputs, problems, difficulties, and challenges in the provision of urban-scale support services begin to be discovered. Monitoring occurs at the tactical and operational levels, whereas evaluation is carried out at the tactical and strategic levels of UHFM. The majority of respondents' understanding of the differences between assessment, monitoring, and evaluation suggests that they are highly aware of and committed to flexible and adaptive urban heritage facility management practices. It is presumed that these respondents and their institutions have included monitoring and evaluation in their daily practices, thereby improving the general efficiency of urban-scale support services in preserving the OUV and integrity of the WH sites from time to time.

5. Conclusions

The urban heritage facility management (UHFM) framework reveals a deep comprehension of the complex dynamics that govern the delivery of support services on a large scale in WH sites. The exploration, driven by the two research questions on the efficient organization of these services and the role of coordination functions in maintaining the WH status, has resulted in detailed observations from three Norwegian World Heritage Sites: Røros, Rjukan, and Notodden. The UHFM framework contains the primary information obtained from interviews and exchanging correspondence with key stakeholders. The cross-sectional table between the seven UHFM steps and the six technical department clusters serves as a navigational tool, streamlining the intricate interactions and responsibilities in managing urban-scale support services. This matrix functions both as a visual representation and a condensed narrative, revealing the complexities of stakeholder engagements and the coordination of support services. The detection of crucial elements absent in the UHFM framework serves as a reflection of the difficulties and gaps in the delivery of support services. The gaps between the theoretical keypoints from the scoping literature review process and the conceptual framework obtained from the studied cases reflect the challenges encountered when trying to balance heritage preservation, authenticity, and modern development. The lack of integration of information modeling tools throughout several UHFM steps is particularly interesting, emphasizing the need for improvement and efficiency in future implementations.

The additional step, monitoring and evaluation, allows the UHFM framework to become a powerful and flexible tool adaptable to all possible social, economic, and environmental changes. The ability of this asset to capture the complex connections among technical departments, governance structures, and stakeholders in providing urban-scale support services while maintaining the OUV, visual quality, authenticity, and significance of the studied WH sites makes it a valuable tool in heritage management, alongside the original HUL approach and other existing heritage conservation frameworks addressing the core business of WH sites. The importance of a collaborative and unified strategy, which involves the integration of heritage preservation, management of urban-scale facilities, and collaboration with stakeholders, is emphasized by this study. The UHFM framework effectively tackles both present challenges and serves as a basis for ongoing enhancement and adaptable strategies in the constantly changing field of urban heritage preservation.

The UHFM organizational framework addresses the challenges of managing facilities and how to effectively organize urban-scale support services in an urban heritage area or World Heritage site. The framework highlights the necessity of simplifying stakeholder interactions between UHFM stakeholders by placing heritage values at the center of urban heritage conservation while providing urban-scale service delivery. Within the World Heritage context, the OUV serves as the foundation for inscribing cultural heritage, making its preservation non-negotiable and must not be compromised for the sake of efficiency, budget, or traditional understandings of effectiveness in facility management. The proposed UHFM framework provides insights into coordinating and orchestrating all urban-scale support services in the urban heritage district. In the newly proposed urban heritage facility management field, the UHFM process flowchart provides the workflow steps that must be taken one after another and the decisions that must be made when providing support services on an urban scale inside heritage areas. The perpetual cycle of monitoring and evaluation enables the necessary modifications predicated on input, guaranteeing the continuous improvement of urban-scale service delivery provision.

The proposed UHFM framework plays a role in engaging and benefiting stakeholders and users by fostering a collaborative and informed approach to urban heritage facility management. The framework's capacity to streamline coordination, improve communication channels, and offer a structured comprehension of urban-scale support services will be beneficial to stakeholders, including the public, private sector, and governing authorities. The clarity offered by the framework ensures that stakeholders can actively contribute to the preservation of heritage values while aligning with contemporary needs. Users, including heritage professionals, municipal authorities, and the community, will benefit from a user-friendly and adaptable tool that facilitates efficient decision-making, resource allocation, and strategic planning. The UHFM framework that enables efficient decision-making, resource allocation, and strategic planning will benefit various stakeholders, such as heritage authorities, technical departments, and the community. The UHFM framework promotes a sense of responsibility for the sustainable management of urban heritage areas by highlighting the importance of heritage significance, authenticity, and visual quality.

This study does not intend to make broad generalizations that can be applicable to all types of technical departments, support services, and different types of World Heritage sites outside of Norway. This study was designed to be an initial umbrella study of urban-scale heritage facility management using Norwegian WH sites as a context, which provides the basis for further research in the realm of Urban FM, urban heritage conservation, and detailed parts of UHFM. Various terms in this study are used interchangeably in English and the Norwegian version due to technical and practical reasons. This study represents a progression in the domain of urban heritage management and Urban FM by introducing a framework that addresses the complexity associated with managing urban heritage facilities, specifically focusing on the Norwegian WH sites, which is in contrast to previous studies that typically examined specific aspects of heritage conservation or facility management of protected buildings only. Furthermore, this study offers a conceptual framework that can be applied to various contexts worldwide. This study serves as an invitation for further academic discussion, research, and implementation of the UHFM framework in order to shape sustainable, resilient, and culturally vibrant urban environments for future generations. The results and findings of this study pave the way for future research to replicate similar studies in other non-WH historic towns and urban heritage districts in Norway, as well as in urban heritage areas and WH sites outside of Norway. This will contribute to a more comprehensive understanding of facility management at an urban scale in urban heritage areas.

Author Contributions: Conceptualization, B.N.P. and A.T.S.; methodology, J.L.; software, B.N.P.; validation, B.N.P., A.T.S. and J.L.; formal analysis, B.N.P.; investigation, B.N.P.; resources, B.N.P.; writing—original draft preparation, B.N.P. and A.T.S.; writing—review and editing, B.N.P., A.T.S. and J.L.; visualization, B.N.P.; supervision, A.T.S. All authors have read and agreed to the published version of the manuscript.

Funding: This research received no external funding.

Data Availability Statement: Data are contained within the article.

Acknowledgments: This study is supported by the Department of Civil and Environmental Engineering, Faculty of Engineering, Norwegian University of Science and Technology (NTNU), the Directorate General of Resources for Science, Technology, and Higher Education, The Ministry of Education, Culture, Research, and Technology of the Republic of Indonesia, and Diponegoro University.

Conflicts of Interest: The authors declare no conflicts of interest.

Appendix A

Table A1. Hard UHFM Support Services.

Tasks/Urban-Scale Support Services	Department/Institution/Organization in Charge		
	Røros	Rjukan	Notodden
District heating and cooling, district/neighborhood heat management (*fjernvarme*) (1, 2, 5)	Ren Røros Strøm AS, Norsk Varme	Statkraft AS, Norsk Varme, Green Mountain (data center excess heat)	Thermokraft AS, Norsk Varme, (owned by Notodden Energi)
Power provider(*strømleverandøren*) (2, 5)	REN Røros Strøm AS	Tinn Energi ASHydro Energi AS Telemark	Notodden Energi Kraft AS
Energy management(*strømnettet*/power grid) (2, 5)	Røros E-Verk Nett	Stannum	Everket AS
Water supply (2, 5)	Røros kommune, Norsk Vann	Tinn kommune (Rjukan vannverks), Norsk Vann	Notodden kommune (Notodden vannverks), Norsk Vann

Table A1. Cont.

Tasks/Urban-Scale Support Services	Department/Institution/Organization in Charge		
	Røros	Rjukan	Notodden
Clean/drinking water system (1, 2, 5)	Røros kommune, Norsk Vann	Tinn kommune, Norsk Vann	Notodden kommune, Norsk Vann
District sewerage system (1, 2, 5)	Røros kommune	Tinn kommune	Notodden kommune
Black water system (1, 2, 5, 6)	Røros kommune, Norsk Vann	Tinn kommune, Norsk Vann	Notodden kommune, Norsk Vann
Neighborhood/district drainage and flood control system (1, 2, 5, 6)	Røros kommune	Tinn kommune	Notodden kommune
Heritage buildings and structures (4)	Byantikvar, Verdensarvkoordinator, Department of cultural heritage	Byantikvar, Verdensarvkoordinator, Department of cultural heritage	Byantikvar, Verdensarvkoordinator, Department of cultural heritage
Core zone and buffer zone (World Heritage sites) (1, 4)	Verdensarvkoordinator, Riksantikvaren (supervised by WHC/UNESCO), Verdensarvrådet	Verdensarvkoordinator, Riksantikvaren (supervised by WHC/UNESCO)	Verdensarvkoordinator, Riksantikvaren (supervised by WHC/UNESCO)
Urban heritage visual quality (3, 4)	Byantikvar, Verdensarvkoordinator, Department of cultural heritage	Byantikvar, Verdensarvkoordinator, Department of cultural heritage	Byantikvar, Verdensarvkoordinator, Department of cultural heritage
Urban heritage street furniture (2, 3, 4)	Røros kommune	Tinn kommune	Notodden kommune
Outdoor and public lighting (1, 2, 6)	Røros kommune, *Statens vegvesen* (The Norwegian Public Roads Administration)	Tinn kommune, *Statens vegvesen* (The Norwegian Public Roads Administration)	Notodden kommune, *Statens vegvesen* (The Norwegian Public Roads Administration)
Street and road infrastructures and maintenance (1, 2, 6)	Røros kommune, Trøndelag fylkeskommune, *Statens vegvesen* (The Norwegian Public Roads Administration)	Tinn kommune, Vestfold og Telemark fylkeskommune, *Statens vegvesen* (The Norwegian Public Roads Administration)	Notodden kommune, Vestfold og Telemark fylkeskommune, *Statens vegvesen* (The Norwegian Public Roads Administration)
Telecommunication infrastructures (1, 2)	Infonett Røros AS (cable-based telecommunication), Telenor, Telia	Telenor, Telia and ICE	Telenor, Telia and ICE

Clusters of departments: (1) PLZ = planning and zoning, (2) PWI = public works and infrastructure, (3) TOU = tourism, (4) CCH = conservation and cultural heritage, (5) ESU = environment and sustainability, (6) USS = urban safety and security.

Table A2. Soft UHFM Support Services.

Tasks/Urban-Scale Support Services	Department/Institution/Organization in Charge		
	Røros	Rjukan	Notodden
Neighborhood/district cleaning/hidden trash containers (1, 2)	Røros kommune	Tinn kommune	Notodden kommune
The traditional seasonal market, tourist-oriented shop/retailer, town events (3)	*Rørosmartnan* (Christmas market), *Destinasjon Røros*	*Høstmarked/Bygdas dag* (Autumn market), Rjukan Matfestival, *Solfesten* (Sun Festival), Rjukan Turistkontor, visitRjukan AS	*Høstmarked*, Notodden Vårmarked, Notodden Bluesfestival, Tinfosløpet, *Kjentmannsmerket*

Table A2. Cont.

Tasks/Urban-Scale Support Services	Department/Institution/Organization in Charge		
	Røros	Rjukan	Notodden
Conservation law enforcer, municipal police (4, 6)	Røros kommune	Tinn kommune	Notodden kommune
Post office (2)	Posten Bring AS	Posten Bring AS	Posten Bring AS
The main square (1, 2, 3)	Røros kommune	Tinn kommune	Notodden kommune
District command center (6)	-	-	-
Electrical panel, underground electricity distribution (2)	Røros E-Verk Nett, Røros kommune	Stannum, Tinn kommune	Everket AS, Notodden kommune
Conservation helpdesk (3)	The Røros Museum Call Centre, Røros kommune, *Servicetorget*	Vestfold og Telemark fylkeskommune, Tinn kommune, *Servicetorget*	Vestfold og Telemark fylkeskommune, Notodden kommune, *Servicetorget*
Protected heritage park, garden, void, cemetery (1, 2, 3, 4, 5)	*Kjerkgata* (Harald Sohlberg corridor), *Røros Kirke*, *Slegghaugan* (the slag heaps of Røros)	Rjukan kirke, Rjukan torg	Notodden kirke, Notodden torv, Admini Notodden
Connection with the general transportation system (1, 2)	Røros Airport, Røros Station/*Jernbanedirektoratet* (Norwegian Railway Directorate), Røros bus terminal	Rjukan station/Norwegian Railway Directorate, Rjukan bus stop	Notodden station/Norwegian Railway Directorate, Notodden *skysstasjon* (public transport terminal)
Heritage funicular, travelator, shuttle/site transportation (1, 2, 3, 4)	-	Krossobanen, Gaustabanen	-
Preservation-oriented parking lot (1, 2)	Røros kommune	Tinn kommune	Notodden kommune

Clusters of departments: (1) PLZ = planning and zoning, (2) PWI = public works and infrastructure, (3) TOU = tourism, (4) CCH = conservation and cultural heritage, (5) ESU = environment and sustainability, (6) USS = urban safety and security.

Table A3. Other UHFM Support Services.

Tasks/Urban-Scale Support Services	Department/Institution/Organization in Charge		
	Røros	Rjukan	Notodden
Heritage environmental management (4, 5)	KLD, *Trøndelag fylkeskommune*, Røros kommune	KLD, Vestfold og Telemark fylkeskommune, Tinn kommune	KLD, Vestfold og Telemark fylkeskommune, Notodden kommune
Urban heritage health and safety (5, 6)	Department for culture and public health (*Avdeling for kultur og folkehelse*), *Sosial og helsedirektoratet*, fylkeskommune, Røros kommune	Department for culture and public health, *Helse og omsorgsdepartementet*, *Sosial og helsedirektoratet*, fylkeskommune, Tinn kommune	Department for culture and public health, *Helse og omsorgsdepartementet*, *Sosial og helsedirektoratet*, fylkeskommune, Notodden kommune
Heritage documentation, archiving, digitization, digitalization (4)	The Røros Museum, Røros kommune (*arkiv*/archive)	Norsk Industri-Arbeidermuseum (NIA), Tinn kommune	Norsk Industri-Arbeidermuseum (NIA), Notodden kommune
Urban heritage preservation, restoration, reconstruction, adaptation (2, 4)	Department of cultural heritage (*Avdeling for kulturminner*), *Byantikvar*, *Verdensarvkoordinator*, *Riksantikvaren*	Department of cultural heritage, *Byantikvar*, *Verdensarv-koordinator*, *Riksantikvaren*	Department of cultural heritage, *Byantikvar*, *Verdensarv-koordinator*, *Riksantikvaren*

Table A3. Cont.

Tasks/Urban-Scale Support Services	Department/Institution/Organization in Charge		
	Røros	Rjukan	Notodden
Urban heritage design guidelines comply with the HUL approach (4)	Department of cultural heritage, *Byantikvar*, *Verdensarvkoordinator*, *Riksantikvaren*	Department of cultural heritage, *Byantikvar*, *Verdensarv-koordinator*, *Riksantikvaren*	Department of cultural heritage, *Byantikvar*, *Verdensarv-koordinator*, *Riksantikvaren*
Strategic heritage plan (SHP) (4)	Department of cultural heritage, *Byantikvar*, *Verdensarvkoordinator*, *Riksantikvaren*	Department of cultural heritage, *Byantikvar*, *Verdensarv-koordinator*, *Riksantikvaren*	Department of cultural heritage, *Byantikvar*, *Verdensarv-koordinator*, *Riksantikvaren*
Heritage/tourist-friendly waste management system (2, 5)	Røros kommune	Tinn kommune	Notodden kommune
HBIM, UHIM, HCIM (1, 2)	-	-	-
Heritage-friendly public facilities (2)	Røros kommune	Tinn kommune	Notodden kommune
Customized universal design and accessibilities (2)	Røros kommune	Tinn kommune	Notodden kommune
Urban heritage-related CSR, PPP, and PPPP (N/A)	Trøndelag fylkeskommune, Røros kommune	Vestfold og Telemark fylkeskommune, Rjukan Næringsutvikling AS, Tinn kommune	Vestfold og Telemark fylkeskommune, Notodden kommune
Search and Rescue (6)	The Norwegian SAR/ The Rescue and Emergency Planning Department, Directorate for Civil Protection and Emergency Planning (*Direktoratet for samfunnssikkerhet og beredskap/DSB*)	The Norwegian SAR/The Rescue and Emergency Planning Department, DSB	The Norwegian SAR/The Rescue and Emergency Planning Department, DSB
Emergency preparedness (6)	The Norwegian SAR/ The Rescue and Emergency Planning Department, DSB, Trøndelag fylkeskommune, Notodden kommune	The Norwegian SAR/The Rescue and Emergency Planning Department, DSB, Vestfold og Telemark fylkeskommune, Notodden kommune	The Norwegian SAR/The Rescue and Emergency Planning Department, DSB, Vestfold og Telemark fylkeskommune, Notodden kommune
Tourism (3)	*Destinasjon Røros*, Trøndelag fylkeskommune, Røros kommune	VisitRjukan, Vestfold og Telemark fylkeskommune, Tinn kommune	Vestfold og Telemark fylkeskommune, Notodden kommune
Heritage Education (4)	The Røros Museum, Røros kommune	Norsk Industri-Arbeidermuseum (NIA), Tinn kommune	Norsk Industri-Arbeidermuseum (NIA), Notodden kommune
Interpretation of heritage for public/general audience (4)	The Røros Museum, Røros kommune, Røros World Heritage Foundation (*Røros Verdensarv*)	Norsk Industri-Arbeidermuseum (NIA), Tinn kommune, Norwegian Industrial Heritage Foundation (*Stiftelsen Norsk Industriarbeidermuseum*)	Norsk Industri-Arbeidermuseum (NIA), Notodden kommune, Norwegian Industrial Heritage Foundation (*Stiftelsen Norsk Industriarbeidermuseum*)

Clusters of departments: (1) PLZ = planning and zoning, (2) PWI = public works and infrastructure, (3) TOU = tourism, (4) CCH = conservation and cultural heritage, (5) ESU = environment and sustainability, (6) USS = urban safety and security.

Appendix B

Table A4. List of Document Studies Resources.

Properties	Documents	Year/Date	Institution
Røros Mining Town	Justification for inclusions in the World Heritage list	16 May 1978	Government of Norway
	Advisory body evaluation	15 November 1978	ICOMOS
	Cultural Heritage Act	1978	Government of Norway
	Decision from World Heritage Committee	29 September 1980	WHC—UNESCO
	Planning and Building Act	1985	Government of Norway
	State of Conservation—Bureau of the World Heritage Committee 18th session	26 May 1994	WHC—UNESCO
	Decision's context	26 May 2006	Presentation of the periodic report for sections I and II of Europe
	Decisions adopted at the 30th session of the World Heritage Committee (Vilnius, 2006)	23 August 2006	WHC—UNESCO
	Periodic Reporting—State of Conservation of World Heritage Properties in Europe	2006	WHC—UNESCO
	Advisory Body Evaluation	17 March 2010	ICOMOS
	Advisory Body Evaluation	May 2010	IUCN
	Report of the decisions adopted by the World Heritage Committee at its 34th Session	3 September 2010	WHC—UNESCO
	Decision's context—Evaluations of Cultural Properties—34th ordinary session(25 July–3 August 2010), Brasilia (Brazil)	2010	WHC—UNESCO
	Decision's context—Establishment of the World Heritage List and of the List of World Heritage in Danger	31 May 2010	WHC—UNESCO
	Periodic Report—Second Cycle	19 May 2014	Government of Norway
Rjukan-Notodden Industrial Heritage Sites	Cultural Heritage Act	1978	Government of Norway
	Planning & Building Act	2008	Government of Norway
	Cultural Heritage Act (Amended)	2009	Government of Norway
	Rjukan—Notodden Industrial Heritage Site—Nomination Dossier	2015	Government of Norway
	Advisory body evaluation	12 March 2015	ICOMOS
	Decisions adopted by the World Heritage Committee at its 39th session (Bonn)	8 July 2015	WHC—UNESCO

Table A4. *Cont.*

Properties	Documents	Year/Date	Institution
Rjukan-Notodden Industrial Heritage Sites	Decisions context—Establishment of the World Heritage List and of the List of World Heritage in Danger (Bonn, Germany, 28 June–8 July 2015)	15 May 2015	WHC—UNESCO
	Decision context—Establishment of the World Heritage List and of the List of World Heritage in Danger (Corrigendum)	22 May 2015	WHC—UNESCO
	Decision context—Evaluation of nominations of cultural and mixed properties to the World Heritage list (ICOMOS report for the World Heritage Committee)	April 2015	ICOMOS

References

1. Elliott, M.A.; Schmutz, V. World heritage: Constructing a universal cultural order. *Poetics* **2012**, *40*, 256–277. [CrossRef]
2. Jokilehto, J.; Cameron, C.; Parent, M.; Petzet, M. *The World Heritage List. What Is OUV? Defining the Outstanding Universal Value of Cultural World Heritage Properties*; Hendrik Bäßler Verlag: Berlin, Germany, 2008; Volume 16, ISBN 3930388510.
3. Tucker, H.; Carnegie, E. World heritage and the contradictions of 'universal value'. *Ann. Tour. Res.* **2014**, *47*, 63–76. [CrossRef]
4. Labadi, S. The World Heritage Convention at 50: Management, credibility and sustainable development. *J. Cult. Herit. Manag. Sustain. Dev.* **2022**. [CrossRef]
5. Khalaf, R.W. Integrity: Enabling a future-oriented approach to cultural heritage. *Hist. Environ. Policy Pract.* **2022**, *13*, 5–27. [CrossRef]
6. Prabowo, B.N.; Temeljotov Salaj, A.; Lohne, J. Identifying Urban Heritage Facility Management Support Services Considering World Heritage Sites. *Urban Sci.* **2023**, *7*, 52. [CrossRef]
7. Shrestha, C.B.; Banskota, B. Strengthening the National Capacity for Conservation of National Heritage Monuments and Sites. *Vikas A J. Dev.* **2021**, *1*, 1–13.
8. Cristina Heras, V.; Wijffels, A.; Cardoso, F.; Vandesande, A.; Santana, M.; Van Orshoven, J.; Steenberghen, T.; Van Balen, K. A value-based monitoring system to support heritage conservation planning. *J. Cult. Herit. Manag. Sustain. Dev.* **2013**, *3*, 130–147. [CrossRef]
9. Prabowo, B.N.; Salaj, A.T.; Lohne, J. Urban Heritage Facility Management: A Scoping Review. *Appl. Sci.* **2021**, *11*, 9443. [CrossRef]
10. Collins, D.; Senior, C.; Jowkar, M.; Salaj, A.; Facilities, A.J. The impact of an urban facilities management summer school on the participants. *Facilities* 2021, ahead-of-print. [CrossRef]
11. Temeljotov Salaj, A.; Lindkvist, C.M. Urban facility management. *Facilities* **2021**, *39*, 525–537. [CrossRef]
12. Ginzarly, M.; Houbart, C.; Teller, J. The Historic Urban Landscape approach to urban management: A systematic review. *Int. J. Herit. Stud.* **2019**, *25*, 999–1019. [CrossRef]
13. Rey-Pérez, J.; Pereira Roders, A. Historic urban landscape: A systematic review, eight years after the adoption of the HUL approach. *J. Cult. Herit. Manag. Sustain. Dev.* **2020**, *10*, 233–258. [CrossRef]
14. Van Oers, R.; Pereira Roders, A. Road map for application of the HUL approach in China. *J. Cult. Herit. Manag. Sustain. Dev.* **2013**, *3*, 4–17. [CrossRef]
15. Wilson, D. *Strategic Facility Management Framework*, 1st ed.; The Royal Institution of Chartered Surveyors (RICS): London, UK; International Facility Management Association (IFMA): Houston, TX, USA, 2018; ISBN 978 1 78321 235 4.
16. Prabowo, B.N.; Salaj, A.T. Urban heritage and the four pillars of sustainability: Urban-scale facility management in the World Heritage sites. In *Proceedings of the IOP Conference Series: Earth and Environmental Science*; IOP Publishing: Bristol, UK, 2023; Volume 1196, p. 12105.
17. *ISO 41011: 2017*; Facility Management–Vocabulary. ISO: Geneva, Switzerland, 2017.
18. Modu, M.A.; Sapri, M.; Abd Muin, Z. Towards facilities management practice within a different environment. *J. Infrastruct. Facil. Asset Manag.* **2021**, *3*. [CrossRef]
19. Nijkamp, J.E.; Mobach, M.P. Developing healthy cities with urban facility management. *Facilities* **2020**, *38*, 819–833. [CrossRef]
20. Chizzoniti, D. The nature of cities. In *Cities' Identity Through Architecture and Arts*; Routledge: London, UK, 2018; pp. 297–308. ISBN 1315166550.
21. UNESCO World Heritage Convention. *UNESCO Recommendation on HUL*; UNESCO: Paris, France, 2011; Volume 25.
22. Shah, A.A.; Chanderasekara, D.P.; Naeem, A. Preserving the Past and Shaping the Future: An Articulation of Authenticity of Heritage within Urban Development. *J. Int. Soc. Study Vernac. Settl.* **2023**, *10*. [CrossRef]

23. Erkan, Y. The Way Forward with Historic Urban Landscape Approach towards Sustainable Urban Development. *Built Herit.* **2018**, *2*, 82–89. [CrossRef]
24. González Martínez, P. Built heritage conservation and contemporary urban development: The contribution of architectural practice to the challenges of modernisation. *Built Herit.* **2017**, *1*, 14–25. [CrossRef]
25. Jiang, J.; Zhou, T.; Han, Y.; Ikebe, K. Urban heritage conservation and modern urban development from the perspective of the historic urban landscape approach: A case study of Suzhou. *Land* **2022**, *11*, 1251. [CrossRef]
26. Otero, J. Heritage conservation future: Where we stand, challenges ahead, and a paradigm shift. *Glob. Chall.* **2022**, *6*, 2100084. [CrossRef]
27. Borgos, M. Managing the World Heritage Site Røros Mining Town and the Circumference. *Adapt. Hist. Places Clim. Chang.* 41. Available online: https://whc.unesco.org/en/list/55/ (accessed on 10 February 2024).
28. Guttormsen, T.S.; Fageraas, K. The social production of "attractive authenticity" at the World Heritage Site of Røros, Norway. *Int. J. Herit. Stud.* **2011**, *17*, 442–462. [CrossRef]
29. Sesana, E.; Gagnon, A.S.; Bonazza, A.; Hughes, J.J. An integrated approach for assessing the vulnerability of World Heritage Sites to climate change impacts. *J. Cult. Herit.* **2020**, *41*, 211–224. [CrossRef]
30. Taugbøl, T.; Andersen, E.M.; Grønn, U.; Moen, B.F. *Rjukan-Notodden Industrial Heritage Site*; Nomination to the UNESCO World Heritage List; Riksantikvaren: Oslo, Norway, 2014.
31. Yin, R.K. *Case Study Research*; SAGE Publications: London, UK, 2014.
32. Harris, J. The correspondence method as a data-gathering technique in qualitative enquiry. *Int. J. Qual. Methods* **2002**, *1*, 1–9. [CrossRef]
33. Parris, M. Email Correspondence: A Qualitative Data Collection Tool for Organisational Researchers. 2008. Available online: https://www.anzam.org/wp-content/uploads/pdf-manager/1390_PARRIS_MELISSA-433.PDF (accessed on 10 February 2024).
34. Van Raemdonck, B.; Vanhoutte, E. Editorial theory and practice in Flanders and the Centre for Scholarly Editing and Document Studies. *Lit. Linguist. Comput.* **2004**, *19*, 119–127. [CrossRef]
35. Miles, M.B.; Huberman, A.M. *Qualitative Data Analysis: An Expanded Sourcebook*; Sage: London, UK, 1994; ISBN 0803955405.
36. Franklin, C.; Ballan, M. Reliability and validity in qualitative research. In *The Handbook of Social Work Research Methods*; Sage: London, UK, 2001; Volume 4.
37. Firmansyah, F.; Fadlilah, K.U. Improvement of involvement society in the context of smart community for cultural heritage preservation in Singosari. *Procedia-Soc. Behav. Sci.* **2016**, *227*, 503–506. [CrossRef]
38. Li, Y.; Hunter, C. Community involvement for sustainable heritage tourism: A conceptual model. *J. Cult. Herit. Manag. Sustain. Dev.* **2015**, *5*, 248–262. [CrossRef]
39. Senior, C.; Temeljotov Salaj, A.; Johansen, A.; Lohne, J. Evaluating the Impact of Public Participation Processes on Participants in Smart City Development: A Scoping Review. *Buildings* **2023**, *13*, 1484. [CrossRef]
40. Chi, C.G.; Zhang, C.; Liu, Y. Determinants of corporate social responsibility (CSR) attitudes: Perspective of travel and tourism managers at world heritage sites. *Int. J. Contemp. Hosp. Manag.* **2019**, *31*, 2253–2269. [CrossRef]
41. Xue, Y.; Temeljotov-Salaj, A.; Lindkvist, C.M. Renovating the retrofit process: People-centered business models and co-created partnerships for low-energy buildings in Norway. *Energy Res. Soc. Sci.* **2022**, *85*, 102406. [CrossRef]
42. Della Torre, S.; Boniotti, C. Innovative funding and management models for the conservation and valorization of public built cultural heritage. In *Eresia ed Ortodossia nel Restauro: Progetti e Realizzazioni*; Arcadia Ricerche: Venice, Italy, 2016; pp. 105–114.

Disclaimer/Publisher's Note: The statements, opinions and data contained in all publications are solely those of the individual author(s) and contributor(s) and not of MDPI and/or the editor(s). MDPI and/or the editor(s) disclaim responsibility for any injury to people or property resulting from any ideas, methods, instructions or products referred to in the content.

Article

Conservation and In Situ Enhancement of Earthen Architecture in Archaeological Sites: Social and Anthropic Risks in the Case Studies of the Iberian Peninsula

Sergio Manzano-Fernández *, Camilla Mileto, Fernando Vegas López-Manzanares and Valentina Cristini

Centro de Investigación en Arquitectura, Patrimonio y Gestión para el Desarrollo Sostenible (PEGASO), Universitat Politècnica de València, 46022 Valencia, Spain; cami2@cpa.upv.es (C.M.); fvegas@cpa.upv.es (F.V.L.-M.); vacri@cpa.upv.es (V.C.)
* Correspondence: sermanfe@upv.es

Abstract: Archaeological sites constitute one of the main tourist attractions in the heritage offerings of most populations. Their ability to convey the ways of life and construction techniques of past societies through physical remains positions them as a culturally significant alternative for visitors. However, their physical conservation, essential for efficiently ensuring information with precision, poses a serious challenge for the various professionals involved, as numerous social and anthropic risks threaten long-term preservation for the enjoyment of future generations. Of all traditional building materials, earth is undoubtedly one of the most fragile and sensitive to loss in the absence of the original protection systems, so that a precise assessment of its threats is essential to minimizing the destruction of these non-renewable assets. The objective of this study is to evaluate the most determining human risk factors within the territorial scope of the Iberian Peninsula, including aspects such as its musealization, suitable interpretation, visit planning, agricultural land use, vandalism and rural depopulation. This is achieved through a literature review and on-site data collection from 85 archaeological sites, as well as the development of an analysis tool to assess the degree of vulnerability, aiming to develop prevention measures.

Keywords: threats; preservation; tourism; architectural vulnerability; heritage; traditional construction; risk assessment; durability; adobe; rammed earth

1. Introduction

In past societies, earthen construction was one of the most widely developed construction systems [1], as the material is easily obtained and handled and can be found in abundance in any type of habitat. This has resulted in a broad spectrum of solutions derived from refined techniques, responses to needs, and construction cultures, serving monumental, residential, defensive, productive, and funerary purposes. Broadly speaking, four major construction groups have been identified internationally (mixed structures, cob, adobe and rammed earth), although each of them has given rise to a high number of subvariants in different latitudes, with identities defined by their connection to different communities. Their status as heritage of great interest is firmly cemented thanks to the architectural, historical and ethnological information they transmit.

At present, this legacy, which is still actively used for housing by up to a fifth of the world population, is a rich and highly valued international archaeological heritage. Increasing interest both in terms of cultural landscape (Devon, 2000) [2] and conservation (Lyon, 2016) [3] has attracted greater attention at international events such as the TERRA World Conference, as well as from international organizations such as UNESCO [4] and ICOMOS-ISCEAH. More attention has also been paid to intervention [5–7], promotion, and display to the wider public through different musealization strategies.

1.1. The Context of the Iberian Peninsula

In the Iberian Peninsula, the presence of this type of heritage has gradually been confirmed in past societies, including prehistory [8], protohistory [9] and the Roman period [10], confirming that this territory was particularly prolific for its development. This culture is currently transmitted to the wider public through the conservation of numerous archaeological sites, found in varying degrees in Spain and Portugal. Although the use of these systems has been copiously documented throughout Portugal [11,12], the levels of in situ preservation and display are far more limited.

The different studies carried out in Spain have revealed a rich representation of all sorts of construction techniques, although these are far more limited among the more vulnerable typologies, such as domestic, productive and funerary constructions, given their characteristics and size. These vulnerable groups are of greater interest as case studies, as they feature different remains in cob (Figure 1a), adobe (Figure 1b), and rammed earth (Figure 1c), as well as mixed structures in conjunction with wood. However, that combination cannot be directly observed in their original execution on site due to the biological nature of wood, which facilitates its decomposition over time, and requires interpretative reconstructions for the purposes of education (Figure 1d). The most commonly preserved cases until the Roman period are stabilized adobe structures with vegetation, complemented with stone masonry at the base [13]. In contrast, rammed earth appears to have been standardized at a later stage [14], while preserved examples of cob are identified less consistently.

(a) (b) (c) (d)

Figure 1. Earthen architecture families identified in the study area: (**a**) Cob wall at the archaeological site of the Roman villa La Olmeda (Pedrosa de la Vega, Palencia); (**b**) Adobe wall with alternating courses at the Roman domus in sector 18 of Libisosa (Lezuza, Albacete); (**c**) Rammed earth and plaster wall in the Mezquita del Cortijo del Centeno (Lorca, Murcia); (**d**) Modern reconstruction of a mixed wall with woven reeds at Castellón Alto (Galera, Granada).

1.2. The Cultural Challenge

The suitability of heritage for cultural and tourist use is seen as a complex activity from all angles, as it requires conscious interventions from multidisciplinary viewpoints in order to ensure, to an equal extent, the preservation over time of the found remnants and the proper transmission of information to the public. This scenario presents varying degrees of difficulty depending on the specific heritage, as for earthen constructions, which are inherently affected by abandonment processes brought about by changes in lifestyle which strip them of their original protections. Moreover, the absence of use, as observed in archaeological sites, where the dominant cultural function prevails at the expense of residential use, compromises the maintenance of the heritage.

Earthen architecture faces a number of specific threats [15] beyond natural issues, including social stigma associated with poverty, association with pre-industrial societies, and perceptions of disease [5]. This sense of vulnerability, heightened following catastrophes such as that in Bam (Iran), and coupled with a lack of professional recognition during the 20th century [8], poor preservation practices, and a general lack of interest in conservation, has led to its underrepresentation in archaeological sites worldwide, most notably in Spain. In archaeological terms, the traditional association of earthen architecture and lower status, stemming from the preference for alternative materials in large public constructions, is still found in contemporary society and poses challenges such as the decentralized nature of protection [16] and the lack of dedicated resources [17].

However, these scenarios are not the only ones contributing to the physical loss of remnants and their valorisation. These are exacerbated by the ongoing development of human activities such as material recycling or land use, observed since ancient societies. This issue, first addressed by Spanish institutions on 7 July 1911 [18] through the initial archaeological regulations, has gained great importance since then because of modern fieldwork systems and tools, which can remove artefacts, alter habitats, and destroy earthen structures in situ. While these heritage complexes have seen a major reduction and even disappearance of quarrying activities in recent years, challenges such as looting, vandalism, and new strategies for museum display, research, life, and visits continue to be relevant today.

These factors help establish the proper conservation and valorisation of earthen architecture in archaeological sites as one of the most complex scenarios for its enhancement. The human aspect should therefore be assessed through the observation and identification of existing issues and their origins, in order to minimize the resulting damages and optimize sustainable safeguarding. Proportionally, this challenge represents an added scientific and touristic value for those who successfully address the preservation and enhancement of the remains found, through sensitive and appropriate interventions.

While in recent times the preservation of these remains was predominantly addressed in response to natural threats, including quantitative assessments of cultural heritage in specific relation to those associated with climate change [19–22], the demand for new awareness of human actions in this field began to emerge at the end of the last century [23]. This aimed to encompass the entire process, from the survival of remains in the natural substrate despite agriculture or soil material extraction; uncovering and rescue measures; management, physical protection, and legal safeguards. Evaluating associated human factors can serve as a starting point for proposing guidelines and strategies in the adaptation of earthen archaeological sites to minimize effects, to contribute to the preservation of remains for future generations, as well as to promote the proper transmission of non-renewable historical, construction, and ethnological knowledge they hold (Figure 2).

Figure 2. Earthen structures preserved and protected in the archaeological site of La Celadilla (Ademuz, Valencia).

2. Methodology

The vulnerability assessment is developed based on the documentation compiled for the research, which integrates current characteristics in selected case studies according to social and anthropic factors. Two phases are undertaken to obtain values: the degree of vulnerability, understood as an index linked to exposure and sensitivity, allowing representation of how susceptible it is to potential loss against various threats considered in an abstract context; and the risk level, geographically locating the sites to verify the existence of real issues, aiming to provide a comprehensive overview of urgency.

2.1. Case Studies

Bibliographical reviews of case studies conserved in the field of interest are essential to establishing a solid foundation for subsequent analysis—a methodology often observed in archaeological assessments [24]. This publication is part of a research endeavour dedicated to globally analysing the risks associated with this type of architecture and context within the Iberian Peninsula [25], preselecting a total of 170 archaeological sites with characteristics of interest for this research. Priority is given to domestic, productive, and funerary architecture from the prehistoric, protohistoric, and Roman periods, as these are potentially more vulnerable in contrast with larger-scale, defensive constructions, and those closer to the medieval period. Out of the total sites, 85 are selected for their special documentary interest, based on the convergence of characteristics, relevance, or geographical dispersion (Figure 3), with a view to conducting subsequent assessments of social and anthropic vulnerability.

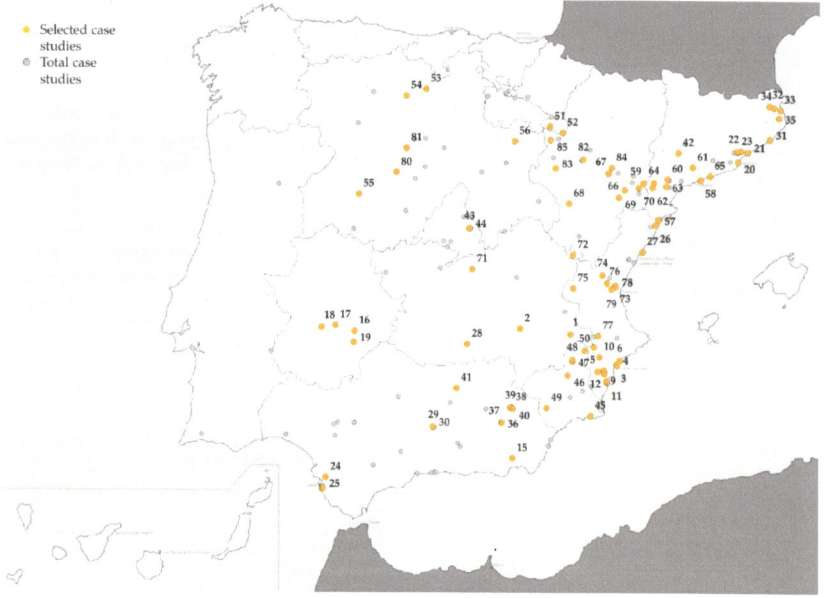

Figure 3. Distribution of total and selected case studies for the analysis of human vulnerability in the Iberian Peninsula, characterized by the presence of earthen architectural structures.

This database, which includes information on various specific factors, is used to statistically identify the most affected and recurring issues, as well as to assess qualitatively and quantitatively vulnerability through a tool that combines these factors, while also allowing for subsequent data cross-referencing in GIS environments. This collective effort provides insights into the aspects of the greatest urgency and attention, as well as possible correlations and origins through related national statistics.

2.2. Vulnerability Level

2.2.1. Human Factors in Vulnerability Assessment

Various characteristics and factors have been identified through the sample of case studies, based on their social or anthropic origin. In terms of analysis, a distinction is made between social vulnerability, referring to the ability to manage, protect, transmit, and appropriately value the structures, conveying their position in society to the public; and anthropic vulnerability, which would encompass potential damages derived from human activity.

Social vulnerability, exposure and dissemination are jointly considered criteria that are vital to this assessment. The level of accessibility granted brings this heritage closer to the wider public and must be provided in greater measure [26]. However, this may not always be optimal due to its precarious nature [23] or its location in inhospitable or remote settings (Figure 4a), with an increasing anthropic risk in contrast to the absence of human interaction while the archaeological remains are still underground [27]. Exposure is also dependent on the degree of visibility of the original materials and technique, so that the reburial of structures limits access for visitors [5], categorizing it as a strategy with inverse input depending on the perspective from which it is analysed.

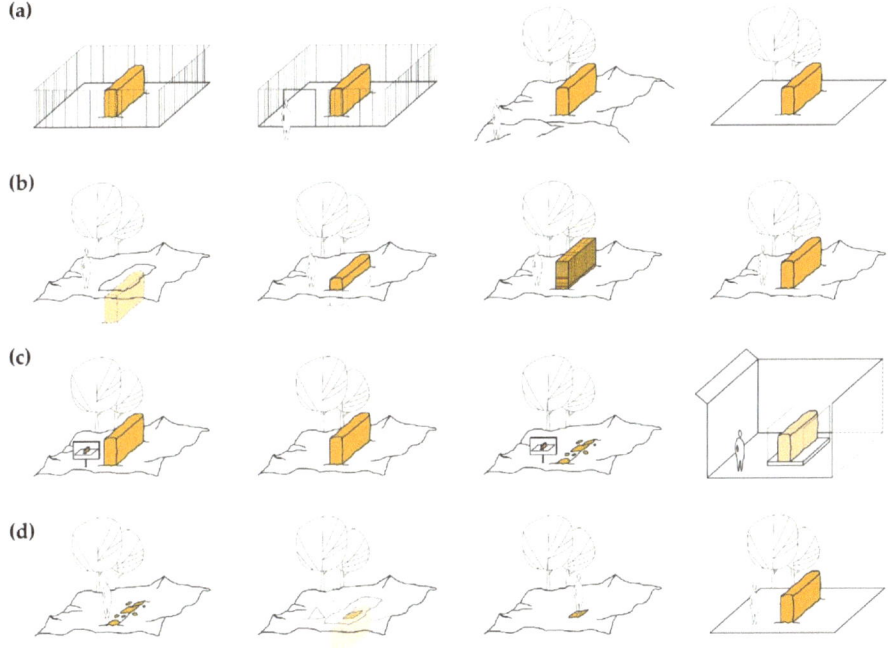

Figure 4. Schematic examples of analysed features: (**a**) Restricted access, limited, obstructed, or completely free; (**b**) Limited exposure due to burial, partially visible, visible with modern materials or original materials; (**c**) Visible with information, visible without information, through signage, or showcased in a museum; (**d**) Destruction, looting, walking over or without vandalism.

In addition, some relevant factors should be considered. These include legal protection, where inclusion in Catalogues of Heritage Protection or Master Plans is strongly recommended to encourage survival and equal intervention in keeping with guidelines [28]; vigilance from looting, which could be considered exhaustive and, therefore, more appropriate, after the installation of cameras or security equipment; daily use of enclaves following musealization can also be encouraged to counteract abandonment and subsequent plundering [29]; visual harmony, already damaged by urbanistic and architectural abuse [30] and its decontextualization, as modern intervention increases its invasiveness into structures, potentially causing issues for the interpretation of original materials and techniques, as well as undesired homogenizations [5] (Figure 4b); or musealization efforts (in situ or ex situ), where the selection of signage content linked to these tasks has been under discussion [31], limiting the visitor's experience of this type of architecture, as the signage of conservation of original materials is considered optimum, with the capacity of transmission decreasing depending on which of these is lacking [23], until its decontextualization due to transfer to a museum (Figure 4c) or the destruction and rendering hidden of remains.

In terms of anthropic vulnerability, it is worth highlighting the issues stemming from the use of earth in agriculture, the extent of which can vary depending on its level of use, contribution of humidity to the soil due to irrigation before excavation of the land where the site is found [32] or any sites adjoining it, along with the use of fertilizers [5]; as well as material extraction, where damage is extensive in active cases [33], although this threat is not as prevalent as it was in the 20th century. It was decided to ignore destruction due to cultivation of the land, as this prior action would have no effect on current vulnerability.

Although these factors have the potential to cause major damage, once the archaeological enclaves are revealed they become subject to even more factors, such as vandalism that is more serious and destructive [34] or plundering [30]; that derived from the opening to the public, such as trampling or carving on structures, which has a more progressive effect over time [5] (Figure 4d); or the absence of maintenance plans and emergency measures, thus risking potential destruction in short periods of time [6]. However, the risk from pollution is very low and is limited to chromatic variations on the surface, with a maximum damage of black crust associated with stone but not with earthen materials. As with the cultivation of the land, this evaluation has also ignored full dismantling for documentary purposes or as a result of urbanistic actions, as these actions have already been completed, although they will be examined and analysed statistically throughout the publication.

The response of each factor is assigned according to the greater or lesser impact of the factors based on the range of possibilities, graded from very low to very high, associated with a Response Value (RV) and a scale from 1 to 5, with the lower value corresponding to lower vulnerability and the higher value indicating a poorer response to the action of external agents. These scales are common in tests for heritage assessment [19] and reflect observations in the reviewed bibliography, adapted to the situation of the case studies selected during data collection. Given that a characteristic may exhibit its maximum level of impact but not influence to risks to the same extent as others, these values are then multiplied by factors between 0 and 1, referred to as Influence Value (IV), based on similar systems tested at environmental level for earthen architectural heritage [15].

2.2.2. Assessment Matrix

This process correlates the factors and values (Table 1) mentioned above through a vulnerability assessment table, using a Leopold matrix [35] for reflecting effects and causes. In recent years, different approaches have been proposed for the quantitative assessment of vulnerability of the architectural and archaeological heritage, including through the defining parameters of exposure, sensitivity and adaptive capacity [19], or degradation [36], as well as a reduction in risk from climate change [20,37]. Given that the current assessment is part of a larger study for the assessment of environmental and natural impact on this type of architecture [25], in this regard the proven accuracy of the Leopold Matrix [38] has been vital to the definitive and unified selection instead of the methodologies mentioned and others within the Multi Attribute Value Approach (MAVA) and extends to the social and anthropic aspects. In addition, its structured and systematic approach, as well as the ease for ruling out any factors which, in general, were difficult to access during data collection, add to the potential of this selection.

Table 1. Response (regular) and importance (bold) values assigned to human vulnerability factors identified. The response values are associated with a scale from 1 to 5, while the importance values are determined between 0 and 1.

Risk Factor	Social Value	Anthropic Value	Risk Factor	Social Value	Anthropic Value
Access	**1.0**	**0.7**	**Vandalism**	**-**	**0.9**
Not accessible	5	1	Not present	-	1
Limited access	3	3	Walked upon/Carving	-	3
Open access (obstacles)	2	5	Looting	-	4
Open access	1	5	Destruction	-	5
Exposure	**0.7**	**0.7**	**Agricultural activity**	**-**	**0.6**
Buried	5	1	Not present	-	1
Reburied	4	2	Previously	-	3
Partially visible	3	3	In adjacent plot	-	5
Visible (covered)	3	5	**Extractive activity**	**-**	**1.0**
Visible (non-original)	3	5	Not present	-	1
Visible	1	5	Present	-	5
Legal protection	**1.0**	**-**	**Maintenance plan**	**-**	**0.9**
Not present	5	-	Not present	-	5
Present	1	-	Present	-	1
Enhancement interventions	**0.7**	**-**	**Pollution**	**-**	**0.2**
Not present	5	-	Not present	-	1
In a museum	3	-	Present	-	5
On panels	3	-	**Aesthetic harmony**	**0.7**	**-**
Preserved without panels	2	-	Completely covered	4	-
Preserved with panels	1	-	Modern reconstruction	3	-
Surveillance	**0.7**	**-**	Encapsulation	3	-
Not present	5	-	Capping	2	-
Occasional	3	-	Traditional reconstruction	2	-
Exhaustive	1	-	Original remains	1	-

In this way, the sum of values is conditioned by the number of known characteristics given that in certain sites it may be impossible to know all chosen factors because of transparency, dissemination, or limited communication due to a low profile. Other complementary strategies implemented include the annulment of factors that are difficult to identify in most case studies and the assignment of high values to those that are particularly challenging. Additionally, consideration is given to the most unfavourable protection system, and high or reduced values are assigned in case of difficult recognition or reburial of the structures, respectively.

Therefore, the level of vulnerability is represented by the following formula:

$$VI_x = \frac{\sum(iv_x \times rv_x)}{\sum iv_x} \quad (1)$$

where: VI = Vulnerability index; iv = Importance value; rv = Response value.

The resulting indices have been tentatively classified into broad groups from 1 to 5, establishing ranges corresponding to low vulnerability (0.00–1.80), low–medium (1.80–2.60), medium (2.60–3.40), medium–high (3.40–4.20), and high (4.20–5.00) vulnerability. While the definition of these groups is practical for the purposes of dissemination, it is considered far more useful for comparison, as it allows for grading the resulting risk through evaluation with common methodologies.

2.3. Level of Risk

2.3.1. Database Creation

Vulnerability indices are contrasted using national demographic documentation obtained from various territorial institutions. The risk maps used correspond to population density, extracted from the Atlas Climático Ibérico (2011) [39] and the Atlas Nacional de España (2019) [40]; municipalities at risk of depopulation based on population density, from the Diagnóstico general del Reto Demográfico (2018) [41], considering threats in those with densities lower than 12.5 inhabitants/km^2; and demographic risk, drawn up by the Red SSPA—Mapa 174 (2020) [42], which integrated new complementary indicators to increase precision such as predictions of population growth or reduction, the physical environment, and demographic evolution over time.

2.3.2. Creation of Risk Maps

Risk maps have been created combining the national information mentioned above with georeferenced vulnerability indices in GIS environments. This has been carried out individually for social and anthropic factors, given their different nature, in order to observe possible unique interrelationships for each of these. The first is overlaid onto depopulation values, as the possible lack of means may result in increased risk. At the same time, anthropic vulnerability is cross-referenced in terms of population density, given the potential for a higher volume of visits and exploitation. This also reduces the need for attention in cases of high vulnerability but low demographic risk.

3. Results

The compilation of information regarding the various vulnerability factors has highlighted the most recurrent issues for this type of heritage site, which on the one hand can compromise the architectural richness offered at tourist level and, on the other, can hinder the feasibility of land use for both professionals and the general public.

3.1. Exposure and Dissemination

Enhancement is inherent to the conservation of archaeological sites with preserved earth structures, which aims to transform them to a greater or lesser extent to facilitate their interpretation and ensure their survival. In this regard, numerous factors can influence the outcome of achieving the quality required by such unique remains.

3.1.1. Accessibility and Visits

The issue of accessibility often cannot be separated from the dimension of archaeological sites, as a large part of these is located in isolated contexts where it is not always possible to provide roads for all types of visitors. Paved roads, for example, will undoubtedly facilitate arrival and promotion, while proper signage and preparation of the paths which make it accessible will also contribute to a lesser extent.

In this regard, the records from Spain present a high number of sites accessible via paved roads with varying degrees of suitability, making up 82% of the sample. However, there are different situations, such as the lack of paved accessibility, as in the case of El Oral (San Fulgencio, Alicante) (Figure 5a), complicating future valorisation efforts. Other examples include the absence of arrival signage and the use of agricultural roads, as seen in Tossal del Moro (Pinyeres, Tarragona) (Figure 5b); the construction of parking lots connected by pedestrian pathways to the archaeological site in Puntal dels Llops (Olocau, Valencia) (Figure 5c) or adjacent to the archaeological site itself, as in the cases of Vilars d'Arbeca (Arbeca, Lleida) and Los Millares (Santa Fe de Mondújar, Almería) (Figure 5d). This implies that one fifth of the sample would clearly be difficult for the general public to access, although it could also potentially have a positive effect by hindering illegal prospecting for looting or vandalism.

(a) (b) (c) (d)

Figure 5. Examples of access and visits in different archaeological sites: (**a**) Restricted access, lack of a road and musealization efforts at El Oral (San Fulgencio, Alicante); (**b**) Full-time open access through rural roads in Tossal del Moro (Pinyeres, Tarragona); (**c**) Roadside parking with a pedestrian path to Puntal dels Llops (Olocau, Valencia); (**d**) Vehicle parking adjacent to the site in Los Millares (Santa Fe de Mondújar, Almería).

Similarly, non-limited public openings can influence the offer of heritage transmission. In this regard, there is a wide variety of possibilities, including unrestricted access without any fencing in 27% of situations, temporary fencing in 24%, and permanent fencing in 49% (Figure 6). In general, less strict limitations will improve social dissemination, so that providing the most flexible system possible will minimize the risk of invisibility.

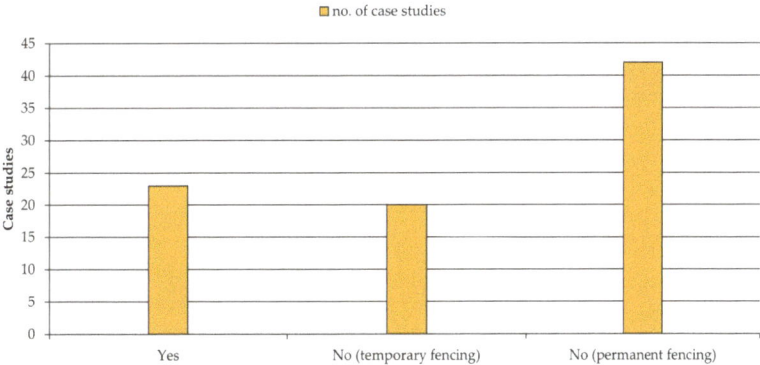

Figure 6. Access: list of case studies and availability for tourist visits based on their temporary or permanent perimeter restriction.

However, this factor presents an opposite risk from the anthropic perspective, as unlimited tourist access will proportionally increase the possibility of destruction or damage by humans, so that surveillance systems should be used to balance this vulnerability for optimal conservation. In this regard, 73% of the sites control access to the interior, although often such protection is illegally breached for looting or plundering purposes, and is therefore not a definitive conservation strategy.

Among those which allow visits, 22% allow unlimited access; 38% for more than 8 days per month; 6% between 4 and 8 days per month; 9% between 1 and 4 days per month; and

25% have not allowed public visits during the drafting of this research (Figure 7). This indicates that a quarter of the sites are either still undergoing excavation and research, exhibiting some degradation that jeopardizes the physical integrity and safe visits to the site, or are waiting for a musealization intervention. In contrast, 60% offer numerous visitation facilities beyond weekly holidays, while 15% limit access to holidays or opt for openings on the first, second, third, or last Sunday of each month.

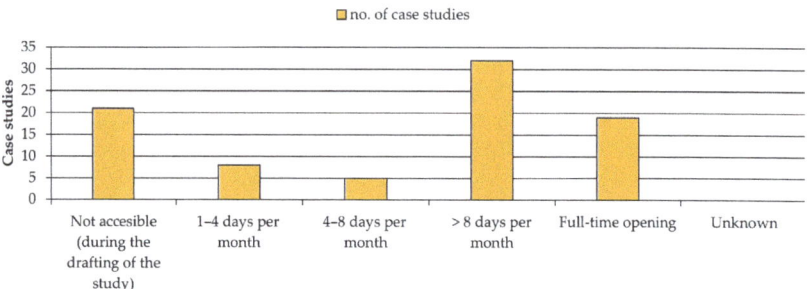

Figure 7. Opening days: connection between case studies and tourist visit options.

Meanwhile, a variety of security measures are employed, including preventive video surveillance in sites with limited access, such as Turó d'en Roïna/Can Taco (Montornès del Vallès, Barcelona) or Los Torrejones (Yecla, Murcia). Moreover, human security personnel regularly monitor certain archaeological complexes that are open continuously, such as Castellet de Banyoles (Tivisa, Tarragona) on the Ruta dels Íbers, reducing the incidents of illicit prospecting. In total, 36% of the sample had regular surveillance; 4% occasional; and 16% lacked these systems (Figure 8). Given the characteristics of the factor, i.e., whether active surveillance exists or is absent, could not be conclusively ascertained in 42% of cases.

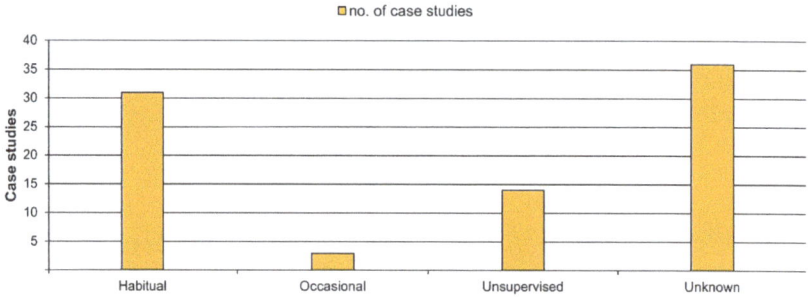

Figure 8. Surveillance: connection between case studies and their surveillance systems.

3.1.2. Musealization, Architectural Legibility, and Aesthetic Harmony

Another essential dimension of enhancement for tourist promotion is the transition from exposed remains to heritage sites adapted for visits. This subsequent cultural and exhibition use must reconcile preservation with research, encouraging promotion among the general public, showcasing as many typologies of musealization as there are museums, and requiring multidisciplinary collaboration to achieve optimum possible objectives. Broad distinctions can be established between in situ and ex situ musealization of archaeological sites.

In situ musealization involves the processes of incorporating materials, structures, and accessories for survival and visitation, ensuring physical preservation, adaptation of routes, and explanatory activities (Figure 9). As this research aims to highlight the human dimension, it is worth reflecting more deeply on the latter. The common use of signage and access paths introduces a physical limitation of informational space that affects content selection, potentially prioritizing other areas (historical, anthropological, etc.) over earthen construction (Figure 10a), without making up for the omitted information through QR codes or websites. This can result in information silence, further exacerbating the already diminished social recognition of earthen architecture by the general public, as seen in La Olmeda (Pedrosa de la Vega, Palencia), where reconstruction videos do not explore the material architectural characterization.

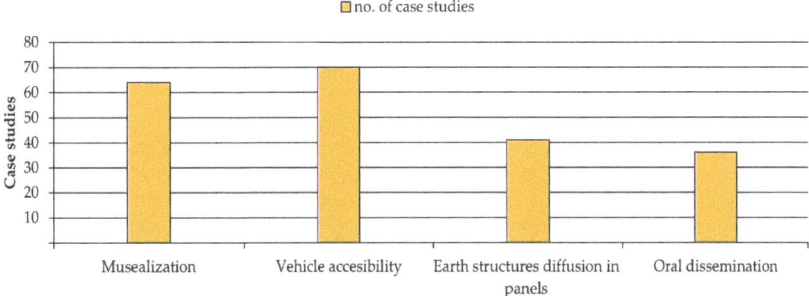

Figure 9. Social enhancement factors: study cases and availability of tourist facilities.

Figure 10. Examples of musealization and dissemination in different archaeological sites: (**a**) Selection of construction content excluding earthen techniques at the Roman villa of La Olmeda (Pedrosa de la Vega, Palencia); (**b**) Guided tours with a focus on adobe construction at Tossal del Moro (Pinyeres, Tarragona); (**c**) Exhibition of movable assets (adobes) extracted from the excavation at the museum of Cerro de la Cruz (Almedinilla, Córdoba); (**d**) Ex situ reconstruction of an earthen dwelling at the interpretation centre of the Cerro de las Cabezas (Valdepeñas, Ciudad Real).

In this regard, 48% of the case studies selected have shown some form of on-site dissemination of these construction techniques, implying in turn that half of the sample does not present a complete knowledge transmission process, despite the rise in supporting elements experienced in musealization in recent years [43]. This occurs, for example, in oral communication, where audio guides do not always explore these fields, unlike guided tours, where these issues are more frequently described, providing the site offers such a service with trained technical staff (Figure 10b). In general terms, 42% of the cases engage in oral dissemination, a figure which increases to 63% for the sites that offer this type of service.

A separate matter is ex situ musealization, outside the original site, usually in interpretation centres or municipal libraries. Typically, this involves more or less movable property [31], which could be of interest if it corresponded to construction pieces, such as adobes extracted from the excavation, as seen in the museum at Cerro de la Cruz (Almedinilla, Córdoba) (Figure 10c). However, sometimes full-scale reconstructions of immovable property are carried out for the faithful reproduction of the architectural space, as seen in Cerro de las Cabezas (Valdepeñas, Ciudad Real) (Figure 10d) and Cerro de la Virgen (Orce, Granada).

In total, 75% of archaeological sites feature some form of musealization. The remaining 25% are either currently being excavated, closed due to various risks, or temporarily abandoned. These strategies complicate the transmission to the public, depending on how effectively the construction culture is conveyed to the untrained visitor. Conflicting characteristics such as durability versus traditional harmonious appearance, or protection, context, and landscape, often come into play here. When combined with the wide range of available intervention solutions in the market, these give rise to noticeable alterations in how knowledge of these techniques is transmitted to tourists. The heterogeneous nature of modern materials for protection and preservation plays a crucial role in this debate, as does simple maintenance, which is highly recommended for risk prevention in the physical object, concealing heterogeneous geographical construction under sacrificial protection or coatings.

3.1.3. Demographic Issue

Another factor of relevance when carrying out maintenance, adaptation, and tourist use activities smoothly is the availability of economic and human resources. In this regard, it is worth highlighting the demographic challenges faced by a substantial portion of Spain. These challenges give rise to scenarios of severe depopulation, potentially leading to the abandonment and deterioration of significant sites, stemming from responses that are less efficient, less immediate, and of lower quality. Likewise, the appeal to tourism could face greater challenges due to the lack of visibility in these depopulated areas, given the scarcity of complementary attractions to enhance their appeal.

Although the correlation of this scenario and actual conservation deficiencies has not been evaluated in the present analysis, it is nonetheless interesting to note the number of sites located in these municipalities as a further point for analysis in the social sphere.

In this context, the results concerning studies related to population density would indicate that 21% of cases are at risk. However, considering recent analyses incorporating other characteristics such as the physical environment, demographic trends, and predictions of population growth or reduction, this figure rose to 28% in 1991 and 34% in 2022, with 1% classified as very severe, 11% as severe, and 22% as intermediate (Figure 11).

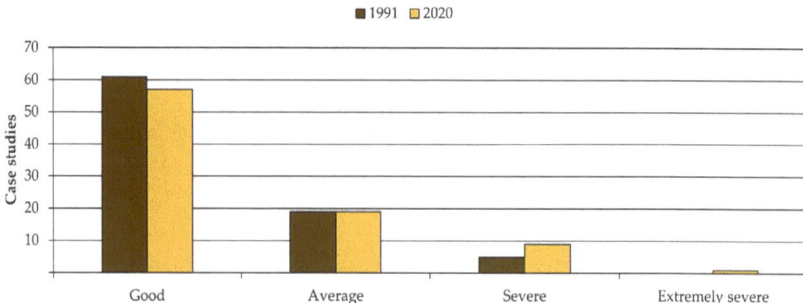

Figure 11. Demographic risk: evolution of depopulation risk in municipalities with tourist offerings of archaeological sites with earthen architecture.

3.2. Legal and Urbanistic Protection

At the management level, legal protection provides the starting point for protecting and planning activities, with comprehensive knowledge that highlights the complexity of the site and its needs. Therefore, there may be a higher risk for sites lacking any protection, such as recognition as a Local Relevance Asset (BRL), Cultural Interest Asset (BIC), the development of Master Plans, or inclusion in Catalogues of heritage protection.

The sample shows results of 82% recognition as BIC and 1% as BRL (Figure 12). This would mean that almost a fifth of the sites exhibited a degree of legal lack of protection and the consequent higher risk of physical loss. Additionally, Master Plans have been accessed for 22% of the case studies, with some of them in various stages of development or final approval.

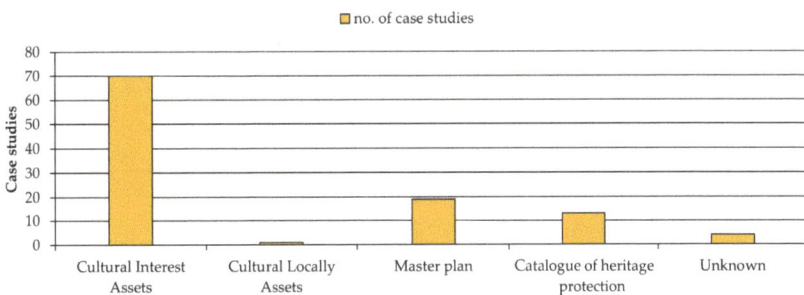

Figure 12. Legal protection: variability of tools identified in the different case studies.

3.3. The Use of the Land

Human utilization of the land can take diverse forms and serve various purposes, with agricultural (Figure 13) and extractive (Figure 14) exploitation of materials emerging as the systems with the most influence on archaeological sites. Although the extractive exploitation of materials focused its destructive impact in the 20th century, erasing sensitive archaeological areas, while it is now inactive in all case studies, agriculture continues to present risks associated with ploughing, tillage, and harvesting (prior to excavation), as well as with indirect factors like soil compaction, irrigation, terrace construction, or compositional modification. These activities can reduce soil permeability, increase moisture (in the case of intensive irrigation in areas adjacent to archaeological sites), or introduce chemical pollutants that may alter groundwater. All of these factors require specific studies that consider soil characterization and, consequently, the real impact and extent based on the particular location in which they occur.

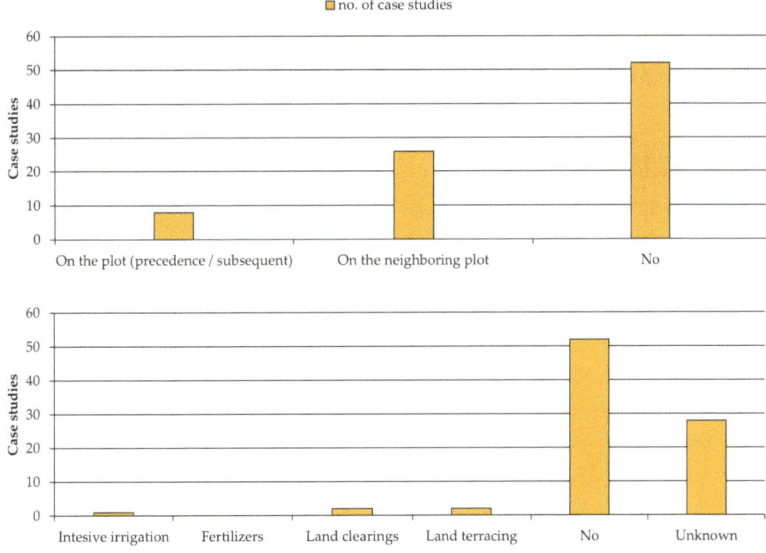

Figure 13. Agricultural activity: recurring scenarios affected by agricultural activity and type of activity conducted.

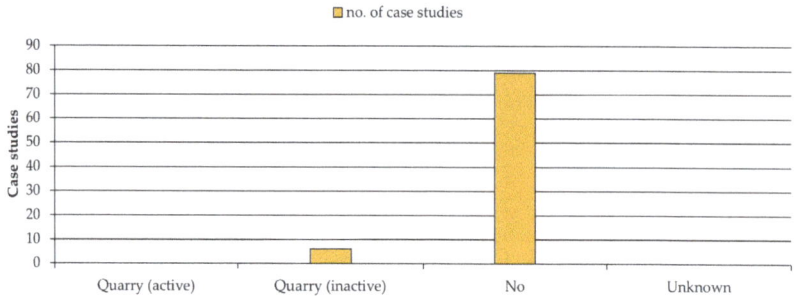

Figure 14. Extractive activity: variability in material extraction activity in the archaeological sites of the sample.

These scenarios have been observed in various archaeological sites under study, such as Cancho Roano (Zalamea de la Serena, Badajoz), where the impact of agriculture has led to the destruction of half of the adobe walls [44] (Figure 15a), which are currently reinterpreted through reconstructions. Another example is the Castellet de Banyoles (Tivissa, Tarragona), where the walls were reduced to a mere 30 cm in height [45] and have now been entirely lost. Other examples of reported intense agricultural activity could include the Castanheiro do Vento complex (Vila Nova de Foz Côa, Guarda), affecting the south, east, and west [46]; Casa del Mitreo (Mérida), in agricultural operation until the 20th century; or the Roman villa Piecordero I (Cascante, Navarra), with potential damage [47]. However, while the relationship between archaeological discoveries and farming activities is common, quantifying the actual extent of loss of structures is challenging once the remains are found. In general terms, up to 61% of the sample does not present a notable current risk due to factors derived from this issue, while 31% of structures are located in neighbouring or adjacent plots, and 9% show explicit agricultural use either before or after excavation (Figure 13).

Figure 15. Examples of events and threats regarding land use: (**a**) In the foreground, reconstructed adobes after loss caused by agricultural activities. In the background, original structures unaffected by these, at Cancho Roano (Zalamea de la Serena, Badajoz); (**b**) Agricultural use near the archaeological site of Alquería de Bofilla (Bétera, Valencia); (**c**) Loss of structures due to gypsum quarrying at Cabezo Redondo (Villena, Alicante); (**d**) Loss of structures due to stone quarrying at Puig de la Nau (Benicarló, Castellón).

Conversely, given the characteristics of the various soils, agricultural impact has been minimized in certain less suitable or fertile sites for the development of these activities, such as in Tossal del Moro [48]. Similarly, mountainous landscapes, like Turó d'en Roina/Can

Taco (Montornès del Vallès), have also contributed to effective preservation of the remains, aided by the increased difficulty of access for curious visitors and enthusiasts.

The destruction caused by material extraction, now inactive, has potentially contributed to the widespread disappearance of structures, as observed in Puig de la Nau (Benicarló, Castellón) [33] (Figure 15d), El Oral (San Fulgencio, Alicante), and Cabezo Redondo (Villena, Alicante) [49] (Figure 15c), making 7% of total case studies; 93% of the samples did not experience these threats (Figure 14). It should also be noted how these earth-moving activities can introduce changes in topography that alter water runoff, disruptions in soil stratigraphy, or complications for tourist or scientific visits.

3.4. The Anthropic Damage

The excavation and enhancement of archaeological remains initiate new processes of degradation which were absent while these were underground, and active interaction with human visitors is one of the most prominent factors. The effects can be classified based on their origin, ranging from voluntary and involuntary vandalism (Figure 16) to the consequences of archaeological documentation actions, including various construction projects unearthing remains and, even in the worst cases, leading to their dismantling, burial, or destruction.

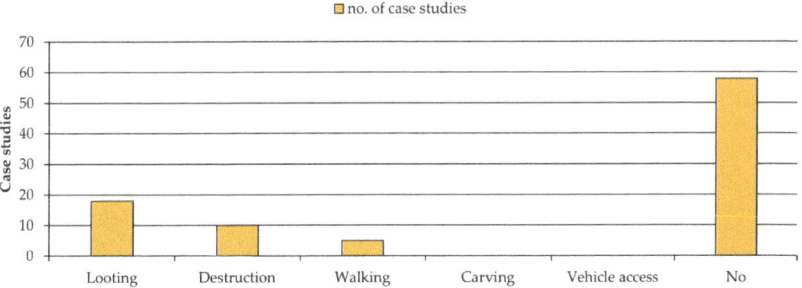

Figure 16. Anthropic damages: recurrence of conservation threats of a human nature after tourist use and enhancement.

Vandalism plays an active role in the destruction of archaeological heritage and can be seen in numerous phases of its life cycle. This includes the indirect damage of looting and pillaging during burial or excavation stages, with a scope of destruction that is hard to determine, as reported in Coimbra del Barranco Ancho (Jumilla, Murcia), where the elevations of four rooms are affected [50], or those at La Lloma de Betxí (Paterna, Valencia) [29], where uncontrolled recurrence prompted their official excavation in 1984. Also notable is the dramatic or gradual destruction of earthen structures due to uncontrolled access by the public, as observed in Medina Siyasa (Cieza, Murcia), or through destruction following illegal trespass, as witnessed in the year 2017 at Los Villares/Kelin (Caudete de las Fuentes, Valencia). It is worth highlighting that, unlike in other types of heritage, the presence of pathologies such as graffiti is virtually non-existent in the examined sample.

However, these degradation effects can also indirectly be caused by the visiting public, either due to poorly planned circulations and routes, an excess of visitors, or other damage derived from poor awareness of value and preservation. This can take the form of walking over or carving on the structures. This damage can range from the complete destruction of earthen structures, as seen in Puntal dels Llops (Olocau, Valencia) [34] (Figure 17a) and La Casa Grande (Alcalá del Júcar, Albacete) [51], the loss of volume, as in the hearths of Puig de la Nau (Benicarló, Castellón) (Figure 17b), or the gradual reduction in thickness, as gradually experienced in the adobe structures of Cerro de la Cruz (Almedinilla, Córdoba) until their definitive enhancement (Figure 17c). Occasionally, the placement of structures in spaces with changes in elevation, such as the cob wall of Caramoro I (Elche, Alicante) [52], could cause future problems due to the lack of signage (Figure 17d). Visitor carving, on the other hand, while documented in various international sites, does not seem to be as much of a concern as vehicle access in the case studies documented.

(a)   (b)   (c)   (d)

Figure 17. Examples of events and effects related to anthropic damage: (**a**) Anthropogenic integral destruction of adobe structures at Puntal dels Llops (Olocau, Valencia); (**b**) Volume loss due to trampling in the hearth at Puig de la Nau (Benicarló, Castellón); (**c**) Reduction in height and thickness in adobe structures at Cerro de la Cruz (Almedinilla, Córdoba), currently restored; (**d**) Potential walking over due to elevation changes at Caramoro I (Elche, Alicante).

Of the total sample, 32% of case studies are related to one of these phenomena, with 21% highlighting looting, 21% involving complete destruction, probably in collaboration with natural agents, and 6% associated with walking over.

The management of archaeological parks often has to face the challenges of the archaeological study and excavation process running parallel to tourism, introducing associated risks (Figures 18 and 19). Much of the destruction in this regard occurred during the 20th century, due to the precarious nature of early excavations, often described as pseudo-clandestine surveys, where these techniques were mostly unknown and safety measures were non-existent.

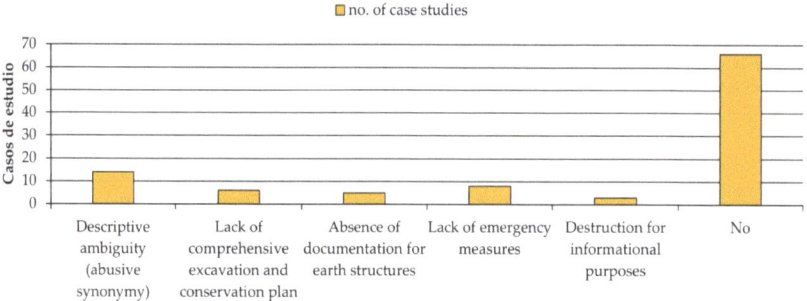

Figure 18. Anthropic excavation or documentation factors: recurrence of conservation threats of a human nature due to excavation activity and documentation of archaeological remains.

Although to some extent these situations have become less frequent, it is still easy to observe destruction for informational purposes. These methodologies involve the voluntary dismantling of earthen structures whose lower stratum becomes inaccessible, aiming to complete the stratigraphic sequence of the ensemble and extract all available historical information. This usually comes at the expense of the disappearance of the original material, whose intrinsic characteristics cannot be recovered, as an inherent process to the ongoing destructive nature of archaeology, and one whose essentiality continues to be justified by professionals as a necessary compromise. These scenarios have been seen in sites such as Castellet de Banyoles (Tivissa, Barcelona), involving Iberian adobe floors and benches (Figure 20b), and in Cástulo (Linares, Jaén), where earthen paving has been removed, revealing Roman mosaics. This has also been observed in cases where preservation is impossible, such as in El Castillar (Mendavia, Navarra), taking advantage of the extraction process for thorough documentation [53].

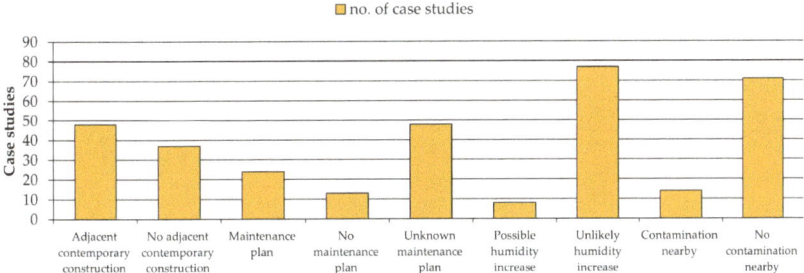

Figure 19. Other anthropic factors: recurrence of other human-related issues in archaeological sites in Spain.

Figure 20. Examples of the impact of other anthropogenic factors: (**a**) Loss of structures due to the absence of relief measures, due to the delayed acquisition of remains at Illeta dels Banyets (El Campello, Alicante); (**b**) Dismantling of pavement and adobe benches for documentary reasons to complete the archaeological sequence at Castellet de Banyoles (Tivissa, Tarragona); (**c**) Abandonment and risk of urbanization over kilns at Tossal de les Basses (Alicante); (**d**) Ventilation of structures in the semi-basement of the container building of Cerro de la Mota (Medina del Campo, Valladolid).

While these factors are challenging to detect retrospectively, 78% of the selected cases do not exhibit this impact, and descriptive ambiguity is noted in 16% of these. The absence of emergency measures accounts for 9%, lack of comprehensive planning in the excavation and intervention process for 7%, absence of documentation in 6% of cases, and physical destruction or dismantling for informational purposes in 4% of the sample.

Another issue associated with human action, although not necessarily with tourism, is related to indirect impacts on the urban or built environment, as well as the execution of infrastructure. It is clear that a significant number of archaeological findings are directly associated with urban development initiatives, which, in favourable instances, document, safeguard, and bury the remnants, or alter their routes to minimize impact. In less favourable scenarios, these initiatives can lead to physical damage, dismantling, or outright destruction of archaeological sites.

In an urban context, some sites are affected by this issue. This can be seen in the kilns of El Arsenal (Elche, Alicante), affected by the proposed urban development plan for the area [54] or Tossa de les Basses (Alicante), at risk of destruction due to channelling of the ravine, expansion, or renovation of the railway (Figure 20c). Regarding infrastructure, there are various clear examples, such as the slight deviation of the A-4 motorway brought about by the discovery of the Cerro de las Cabezas (Valdepeñas, Ciudad Real); the impact of the A-60 on the Lancia kilns, protected and buried in 2012 [55]; the destruction of Sitjar Baix (Onda, Castellón) due to the CV-10; or the dismantling of the L'aumedina kiln (Tivissa, Tarragona), adobe by adobe [56]. In special cases, this impact has been mitigated to prevent destruction through relocation interventions to a museum, such as the Arrollo Villalta kiln (Bobadilla, Málaga) to the Antequera Museum [57]; or kiln 5 of the El Ruedo pottery complex (Almedinilla, Córdoba) [58].

Finally, a range of minor factors arising from activities and circumstances generated by human intervention that could worsen conservation have also been considered. Among these factors are the potential to increase ambient humidity due to excess visitors in confined and unventilated spaces, the absence of maintenance plans, or the existence of elevated levels of environmental pollution. In some cases, such as the presence of noise pollution due to proximity to roads with heavy traffic (evident in sites like Tos Pelat in Moncada, Valencia), the tourist experience may be worsened, although they do not play an active role in the physical preservation of archaeological remains.

From a statistical perspective, these are anecdotal observations; 91% do not have containment structures that would favour the emergence of microclimates, with the remaining 9% being equipped with adequate ventilation systems; 84% of cases are not located in areas affected by pollution (industrial zones, large cities); and 56% are in direct contact or adjacent to contemporary structures.

3.5. Quantitative Risk Assessment

The numerous trade-offs presented in the in situ conservation of archaeological remains in relation to tourism, scientific exploration, and land use demand intervention measures—as well as prevention and organization efforts—to minimize their occurrence and development. In this regard, assigning precise values according to the vulnerability assessment methodology described can establish a gradation of urgency and a review of the needs of such sites in the territorial landscape of Spain (Table 2).

Table 2. Vulnerability indices resulting from 0 to 5 after assigning values to different human vulnerability factors and the 85 study cases with earthen architecture from Spain.

Archaeological Site	Social	Anthropic	Archaeological Site	Social	Anthropic
1. El Amarejo	2.46	3.32	44. Casa de Hippolytus	1.85	2.72
2. Libisosa	2.15	1.84	45. El Molinete	2.00	2.00
3. Tossa de les Basses	3.00	2.80	46. Medina Siyasa	1.85	3.28
4. Tossal de Manises	1.85	2.54	47. Coimbra del barranco ancho	2.17	3.38
5. Peña Negra	2.17	2.84	48. Villa de Los Cipreses	3.00	2.84
6. Illeta dels Banyets	1.85	1.84	49. Mezquita cortijo del centeno	3.00	2.28
7. El Arsenal	4.83	1.86	50. Villa romana de Los Torrejones	1.85	3.04
8. Caramoro I	3.12	2.06	51. Villa Romana Piecordero I	3.35	3.58
9. La Alcudia	2.15	2.32	52. Alto de la Cruz	3.68	2.20
10. El Monastil	1.85	2.56	53. Horno La Jericó	3.1	2.52
11. La Fonteta	2.15	1.42	54. Villa romana La Olmeda	2.00	2.08
12. Rábita Califal	1.56	1.84	55. Cerro de San Vicente	1.85	2.56
13. El Oral	3.85	1.62	56. Numancia	1.71	1.84
14. Cabezo Redondo	2.15	2.56	57. Moleta del Remei	2.00	2.56
15. Los Millares	1.85	2.56	58. Villa romana Els Munts	1.71	1.84
16. La Mata	2.15	2.56	59. Tossal del Moro	2.38	3.32

Table 2. Cont.

Archaeological Site	Social	Anthropic	Archaeological Site	Social	Anthropic
17. Casas del Turuñuelo	2.56	2.04	60. Calvari el Molar	1.88	3.32
18. Casa del Mitreo	1.85	2.24	61. Horno de Fontscaldes	2.85	2.76
19. Cancho Roano	1.85	2.44	62. Coll del Moro	2.71	2.52
20. Domus Avinyó	1.85	2.00	63. Castellet de Banyoles	1.88	3.80
21. Ca L'Arnau y Can Rodón	2.29	2.80	64. Turó del Calvari	2.42	2.84
22. Turó d'en Roïna/Can Taco	1.71	2.00	65. Ciutat Ibèrica de Calafell	1.56	2.56
23. Horno Camp d'en Ventura de l'Oller	2.42	1.72	66. El Palao	2.46	3.38
24. Doña Blanca	1.56	3.28	67. Cabezo de Alcalá	3.02	2.14
25. Horno de la Torrealta y Camposoto	1.85	2.72	68. La Caridad	3.29	2.76
26. Puig de la Nau	1.85	2.20	69. Hornos Mas de Moreno	3.54	2.76
27. Orpesa la Vella	3.85	1.14	70. San Cristóbal	1.29	2.84
28. Cerro de las cabezas	1.71	3.76	71. Plaza de los moros	1.88	3.32
29. Cerro de la Cruz	2.15	2.86	72. La Celadilla	1.88	3.38
30. Horno villa romana El Ruedo	1.71	3.04	73. Alquería de Bofilla	2.44	2.32
31. Turó Rodó	2.15	2.56	74. Castellet de Bernabé	2.56	3.00
32. Mas Castellar	2.38	3.38	75. Los Villares/Kelin	2.29	3.76
33. Ampurias	1.56	1.84	76. Tossal de Sant Miquel-Edeta	1.71	2.56
34. Horno Clos Miquel	1.42	2.00	77. Bastida de les Alcusses	1.85	2.38
35. Illa d'en Reixac	3.68	2.48	78. Tos Pelat	2.00	3.28
36. Cerro Santuario/Basti	2.00	3.04	79. Lloma de Betxí	1.73	4.02
37. Cerro Cepero/Basti	3.02	2.62	80. Cerro de La Mota	1.85	2.56
38. Necrópolis de Tútugi	1.56	3.10	81. Soto de Medinilla	3.85	2.50
39. Castellón Alto	2.15	2.56	82. Contrebia Belaisca	2.85	2.98
40. Cerro de la Virgen	2.85	2.56	83. Bílbilis	2.17	1.84
41. Cástulo	2.00	3.10	84. Lépida Celsa	2.17	2.12
42. Vilars d'Arbeca	1.56	2.32	85. La Oruña	1.94	3.32
43. Casa de los grifos	1.85	2.72			

These vulnerability indices, with a high degree of abstraction, have been overlaid in GIS environments using national information of human relevance, such as population density or demographic risk. This layer adds an additional dimension of specificity that lays the groundwork for studies reflecting on potential correlations between conservation and the degree of exploitation of heritage sites.

In terms of social risk, the vulnerability levels obtained show a distribution of 18% for the low category, 59% for the low–medium, 15% for the medium, 7% for the medium–high, and 1% for the high category. Noteworthy cases include El Arsenal (Elche, Alicante) for urbanistic vulnerability, the Mas de Moreno kilns (Foz-Calanda, Teruel), and Illa d'en Reixac (Ullastret, Girona) for difficulty of access, isolation, and lack of enhancement, respectively. In addition, the reburied sites of El Oral (San Fulgencio, Alicante), Alto de la Cruz (Cortes, Navarra) and Soto de Medinilla (Valladolid) stand out for their invisibility to the general public.

After data analysis, a higher demographic risk would be observed in municipalities such as La Oruña (Vera de Moncayo, Zaragoza), Lépida Celsa (Velilla de Ebro, Zaragoza), La Celadilla (Ademuz, Valencia), Cabezo de Alcalá (Azaila, Teruel), Cerro de la Virgen

(Orce, Granada), Libisosa (Lezuza, Albacete), El Amarejo (Bonete, Albacete), Castellón Alto, and the necropolis of Tútugi (Galera, Granada). The case of La Oruña stands out as very severe, and El Amarejo, Tútugi, Castellón Alto, Lépida Celsa, and Cerro de la Virgen show a severe risk [42] (Figure 21).

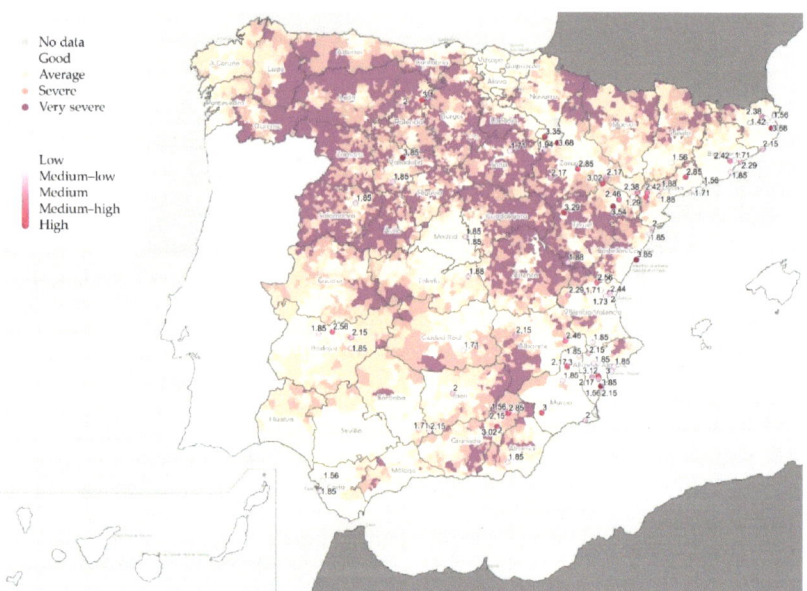

Figure 21. Data cross-referencing in case studies from Spain: social vulnerability indices (low–high) and demographic risk (good–very severe).

This reveals a significant concurrence between human vulnerability and depopulation, especially in intermediate levels, with up to seven case studies showing medium vulnerability in municipalities with intermediate risk levels (Figure 22). However, a more in-depth study is required to determine a direct correlation between both factors, as this could be due to the broader spectrum in these vulnerability levels.

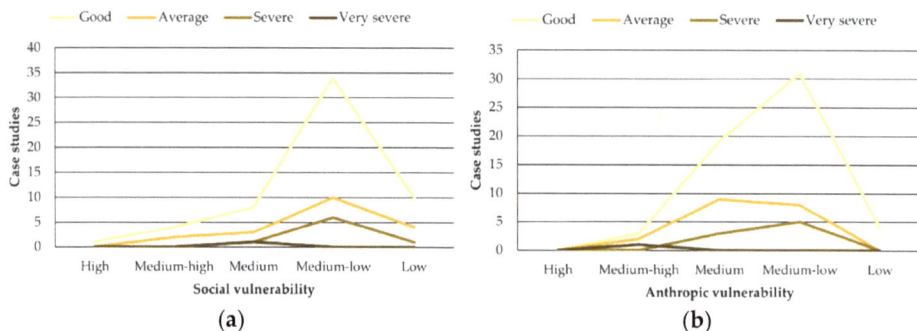

Figure 22. Data cross-referencing in case studies from Spain: demographic risk (good, average, severe, very severe) compared to results of social vulnerability (**a**) and anthropic vulnerability (**b**) (low, medium-low, medium, medium-high, high) of the archaeological sites.

In terms of anthropic risk, the vulnerability levels obtained are slightly higher, with 5% for the low category; 52% for low–medium; 36% for medium; and 7% for medium–high. Particularly vulnerable cases stand out, including La Oruña, inserted in agricultural landscapes, the Roman villa Piecordero I (Cascante, Navarra), La Lloma de Betxí (Paterna, Valencia), Castellet de Banyoles and Cerro de las Cabezas (Figure 23).

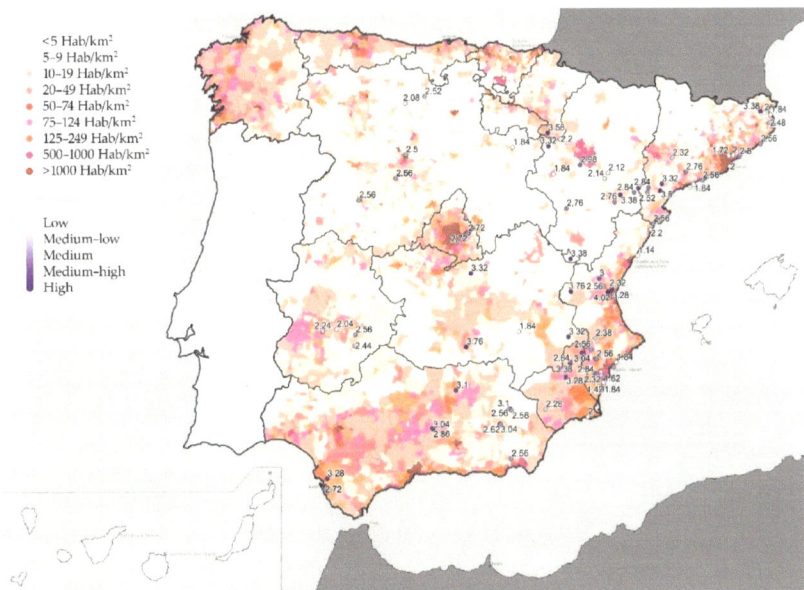

Figure 23. Data cross-referencing in case studies from Spain: anthropic vulnerability indices (low–high) and population density (from <5 inhabitants/km^2 to >1000 inhabitants/km^2).

In the data cross-referencing, a higher accumulation of case studies with elevated vulnerability levels is observed in municipalities with a population density of <75 inhabitants/km^2, particularly in terms of social vulnerability, while a similar accumulation of risk is also present in densely populated cities with densities exceeding 1000 inhabitants/km^2 (Figure 24). Although further study is needed to determine direct correlations between factors, similar to the demographic data analysis, this concentration in environments with smaller populations can cause greater real impact due to limited economic or human resources for dissemination, conservation, and tourist use.

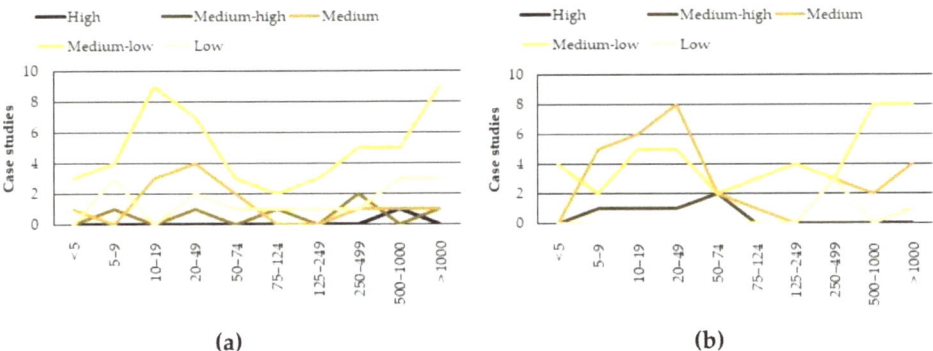

Figure 24. Data cross-referencing in case studies from Spain: municipal density <5 to >1000 inhab/km^2 compared to results of social vulnerability (**a**) and anthropic vulnerability (**b**) (low, medium-low, medium, medium high, high) of the archaeological sites.

4. Discussion

The qualitative reflection, statistical compilation, and quantitative evaluation lead to several noteworthy insights regarding the social aspect of tourist use, valorisation, legal protection, and associated threats. Overall, the level of vulnerability observed corresponds to a medium-low and medium value for the social domain and is slightly higher for the anthropic domain, with a predominance of medium indices (Figure 25).

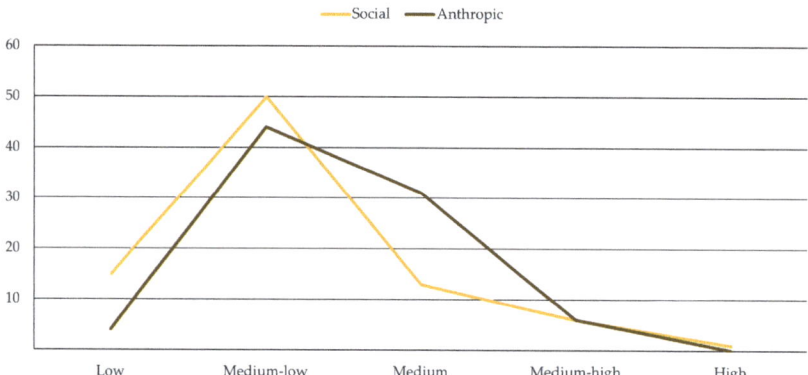

Figure 25. Levels of vulnerability for the social and anthropic aspects.

Regarding this last point, it is worth noting that a quarter of the case studies cannot be visited, although those accessible generally allow visits for more than eight days a month or freely and permanently. In addition, a vast majority are musealized and legally protected through at least one available tool. In contrast, only half have provided information about earthen structures on their on-site informational signage, and approximately half lack comprehensive surveillance measures (or these are unknown).

The aspect of accessibility, positively oriented towards dissemination and tourist visits, is inversely valued regarding anthropic risk. This may have contributed to its overall higher vulnerability, exposing the remains to increased human interaction, potentially resulting in more walking over or degradation as visitors lean on or touch the upper sections of low structures, such as homes and walls, as well as pavements. In contrast, as the impact of carving is moderate or non-existent in the case of carving, it should be addressed through preventive social recognition but is not considered so urgent.

Regarding the remaining anthropic factors, in many cases there has been a drop in activity in the previous agriculture and quarrying, so that many of them are currently inactive. However, it is important to consider the presence of intensive irrigation adjacent to the conservation and tourist valorisation project, as water control in the area can be a relevant factor when proposing solutions.

Destruction due to looting, vandalism, and scientific factors should be reflected on. In this regard, it is vital to establish cross-disciplinary points shared by archaeology, architecture and conservation specialists, in order to unify their opposing perspectives, allowing them to meet demands with the means currently available. In keeping with this, and in cases where non-invasive technology is insufficient for documentation, it is advisable not to rule out proposals for selective excavations in agreed areas where extensive dismantling is not required. Equally, whenever possible, knowledge of specific spaces should be inferred or discarded based on the sequences of adjoining spaces with no earthen structures preserved. As this issue affects one in every fifteen enclaves of these characteristics in the Iberian Peninsula, an updated and thorough debate in contemporary society is required for international impact.

5. Conclusions

The domestic, productive and funerary archaeological heritage executed in earth in Spain is a valuable resource, found in limited amounts, which can lead to a vigorous promotion of tourism around its locations and surroundings (Figure 26). However, there are numerous vulnerability factors which can minimize the social scope and increase the physical loss of remains when the site is in operation.

Figure 26. Enhancement and protection through permanent covers in La Mata (Campanario, Badajoz), as a focal point in municipal tourism promotion.

The scope of anthropic influence generates problems of relative severity following both excavation and exposure, in combination with other natural risks. Invisibilisation is also an issue, as it can deny society the enjoyment of the knowledge contained. The conditions imposed for cultural use and dissemination in cases of states of partial collapse, which require strategies for preservation in the event of the general natural risks affecting heritage, inevitably have a bearing on the maintenance of the original setting and the excessive visual homogenization of the spectrum of construction techniques. The general lack of master plans highlights the heterogeneous interventions currently observed in the professional community, imposing confusions which can discourage tourist interest and facilitate erroneous associations of systems. The selection of content to be transmitted through traditional signage and new technologies prioritizes the historical aspect, while succinctly addressing that of construction, with a more informative than educational approach, and not always explicitly displaying precision in terms of the material component, thus prolonging its limited recognition. A compromised assessment encourages the recurrence of acts of vandalism, both voluntary and involuntary, giving rise to the need for increasingly invasive interventions linked to the remaining factors.

Designing interventions to address this issue requires consideration of the particular characteristics of each archaeological site, achievable through studies initiated via Master Plans and multidisciplinary observation involving various fields throughout the entire process. While issues such as vandalism or looting should be addressed through educational exercises, administrative actions can provide the necessary surveillance and legal protection. The current process for ascertaining recurrence and urgency can preventively locate focal points of interest, minimize threats which arise when excavating the original structures, and facilitate the safeguarding of the remains throughout their cultural life cycle.

Author Contributions: Conceptualization, S.M.-F., C.M., F.V.L.-M. and V.C.; methodology, S.M.-F., C.M. and F.V.L.-M.; software, S.M.-F.; validation, S.M.-F.; formal analysis, S.M.-F.; investigation, S.M.-F.; resources, S.M.-F., C.M., F.V.L.-M. and V.C.; data curation, S.M.-F.; writing—original draft preparation, S.M.-F.; writing—review and editing, S.M.-F., C.M., F.V.L.-M. and V.C.; visualization, S.M.-F.; supervision, C.M., F.V.L.-M. and V.C.; project administration, C.M. and F.V.L.-M.; funding acquisition, C.M. and F.V.L.-M. All authors have read and agreed to the published version of the manuscript.

Funding: This work is part of the research project "Risk-Terra: La arquitectura de tierra en la Península Ibérica: Estudio de los riesgos naturales, sociales y antrópicos y estrategias de intervención e incremento de la resiliencia" (Ref. RTI2018-095302-B-I00; main researchers: Camilla Mileto and Fernando Vegas), funded by the Spanish Ministry of Science, Innovation and University. This research has been developed within a doctoral thesis funded by the Spanish Ministry of Science, Innovation and University (Ref. PRE2019-089629).

Data Availability Statement: The original contributions presented in the study are included in the article, further inquiries can be directed to the corresponding author.

Acknowledgments: To the Ministry of Science and Innovation of Spain for funding the research that is part of the present study at the Research Centre for Architecture, Heritage, and Management for Sustainable Development (PEGASO) at the Polytechnic University of Valencia. Also, to the guides of the different archaeological sites for sharing their experiences and knowledge in the preservation and dissemination of each of them.

Conflicts of Interest: The authors declare no conflicts of interest. The funders had no role in the design of the study; in the collection, analyses, or interpretation of data; in the writing of the manuscript, or in the decision to publish the results.

References

1. Guillaud, H. An Approach to the Evolution of Earthen Building Cultures in Orient and Mediterranean Regions. What Future for Such an Exceptional Legacy? *Al-Rāfidān J. West. Asiat. Stud.* **2003**, *24*, 1–28.
2. Fidler, J. *Terra 2000 Postprints: 8th International Conference on the Study and Conservation of Earthen Architecture, Torquay, Devon, UK, May 2000*; Routledge: London, UK, 2002.
3. Joffroy, T.; Guillaud, H.; Sadozaï, C. *Terra Lyon 2016: Articles Sélectionnés Pour Publication En Ligne/Articles Selected for on-Line Publication/Artículos Seleccionados Para Publicación En Línea*; Editions Craterre: Villefontaine, France, 2018.
4. UNESCO World Heritage Centre. *World Heritage Earthen Architecture Programme 2009/2017*; UNESCO World Heritage Centre: Paris, France, 2008.
5. Cooke, L. *Conservation Approaches to Earthen Architecture in Archaeological Contexts*; British Archaeological Reports; BAR Publishing: Oxford, UK, 2010.
6. Pedelì, C.; Pulga, S. *Conservation Practices on Archaeological Excavations: Principles and Methods*; Getty Conservation Institute: Los Angeles, CA, USA, 2013.
7. Bendakir, M. *Les Vestiges de Mari: La Préservation d'une Architecture Millénaire*; Editions de La Villette: Grenoble, France, 2009.
8. Pastor Quiles, M. *La Construcción Con Tierra En Arqueología. Teoría, Método, Técnicas y Aplicación*; Publicaciones de la Universidad de Alicante: Sant Vicent del Raspeig, Spain, 2017.
9. Belarte Franco, M.C. *L'utilisation de La Brique Crue Dans La Péninsule Ibérique Durant La Protohistoire et La Période Romaine*; de Chazelles, C.-A., Klein, A., Pousthomis, N., Eds.; Editions de l'Espérou: Montpellier, France, 2011.
10. Beltrán Lloris, M. La Casa Hispanorromana. Modelos. *Bolskam* **2003**, *20*, 13–63.
11. Bruno, P. Arquitecturas de Terra nos Espaços Domésticos Pré-Históricos do Sul de Portugal. Sítios, Estruturas, Tecnologías e Materiais, Universidade de Lisboa. Available online: https://repositorio.ul.pt/handle/10451/3475 (accessed on 14 January 2024).
12. Olaio, A. From Earth to Structure: New Insights on the Iron Age Domestic Architecture from Quinta Do Almaraz (Almada, Portugal). In Proceedings of the ARAR Arqueología & Arquitectura. La Arquitectura de Tierra en el Mediterráneo Antiguo: Perspectivas, Estrategias y Metodologías (Guareña-Badajoz, 21–24 de febrero de 2024), Guareña, Spain, 21–24 February 2024.
13. Manzano-Fernández, S.; Mileto, C.; Vegas, F.; Cristini, V. Examination of Earthen Construction in Archaeological Sites of the Iberian Peninsula for Risk Analysis. In *Vernacular Heritage: Culture, People and Sustainability: Heritage 2022 International Conference, Valencia, Spain, 15–17 September 2022*; Mileto, C., Vegas, F., Cristini, V., García-Soriano, L., Eds.; Universitat Politècnica de València: Valencia, Spain, 2022; pp. 401–408. [CrossRef]
14. De Chazelles, C.-A. Les Constructions En Terre Crue d'Empúries a l'époque Romaine. *Cypsela* **1990**, *VIII*, 101–118.
15. Mileto, C.; Vegas, F.; Cristini, V.; García-Soriano, L. Initial Assessment of Multi-Risk Social Vulnerability for Iberian Earthen Traditional Architecture. *Procedia Struct. Integr.* **2020**, *29*, 34–39. [CrossRef]
16. Reeder-Myers, L.A. Cultural Heritage at Risk in the Twenty-First Century: A Vulnerability Assessment of Coastal Archaeological Sites in the United States. *J. Isl. Coast. Archaeol.* **2015**, *10*, 436–445. [CrossRef]

17. Taçon, P.S.C.; Baker, S. New and Emerging Challenges to Heritage and Well-Being: A Critical Review. *Heritage* **2019**, *2*, 1300–1315. [CrossRef]
18. AA.VV. *Las Excavaciones Arqueológicas y Sus Problemas. Legislación*; Institución Fernando el Católico: Zaragoza, Spain, 1980.
19. Nakhaei Ashtari, M.; Correia, M. Assessment of Vulnerability and Site Adaptive Capacity to the Risk of Climate Change: The Case of Tchogha Zanbil World Heritage Earthen Site in Iran. *J. Cult. Herit. Manag. Sustain. Dev.* **2022**, *12*, 107–125. [CrossRef]
20. Daly, C. A Framework for Assessing the Vulnerability of Archaeological Sites to Climate Change: Theory, Development, and Application. *Conserv. Manag. Archaeol. Sites* **2014**, *16*, 268–282. [CrossRef]
21. García, B.M. Resilient Cultural Heritage for a Future of Climate Change. *J. Int. Aff.* **2020**, *73*, 101–120.
22. Bertolín, C.; Camuffo, D. Deliverable 5.2 Climate Change Impacts Movable and Immovable Cultural Heritage throughout Europe. Damage risk assessment, economic impact and mitigation strategies for sustainable preservation of cultural heritage in the times of climate Change. In *Climate for Culture*; The European Climate Adaptation Platform Climate-ADAPT: Washington, DC, USA, 2014.
23. Arias Vilas, F. Sitios Musealizados y Museos de Sitio: Notas Sobre Dos Modos de Utilización Del Patrimonio Arqueológico. *Museo* **1999**, *4*, 39–57.
24. Correia, M. Evaluation Criteria for Earthen Archaeological Sites. In *Conservation des Architectures de Terre sur les Sites Archéologiques. Nouvelles Pratiques et Perspectives*; Gandreau, D., Sadozaï, C., Eds.; Éditions CRAterre: Grenoble, France, 2015; pp. 54–58.
25. Manzano-Fernández, S. Arquitectura de Tierra En Yacimientos Arqueológicos de La Península Ibérica: Estudio de Riesgos Naturales, Sociales, Antrópicos y Estrategias de Intervención, Universitat Politècnica de València. 2023. Available online: https://riunet.upv.es/handle/10251/197994 (accessed on 24 January 2024).
26. Juncà Ubierna, J.A.; Amigo Alvaro, R.; Ruiz Díaz, F.J.; García García-Castro, C.; Zarza Alejo, R. *Accesibilidad Universal Al Patrimonio Cultural. Fundamentos, Criterios y Pautas*; Ministerio de Sanidad, Política social e Igualdad: Madrid, Spain, 2011.
27. de Guichen, G. *Plan de Conservación Preventiva de La Cueva de Altamira*; Instituto del Patrimonio Cultural de España: Madrid, Spain, 2014.
28. Instituto del Patrimonio Cultural Español. La Conservación y Restauración del Patrimonio Cultural Exige un Conocimiento Profundo y Riguroso de Los Bienes Que lo Integran: Su Historia, Significados o Valores, Las Circunstancias, Procesos y Contextos de su Creación y Evolución. Available online: https://ipce.cultura.gob.es/investigacion/conservacion-bienes-culturales/documentacion-estudios.html (accessed on 9 April 2024).
29. De Pedro Micó, M.J.; Fumanal García, M.P.; Ferrer García, C.; Jover Maestre, F.J.; Lopez Padilla, J.A.; Grau Almero, E.; Pérez-Jordà, G.; Sarrión Montañana, I. *La Lloma de Betxí (Paterna, Valencia), Un Poblado de La Edad Del Bronce*; Diputación Provincial de Valencia, Servicio de Investigación Prehistórica: Valencia, Spain, 1998.
30. Melucco Vaccaro, A. La Particularidad Del Problema Arqueológico. In *La Carta de Riesgo: Una Experiencia Italiana Para la Valoración Global de los FACTORES de Degradación del Patrimonio Monumental*; Istituto Superiore per la Conservazione ed il Restauro, Instituto andaluz del patrimonio Histórico, Eds.; Junta de Andalucía, Instituto Andaluz del Patrimonio Histórico: Sevilla, Spain, 1992; pp. 18–21.
31. Lasheras Corruchaga, J.A.; Hernández Prieto, M.Á. Explicar o Contar. La Selección Temática Del Discurso Histórico En La Musealización. In *III Congreso Internacional Sobre Musealización de Yacimientos Arqueológicos. De la Excavación al Público. Procesos de Decisión y Creación de Nuevos Recursos*; de Francia Gómez, C., Erice Lacabe, R., Eds.; Ayuntamiento de Zaragoza, Área de Cultura y Turismo, Servicio de Cultura: Diputación Provincial de Zaragoza; Institución 'Fernando el Católico': Zaragoza, Spain, 2005; pp. 129–136.
32. Gandreau, D. Patrimoine Archéologique En Terre et Développement Local. Enjeux Interdisciplinaires et Perspectives de Formation, Université Grenoble Alpes, ENSAG. 2017. Available online: https://theses.hal.science/tel-01734984 (accessed on 11 December 2023).
33. Gusi Jener, F.; Oliver Foix, A.; Gómez Bellard, F.; Arenal, I.; Pérez-Pérez, A.; Valdés, L.; Cubero, C.; López de Roma, M.T.; Burjachs, F.; Castaños Ugarte, P.; et al. *El Puig de La Nau: Un Hábitat Fortificado Ibérico En El Ámbito Mediterráneo Peninsular*; Diputació de Castelló: Castellón de la Plana, Spain, 1995.
34. Bonet Rosado, H.; Mata Parreño, C. *El Puntal Dels Llops: Un Fortín Edetano*; Diputación Provincial de Valencia, Servicio de Investigación Prehistórica: Valencia, Spain, 2002.
35. Leopold, L.B.; Clarke, F.E.; Hanshaw, B.B.; Balsey, J.R. *Procedure for Evaluating Environmental Impact*; US Department of the Interior: Washington, DC, USA, 1971; Volume 2. [CrossRef]
36. Cappai, M.; Pia, G. Un modelo Fuzzy para el estudio de la cinética de degradación en sitios arqueológicos. In Proceedings of the ARAR Arqueología & Arquitectura. La Arquitectura de Tierra en el Mediterráneo Antiguo: Perspectivas, Estrategias y Metodologías (Guareña-Badajoz, 21–24 de Febrero de 2024), Guareña, Spain, 21–24 February 2024.
37. Forino, G.; MacKee, J.; von Meding, J. A Proposed Assessment Index for Climate Change-Related Risk for Cultural Heritage Protection in Newcastle (Australia). *Int. J. Disaster Risk Reduct.* **2016**, *19*, 235–248. [CrossRef]
38. Verd, J. Recursos Para Las CTMA: La Matriz de Leopold, Un Instrumento Para Analizar Noticias de Prensa de Temática Ambiental. *Enseñanza Cienc. Tierra* **2000**, *8*, 239–246.
39. AA.VV. *Atlas Climático Ibérico: Temperatura Del Aire y Precipitación (1971–2000)*; Agencia Estatal de Meteorología, Ministerio de Medio Ambiente y Medio Rural y Marino, Instituto de Meteorología de Portugal: Lisbon, Portugal, 2011. [CrossRef]
40. AA.VV. *España En Mapas. Una Síntesis Geográfica*; Ministerio de Fomento: Madrid, Spain, 2019. [CrossRef]

41. AA.VV. *Directrices Generales de la Estrategia Nacional Frente al Reto Demográfico*; Ministerio para la Transición Ecológica y el Reto Demográfico: Madrid, Spain, 2023.
42. Zúñiga-Antón, M.; Guillén, J.; Caudevilla, M.; y Bentué-Martínez, C. Mapa 174. Zonificación de los Municipios Españoles Sujetos a Desventajas Demográficas Graves y Permanentes. StoryMap. Available online: https://storymaps.arcgis.com/stories/9dd9b6e20cad403c95e87d4cc493c8fb (accessed on 26 January 2024).
43. Beltrán de Heredia Bercero, J.; Sánchez Montes, A.L.; Rascón Marqués, S. Pasado, Presente y Futuro de La Musealización de Yacimientos En España. *VI Congr. Int. Musealización Yacim. Patrim.* **2010**, *1*, 139–159.
44. Maluquer de Motes, J. *El Santuario Protohistórico de Zalamea de La Serena, Badajoz*; Consejo Superior de Investigaciones Científicas, Universidad de Barcelona: Barcelona, Spain, 1981.
45. Asensio Vilaró, D.; Sanmartí Grego, J.; Jornet Niella, R.; Miró Alaix, M.T. L'urbanisme i l'arquitectura Domèstica de La Ciutat Ibèrica Del Casteller de Banyoles (Tivissa, Ribera d'Ebre). In *Iberos del Ebro. Actas del II Congreso Internacional (Alcañiz-Tivissa, 16–19 de Noviembre de 2011)*; Belarte Franco, M.C., Benavente Serrano, J.A., Fatás Fernández, L., Diloli Fons, J., Moret, P., Noguera Guillén, J., Eds.; Institut Català d'Arqueologia Clàssica (ICAC): Tarragona, Spain, 2012; pp. 173–193.
46. Oliveira Jorge, V. Castanheiro Do Vento, Uma Ruína de Arquitetura de Terra ("Colina Monumentalizada") Pré-Histórica Do Norte de Portugal. In *SOS-Tierra: Restauración y Rehabilitación de la Arquitectura Tradicional de Tierra en la Península Ibérica*; López-Manzanares, F.V., Mileto, C., Eds.; *manuscript in preparation, submitted*.
47. Gómara Miramón, M.; Serrano Arnáez, B.; Bonilla Santander, Ó. Un Torcularium de Los Siglos I a.C.-I d.C. Del Yacimiento Romano Piecordero i (Cascante, Navarra). In *Estudis sobre ceràmica i arqueologia de l'arquitectura. Homenatge al Dr. Alberto López Mullor*; Aquilué Abadías, J., Beltrán de Heredia, J., Caixal Mata, A., Fierro Macià, X., Kirchner, H., Eds.; Diputación Provincial de Barcelona: Barcelona, Spain, 2020; pp. 417–425.
48. Arteaga Matute, O.; Sanmartí Grego, E.; Padró Parcerisa, J. *El Poblado Ibérico Del Tossal Del Moro de Pinyeres (Batea, Terra Alta, Tarragona)*; Institut de Prehistòria i Arqueologia: Santander, Spain, 1990.
49. Hernández Pérez, M.S.; García Atiénzar, G.; Barciela González, V. *Cabezo Redondo (Villena, Alicante)*; Universidad de Alicante: San Vicente del Raspeig, Spain, 2019.
50. García Cano, J.M.; Hernández Carrión, E.; Page del Pozo, V. Excavación de Urgencia En El Conjunto Arqueológico de Coimbra Del Barranco Ancho (Jumilla-Murcia) 1995. In *Séptimas Jornadas de Arqueología Regional*; Lechuga Galindo, M., Sánchez González, M.B., Eds.; Editora Regional de Murcia: Murcia, Spain, 2002; pp. 222–226.
51. Broncano Rodríguez, S.; Coll Conesa, J. Horno de Cerámico Ibérico de La Casa Grande, Alcalá de Júcar (Albacete). *Not. Arqueol. Hispánico* **1988**, *30*, 187–228.
52. Pastor Quiles, M.; Jover Maestre, F.J.; Martínez Monleón, S.; López Padilla, J.A. La Construcción Mediante Amasado de Barro En Forma de Bolas de Caramoro I (Elche, Alicante): Identificación de Una Nueva Técnica Constructiva Con Tierra En Un Asentamiento Argárico. *Cuad. Prehist. Arqueol. Univ. Autónoma Madr.* **2018**, *1*, 81–99. [CrossRef]
53. Fonseca de la Torre, H.J.; Arróniz Pamplona, L.; Calvo Hernández, C.; Cañada Sirvent, L.; Meana Medio, L.; Bayer Rodríguez, X.; Pérez Legido, D. The Problematic Conservation of Adobe Walls in the Open-Air Site of El Castillar (Mendavia, Navarre, Spain). In *Earthen Construction Technology: Proceedings of the XVIII UISPP World Congress, Paris, France, 4–9 June 2018*; Daneels, A., Torras Freixa, M., Eds.; Achaeopress: Oxford, UK, 2021; pp. 109–117.
54. Serrano, P. Cómo Se Ha Llegado Al Desaguisado de El Arsenal de Elche: Cronología de Otro Conflicto Urbanístico. *Elche Plaza*. 5 February 2021. Available online: https://alicanteplaza.es/como-se-ha-llegado-al-desaguisado-urbanistico-de-el-arsenal-de-elche-cronologia-de-otro-conflicto-urbanistico (accessed on 14 April 2023).
55. Gustavo López, D. Sos Lancia (3) ¿También Serán Destruidos Los Hornos de Lancia? *ProMonumenta: Asociación de Amigos del Patrimonio Cultural de León*. 5 June 2017. Available online: http://promonumenta.com/sos-lancia-3-tambien-seran-destruidos-los-hornos-de-lancia (accessed on 21 March 2023).
56. Pérez Suñé, J.M.; Rams Folch, P. *Memòria. Desmuntatge de Les Estructures Arqueològiques Del Jaciment de l'Aumedina Situades Al PK 20+500 de Ka Carretera C-44*; Generalitat de Catalunya: Barcelona, Spain, 2010. Available online: http://hdl.handle.net/10687/425594 (accessed on 21 June 2022).
57. Santos Fernández, J.L. Adif Traslada al Museo de Antequera un Horno Romano Hallado Durante la Ejecución de las Obras de alta Velocidad. *Terrae Antiqvae*. 7 May 2012. Available online: https://terraeantiqvae.com/m/blogpost?id=2043782:BlogPost:205746 (accessed on 5 May 2023).
58. Muñiz Jaén, I. Seguimiento Arqueológico En La Villa Romana de 'El Ruedo' (Almedinilla-Córdoba) II. *Anu. Arqueol. Andal. 1998* **2001**, *3*, 215–223.

Disclaimer/Publisher's Note: The statements, opinions and data contained in all publications are solely those of the individual author(s) and contributor(s) and not of MDPI and/or the editor(s). MDPI and/or the editor(s) disclaim responsibility for any injury to people or property resulting from any ideas, methods, instructions or products referred to in the content.

Article

Community Attachment to AlUla Heritage Site and Tourists' Green Consumption: The Role of Support for Green Tourism Development

Ibrahim A. Elshaer [1,2,*], **Mansour Alyahya** [2], **Alaa M. S. Azazz** [3,4] **and Sameh Fayyad** [2,5]

1. Department of Management, College of Business Administration, King Faisal University, Al-Ahsaa 380, Saudi Arabia
2. Hotel Studies Department, Faculty of Tourism and Hotels, Suez Canal University, Ismailia 41522, Egypt; malyahya@kfu.edu.sa (M.A.); sameh.fayyad@tourism.suez.edu.eg (S.F.)
3. Department of Social Studies, Arts College, King Faisal University, Al-Ahsaa 380, Saudi Arabia; aazazz@kfu.edu.sa
4. Tourism Studies Department, Faculty of Tourism and Hotels, Suez Canal University, Ismailia 41522, Egypt
5. Hotel Management Department, Faculty of Tourism and Hotels, October 6 University, Giza 12573, Egypt
* Correspondence: ielshaer@kfu.edu.sa

Abstract: This study explores the interrelationship between community attachment in AlUla Heritage City (located in Saudi Arabia) and tourists' green consumption practices, testing support for green tourism development as a mediator. The old historical city of AlUla, a significant city experiencing ongoing preservation and tourism development, represents an adequate context for exploring the link between community attachment and green tourism practices. This study employs a quantitative approach, including surveys with 328 local residents of AlUla. A structural equation modeling partial least square (PLS-SEM) analysis is conducted to explore the indirect influence of community attachment on tourists' green consumption through the mediating role of support for green tourism development. The findings indicated a positive path from community attachment in AlUla to tourists' tendencies toward green consumption behavior. Moreover, the mediating effects of support for green tourism development suggested that a deep sense of community attachment improves tourists' support for green practices in the tourism sector. This study adds to the extended body of the literature on place attachment and green tourism by emphasizing the significance of community place attachment in stimulating tourists' green consumption practices. Several practical implications for policymakers seeking to promote green tourism practices in heritage cities like AlUla are explored from the study results.

Keywords: community attachment; place dependence; place identity; affective attachment; sustainable tourism; green consumption

1. Introduction

The Saudi city of AlUla has emerged as a popular tourist destination because of its stunning natural features, unique examples of wildlife, diverse history, and thousands of years old antiquities [1]. AlUla has consequently grown to be one of the most significant ecotourism destinations, which prioritizes environmental preservation and provides tourists with distinctive experiences by promoting sustainable economic growth, preserving natural resources, and supporting residents [2]. Therefore, the responsible authorities in AlUla resorted to adopting strategies to develop green tourism in the city to conserve the ecosystem and preserve the well-being of local residents [3]. According to Dodds and Joppe [4], the concept of green tourism has four components including (1) environmental commitment: preserving and improving nature and the physical environment to guarantee the long-term health of the ecosystems that support life; (2) local economic vitality: to maintain economic sustainability, local economies, enterprises, and communities are supported; (3) cultural diversity: respecting and valuing cultural variety is essential to ensure the long-term survival of the host or local cultures; and (4) experiential richness: offering fulfilling and pleasurable experiences via direct and meaningful engagement with people, places, animals, and cultures. In this context, The Royal Commission for AlUla, to conserve archaeological and heritage sites and involve residents in enhancing the values of the local culture and fostering green tourism, has launched numerous campaigns and initiatives in international educational exchanges and encouraging local engagement with resident communities. For example, The Royal Commission for AlUla's Hammayah program strives to involve residents in managing their natural and cultural heritage to engage and advocate for their community. The Commission has also enlisted the help of numerous local investors to reuse historic buildings, provide residential units that meet local criteria, and provide examples of best restoration practices [5]. Additionally, in 2020, the UNWTO and the G20 Tourism Working Group created a "Framework AlUla for Inclusive Community Development through Tourism" to assist the sector in achieving its potential to donate to and fulfill the development of inclusive communities and the Sustainable Development Goals [6]. There are more initiatives that have established green tourism practices in AlUla, and the goal of these efforts is the complete belief that this environmental framework is the ideal solution to preserve AlUla's special nature (many other initiatives can be viewed on the website of The Royal Commission for AlUla [7]). Accordingly, the present study seeks not to evaluate the green tourism practices that have become established in AlUla as a result of these efforts and initiatives but rather to examine the impact of the community's attachment to AlUla in its support for these efforts and initiatives and its reflection in the behaviors of green AlUla tourists from the residents' perspectives.

The tourism industry has three main parts including visitors, communities, and destinations (attractions and facilities). Additionally, intermediary companies may play a key role between visitors and the destination such as providing transportation and information [8]. From the host community's perspective, tourism success depends on how local residents are attached to their environment and how they perceive and support tourism development [9]. Furthermore, community perception of the advantages generated by tourism can shape their behavior toward its development [10,11]. Hence, community support should be considered to improve green tourist consumption since residents are crucial players in offering tourist quality involvement [12,13]. When the local community experiences a positive impact from tourism, a strong desire will emerge for place attachment and development [14–16]. Place attachment is one of the most dominant non-financial factors that can illustrate why local communities resist or support destination development [17–20]. Place attachment can be conceptualized as the positive psychological and emotional connections built among residents and their environment [21–23]. These ties are essential in planning and tourism development [24] because tourism affects not only the look of local heritages but also the values of heritages and the ties the local community have with others [18,25].

The style of living among the community's citizens may affect the intended changes within the tourism destination taking place as a consequence of the continuing development, such as adjustments in local finances (e.g., [26,27]), social and cultural changes (e.g., [26,28–30]), and environmental changes (e.g., [26,29,31,32]). The efforts to foster green tourist consumption are challenging without the advocacy and involvement of the local residents [33–35].

Research on community attachment has been established in the literature within several domains, including environmental studies, education, green behavior, psychology, management, and tourism (e.g., [36–38]), and significant methodological and theoretical progress has been made by scholars in this area [36]. Several scholars have found a significant link between place attachment and pro-environmental green behaviors (e.g., [37,39,40]). Green behavior can be defined as practices by people that foster or engender the sustainable consumption of destination resources [41,42].

Although the current research paper introduces place attachment as a potentially useful tool to stimulate green behaviors (i.e., tourists' green consumption), outcomes on the link between the two constructs are inconsistent and far from convincing [43]. This can be explained by the various dimensions of place attachment and their interrelationship with green behaviors, which have been investigated in numerous contexts using methodologies with different data analysis techniques (e.g., [44–46]). The conclusion is that few studies (e.g., [47,48]) operationalize community attachment as a multi-dimensional factor, containing three sub-dimensions named (place dependence, place identity, and affective attachment) in a single study. This study aims to fill this gap and employ community attachment as a multidimensional construct and for the first time investigates its impact on tourists' green consumption support through the mediating role of residents' support of green tourism development in one context AlUla Heritage City, Saudi Arabia.

2. The Context of AlUla Heritage City, Saudi Arabia

Saudi Arabia is the main leader in implementing creative practices for a sustainable and environmental post-oil era. The country has undergone substantial economic extension and enlarged its targets to include building new sectors, global investments in the tourism industry, and protecting national heritage [49]. Recently, the government of Saudi Arabia developed strategies that encouraged urbanization by emphasizing the transfer from traditional patterns to modern design and green responsibility. As an example, recent practices were developed to foster tourism effects on environmental destinations for the city of AlUla. Located 1100 km west of Riyadh, AlUla is an archaeological site. After excavating the ancient site consisting of a necropolis, quarries, and settlements (Figure 1), one can now tour the world heritage sites in a rock-cut landscape, which has been sculpted for a million years. AlUla is known for its archaeological ruins and heritage sites, some of which are more than 2000 years old. Previously, this city was famous as the entrance gateway for the merchandise troops traveling from south to north in the Arabic Peninsula [50,51]. Geologically, the city has an oasis and a huge valley, allocated between sandstone highlands (mountains) with exceptional arithmetic shapes [52]. AlUla has fifty-five heritage sites from various eras; Mada'in Salih was one of the UNESCO World Heritage sites, registered in 2008. Furthermore, AlUla's Hegra Heritage Site (Al-Hijr/Mada'in Salih) was the first World Heritage place to be listed in Saudi Arabia, it represents the same outlook as UNESCO's "Memory of the World" (MoW) Program [53] (see Figure 1).

In AlUla, government strategies are ongoing to preserve its heritage landscape and plan for a substantial transformation to foster heritage tourism in Saudi Arabia. The "Royal Commission for AlUla" (RCU), developed in 2017, has the mission of sustaining and developing AlUla's cultural heritage for both current and future generations [55]. Corresponding with Vision 2030 of Saudi Arabia, RCU's initiatives in AlUla spread beyond archaeology and education to include arts and tourism. RCU aims to reinforce AlUla heritage sites as a friendly hub for international visitors. RCU's main goal is to foster cultural exchanges and deliver a glance into the heritage printed in stone, all while being thoughtful of sustainability toward the local community [53].

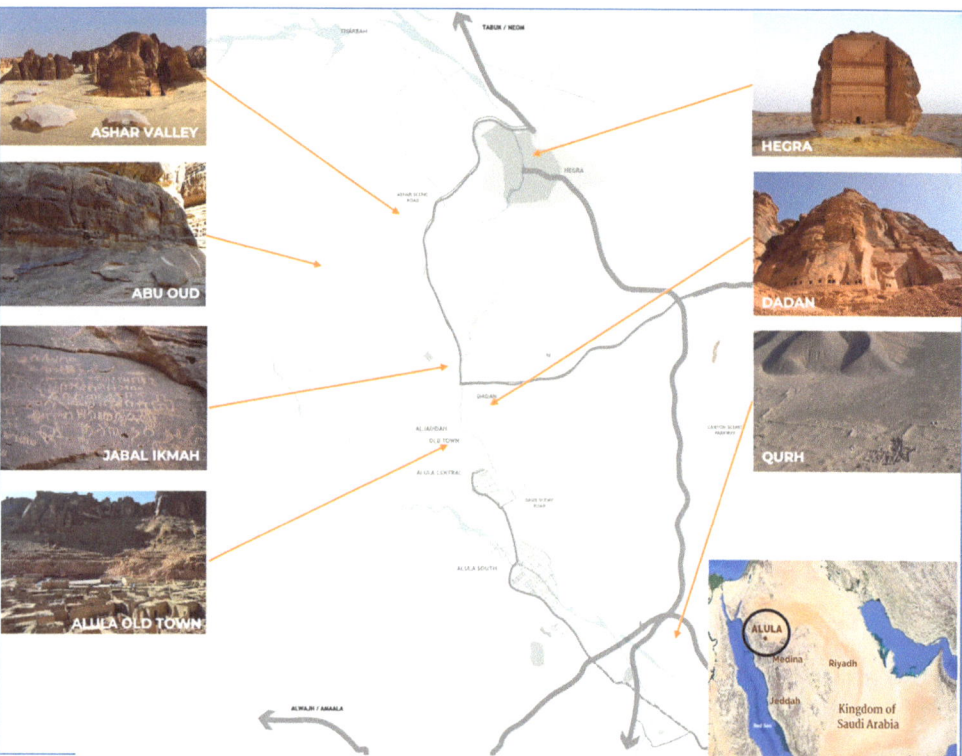

Figure 1. Map of the AlUla oasis and investigated archaeological and natural sites. From: Rock Slope Instabilities Affecting the AlUla Archaeological Sites (KSA) [54].

3. Literature Review and Hypothesis Development

3.1. Community Attachment and Tourists' Green Consumption (TGC)

The literature defines community attachment as the ties residents have with their community founded on affect, meaningfulness, emotion, value, and closeness to the place [33,47]. The community attachment concept stemmed from attachment theory [56]. The construction operationalizations of the place attachment concept vary wildly across multiple disciplines, causing challenges for investigators [47]. However, several studies have conceptualized community attachment through place dependence, place identity, and affective attachment [46,57,58]. These three subdimensions demonstrate that community attachment is characterized as having functional, cognitive, and affective aspects, respectively [46,59].

Regarding place dependence, there are two elements that people and groups use to define place dependency, which is inherent in the place's physical characteristics. The first is the quality of the existing place, and the other is the comparable options' relative qualities [60,61]. That is, a place turned into a resource for achieving objectives, creating, in turn, a dependent relationship [62,63]. This dependence leads to vast interaction with the place, creating a resident–place attachment [64] that boosts their pro-environment behaviors [65]. Meanwhile, residents' eco-friendly behaviors may knock on visitors' attitudes and behaviors about environmental preservation at the destination [66,67]. Nevertheless, prior studies have primarily examined tourists' responsibilities and behaviors, with little attention given to the residents' roles in supporting the destination's sustainability [66,68]. Accordingly, we hypothesize the following:

Hypothesis 1a (H1a). *Place dependence positively affects tourists' green consumption (TGC).*

Place identity pertains to an individual's self-concept shaped by their place of belonging, incorporating aspects associated with the public image of that destination [69,70]. Identity theory asserts that identity encourages and shapes actions [71]. In heritage and natural destinations, an intimate connection with the heritage and nature forms meaningful attachments to and boosts positive values toward these places [72]. This intense contact may shape a resident's environmental identity, leading to favorable environmental attitudes [59]. As such, the more intimate the association between the residents' self-identities and the destination's environment, the greater the likelihood that the locals will stimulate pro-eco-friendly behaviors to safeguard the environment [66]. As a reflection of this identity, several investigations claim that destination residents' identity may motivate tourists to engage in green behaviors [73,74]. Hence, we hypothesize the following:

Hypothesis 1b (H1b). *Place identity positively affects tourists' green consumption (TGC).*

Affective attachment was conceptualized as an emotional connection to a specific environment [57,75]. Studies indicate that affective attachment to a place significantly inspires nature-protective and pro-eco-friendly behaviors [48]. More precisely, affective attachment to nature and heritage destinations improves a resident's empathy toward and sense of connection to it, encouraging green behaviors [40]. However, studies have rarely integrated affective attachment with other attachment dimensions to explore its effect on residents' pro-environmental green behaviors. In general, since residents live in the destination, their affective attachment to it pushes them to protect it and simultaneously stimulates them to encourage visitors to behave in an environmentally sustainable manner [76]. Hence, the below hypothesis is set:

Hypothesis 1c (H1c). *Affective attachment positively affects tourists' green consumption (TGC).*

3.2. Community Attachment and Support for Green Tourism Development (GTD)

Community attachment includes individuals' strong positive emotions, rootedness, and feelings of belonging toward the community [77]. In the tourism and hospitality literature, community attachment has often been used to investigate its influences on residents' perceptions of tourism effects and perspectives concerning support for tourism development [78,79]. The studies' results varied. Some found that the locals who were highly attached to their community saw fewer benefits from tourism development [80]. In contrast, other studies revealed that these residents perceived positive results [10]. However, in the context of green tourism development, there is almost agreement in the literature that a high level of community attachment guides locals toward supporting sustainable tourism development, especially in heritage and nature destinations [33,78,81]. As such, recommendations were made to highlight the role of place identity in apprehending residents' perspectives toward STD [82]. The nature and power of local place identity and the surrounding environment (ecological or heritage) may significantly determine how well they interact with tourism development [83]. Similarly, highly tourism-dependent residents (place dependence) are expected to give more support for sustainable tourism development [84]. Additionally, residents with a high level of affective attachment often strive to support green tourism practices to protect their destination's nature or heritage resources [67]. In sum, with its three sub-dimensions, community attachment may be positively associated with support for GTD. This means that locals who have a more robust community attachment endorse supporting tourism development in ecological and heritage destinations [78]. Hence, the next three hypotheses were developed:

Hypothesis 2a (H2a). *Place dependence positively affects support for GTD.*

Hypothesis 2b (H2b). *Place identity positively affects support for GTD.*

Hypothesis 2c (H2c). *Affective attachment positively affects support GTD.*

3.3. Support for GTD and TGC

Enhancing and preserving the ecological and heritage environment is valuable for protecting the attractiveness of tourism destinations and fulfilling sustainable development. This process must depend on the collaborative endeavors of both local residents and tourists. Therefore, several studies have been conducted to determine how to stimulate residents to display eco-responsible behaviors and encourage tourists' green consumption [13]. In this context, an increasing body of studies confirms that residents can affect tourists' behaviors [85,86]. Accordingly, tourists' green consumption behaviors can potentially be impacted by dealings with residents who support green tourism development. However, empirical research has yet to explore the connections between locals' eco-responsible behaviors and tourists' green consumption thoroughly [13]. Within this framework, we further examine this association between support for green tourism development (GTD) and tourists' green consumption (TGC) as follows:

Hypothesis 3 (H3). *Support for GTD positively affects tourists' green consumption (TGC).*

3.4. The Mediating Role of Support for GTD

Based on previous studies and the above-stated arguments that illustrate the direct connections among the three sub-dimensions of community attachment and tourists' green consumption (TGC) and residents' support for green tourism development (GTD), and GTD and TGC, and in light of attachment theory and social identity theory, the next three hypotheses of mediating impact are presented:

Hypothesis 4a (H4a). *Support for GTD mediates the linkage between place dependence and tourists' green consumption (TGC).*

Hypothesis 4b (H4b). *Support for GTD mediates the linkage between place identity and tourists' green consumption (TGC).*

Hypothesis 4c (H4c). *Support GTD mediates the linkage between affective attachment and tourists' green consumption (TGC).*

Based on the previously mentioned literature and theories, Figure 2 graphically depicts this study's conceptual model.

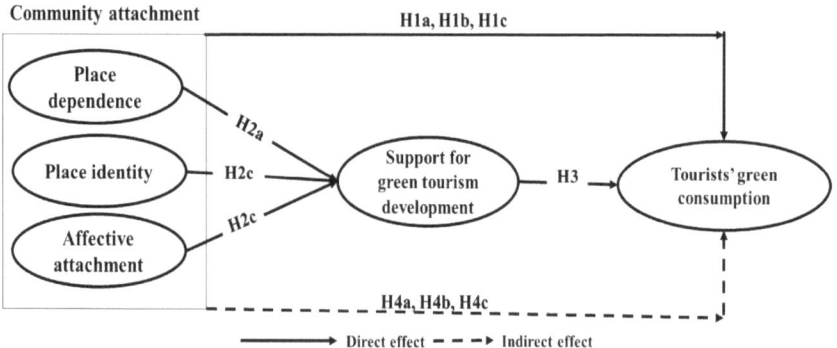

Figure 2. Conceptual model.

4. Methods

4.1. Measures

The scale items of all study variables were adapted based on the previous literature. Ten items from [57,79,87] were used to evaluate the three sub-dimensions of the community attachment construct including place dependence (three variables), place identity (three variables), and affective attachment (four variables). The tourists' green consumption variable was gauged by the 7-item scale suggested by [13]. Finally, five items from Lee's [79] study were adopted to measure the support for the GTD construct (see Appendix A). Seventeen academic experts and executives in sustainable tourism reviewed the survey's validity, and tiny changes were made. Furthermore, Harman's single-factor test was employed to avoid potential bias issues. The findings display that the value of this test is 46.5%, implying no bias in the current study because Harman's single factor score should not exceed 50% [88].

4.2. Collection of Study Data

Local residents in AlUla City, Saudi Arabia, were enrolled in this study, utilizing convenient samples and drop-off and pick-up procedures. AlUla city is situated in the Medina region, northwest of the kingdom, along the historic route used for incense trading. As a case study, AlUla is a popular tourist destination with a remarkable historical and archaeological heritage that led to its designation as Saudi Arabia's first UNESCO site [89]. The landscape, sandstone mountains along an oasis accompanying the ancient trade routes, and carved tombs in the mountains by the ancient Nabataean civilization make AlUla a unique heritage site [52].

The data were collected through an online questionnaire published on the city of AlUla's social media pages. AlUla's city has a population of 40,670 [53]. Residents willingly completed the survey, and their responses remained confidential. The data were gathered from January to April 2024. Around 600 questionnaires were returned, and 328 were found valid after excluding incomplete responses (251) and questionnaires with missing data (21 replies). The sample size of 328 participants is enough to satisfy the needs for the application of PLS-SEM models and it is considered sufficient. The PLS-SEM method entails determining the minimum sample size as a crucial factor. According to the "10 times rule" method suggested by Hair et al. [88], the minimum sample size should be at least 10 times the number of indicators for the most complex construct in the model. Our model consists of 5 latent constructs with 22 reflective indicators.

Consequently, following the rule of 10 times, the least required sample size was 220. Our sample of 349 participants was much more than this minimum recommendation. Furthermore, the sample of 328 exceeded the widely known suggestion that the sample size for SEM should be between 100 and 150 cases to obtain accurate results [88–90]. The sample comprised 212 males (64.6%) and 116 females (35.4%). Ages varied from 26 to 65. Also, most residents had a college degree 272 (82.9%).

The questionnaire structure is classified into four main sections. The first concentrates on demographic facts such as age and gender. Subsequently, the second part gathers data on three sub-dimensions of community attachment, i.e., place dependence, place identity, and affective attachment. The third section handles inquiries about the tourists' green consumption behaviors variable. Finally, the last section includes items regarding mediating variables, explicitly measuring support for the sustainable tourism development variable. All survey items were assessed using five-point Likert scales ranging from 1 (strongly disagree) to 5 (strongly agree).

4.3. Data Analysis

The hypothesized study model was tested by PLS-SEM using Smart PLS 3 for several reasons. First, this technique facilitates the researcher's evaluation of links between constructs in the structural model and their linked latent dimensions in the measurement model. Second, PLS-SEM works well with sophisticated models with moderation and mediation effects. Third,

PLS's graphical user interface is easier to use than those of other path modeling programs, such as AMOS. Previous research has used this approach extensively [89,90]. Using the PLS approach to test the suggested hypotheses of the proposed models requires investigating the validity and reliability of the outer model, also called the measurement, at the first stage and the inner or "structural" model at the second [91].

5. Results
5.1. Measurement Model Outcomes

In this step, the measurement outer model was assessed. Specifically, convergent validity (CV) and discriminant validity (DV) were calculated [92]. This study employed several indices, containing "factor loadings" (λ) (should exceed 0.70), "Cronbach's alpha" (a) (0.70 or greater is preferred), "composite reliability" (CR) (0.70 or greater is preferred), and "Average Variance Extracted" (AVE) (0.50 or greater is acceptable) [93]. Additionally, the AVE of each factor should exceed the squared inter-factor correlations [94], and the "Heterotrait–Monotriat" value (HTMT) should not exceed a value of 0.90 to attain the DV. As shown in Table 1, all the indices provide evidence of a satisfactory CV. Further, Table 2 portrays that an item loading within its variable is greater than any of its cross-loadings with another variable, Table 3 displays that all AVEs are higher than their corresponding squared inter-construction correlations, and Table 4 demonstrates that all HTMTs are <0.90, guaranteeing the DV of the model.

Table 1. Outer model findings.

	λ	(a)	(C.R)	(AVE)
Place dependence		0.816	0.890	0.731
CT1	0.887			
CT2	0.858			
CT3	0.818			
Place identity		0.811	0.889	0.727
CT_4	0.822			
CT_5	0.894			
CT_6	0.840			
Affective attachment		0.840	0.893	0.675
CT_7	0.828			
CT_8	0.809			
CT_9	0.829			
CT_10	0.820			
Tourists' green consumption (TGC))		0.898	0.919	0.619
TGC_1	0.779			
TGC_2	0.765			
TGC_3	0.782			
TGC_4	0.788			
TGC_5	0.773			
TGC_6	0.816			
TGC_7	0.804			
Support for STD		0.819	0.873	0.580
SSTD_1	0.769			
SSTD_2	0.772			

Table 1. Cont.

	λ	(a)	(C.R)	(AVE)
SSTD_3	0.763			
SSTD_4	0.755			
SSTD_5	0.748			

Note: CT1, CT2, and CT3 are measure items for the place dependence variable; CT_4, CT_5, and CT_6 are for the place identity variable; CT_7, CT_8, CT_8, and CT_10 are for the affective attachment variable; TGC_1 to TGC_7 are for the tourists' green consumption variable; and SSTD_1 to SSTD_5 are for the support for green tourism development variable.

Table 2. Factor cross-loadings.

	Place Dependence	Place Identity	Affective Attachment	SSTD	TGC
CT_1	**0.887**	0.517	0.646	0.637	0.661
CT_2	**0.858**	0.417	0.628	0.600	0.625
CT_3	**0.818**	0.438	0.526	0.563	0.466
CT_4	0.510	**0.822**	0.532	0.543	0.550
CT_5	0.443	**0.894**	0.490	0.588	0.556
CT_6	0.418	**0.840**	0.456	0.525	0.500
CT_7	0.566	0.465	**0.828**	0.597	0.596
CT_8	0.513	0.413	**0.809**	0.528	0.510
CT_9	0.595	0.477	**0.829**	0.645	0.584
CT_10	0.634	0.534	**0.820**	0.691	0.594
SSTD_1	0.575	0.538	0.627	**0.769**	0.502
SSTD_2	0.526	0.578	0.653	**0.772**	0.638
SSTD_3	0.470	0.443	0.563	**0.763**	0.532
SSTD_4	0.566	0.453	0.582	**0.755**	0.636
SSTD_5	0.536	0.447	0.432	**0.748**	0.603
TGC_1	0.535	0.528	0.548	0.637	**0.779**
TGC_2	0.551	0.519	0.585	0.631	**0.765**
TGC_3	0.533	0.478	0.474	0.550	**0.782**
TGC_4	0.528	0.450	0.502	0.507	**0.788**
TGC_5	0.501	0.408	0.501	0.581	**0.773**
TGC_6	0.592	0.455	0.619	0.653	**0.816**
TGC_7	0.559	0.604	0.594	0.649	**0.804**

Table 3. "Fornell–Larcker criterion matrix".

	AA	PD	PI	SGTD	TGC
AA	0.822				
PD	0.706	0.855			
PI	0.578	0.536	0.853		
SGTD	0.754	0.703	0.649	0.761	
TGC	0.698	0.691	0.628	0.768	0.787

Note: "Values off the diagonal line are squared inter-construction correlations, while values on the diagonal line are AVEs".

Table 4. HTMT outcomes.

	AA	PD	PI	SGTD	TGC
1-6 AA					
1-6 PD	0.843				
1-6 PI	0.695	0.657			
1-6 SGTD	0.897	0.857	0.791		
1-6 TGC	0.795	0.796	0.731	0.887	

Note: "For an appropriate DV, all HTMT values need to be <0.90".

5.2. Hypothesis Assessment Results

VIF, R2, Q2, and Beta coefficient (β) values were used to validate this study's inner model [93]. VIFs should be <5.0 to avoid multi-collinearity issues among model constructs, R2 in social sciences should be >0.10, β must be significant, and Q2 should be >0.0 [93]. As demonstrated in Table 5, all VIF values were less than 0.5, R2 for TGC was 0.663, R2 for SSTD was 0.672, the Q2 values for the TGC and SSTD variables were more than 0.0, and the β values were significant at $p = 0.01$ (Table 6), proving the fit of this study's structural model.

Table 5. VIF, R^2, and Q^2 findings.

Element	VIF	Element	VIF	Element	VIF	Element	VIF	Element	VIF
CT_1	1.990	CT_6	1.874	SSTD_1	1.898	TGC1	1.976	TGC_6	2.296
CT_2	1.796	CT_7	1.899	SSTD_2	1.798	TGC2	1.915	TGC_7	2.087
CT_3	1.708	CT_8	1.889	SSTD_3	1.725	TGC3	2.014		
CT_4	1.598	CT_9	1.852	SSTD_4	1.720	TGC4	2.132		
CT_5	2.193	CT_10	1.739	SSTD_5	1.637	TGC5	2.039		
Tourists' green consumption						R2	0.663	Q2	0.380
Support for sustainable tourism development						R2	0.672	Q2	0.362

Table 6. Hypothesis results.

Paths	β	t	p	Result
Direct Paths				
H1a—Place dependence → TGC	0.219	3.714	0.000	✔
H1c—Place identity → TGC	0.171	2.723	0.007	✔
H1b—Affective attachment → TGC	0.151	2.065	0.039	✔
H2a—Place dependence → SSTD	0.272	4.810	0.000	✔
H2b—Place identity → SSTD	0.267	7.352	0.000	✔
H2c—Affective attachment → SSTD	0.407	8.146	0.000	✔
H3—SSTD → SSTD	0.389	6.126	0.000	✔
Indirect Mediating Paths				
H4a—Place dependence → SSTD → TGC	0.106	4.204	0.000	✔
H4b—Place identity → SSTD → TGC	0.104	4.610	0.000	✔
H4c—Affective attachment → SSTD → TGC	0.158	4.585	0.000	✔

Further, the below equation can also be used to confirm the PLS-SEM model's Goodness of Fit (GoF). The presented model's GoF is 0.667, meaning a high GoF for the model [95].

$$GoF = \sqrt{AVE_{avy} \times R^2_{avy}}$$

Finally, after confirming the reliability and validity of the outer and inner models, a bootstrapping process with 5000 iterations using Smart PLS 3 was carried out to test this study's hypotheses (see Table 6).

Table 6 and Figure 3 indicate that place dependence positively impacts TGC at $β = 0.219$ and $p = 0.000$ and SSTD at $β = 0.272$ and $p = 0.000$, supporting H1a and H2a. H1b and H2b are also supported because place identity positively influences TGC and SSTD at β = 0.171 and p = 0.007 and at β = 0.267 and p = 0.000, respectively. Similarly, H1c and H2c are confirmed, as affective attachment positively impacts TGC and SSTD at β = 0.151 and p = 0.039 and at β = 0.407 and p = 0.000, respectively. As for H3, the positive effect of SSTD on TGC (β = 0.389, t = 6.126, p = 0.000) is confirmed. In addition, place dependence, place identity, and affective attachment mediate the linkage between SSTD and TGC, all at p = 0.000 and at β = 0.106, β = 0.104, and β = 0.158, respectively, thus proving H4a, H4b, and H4c.

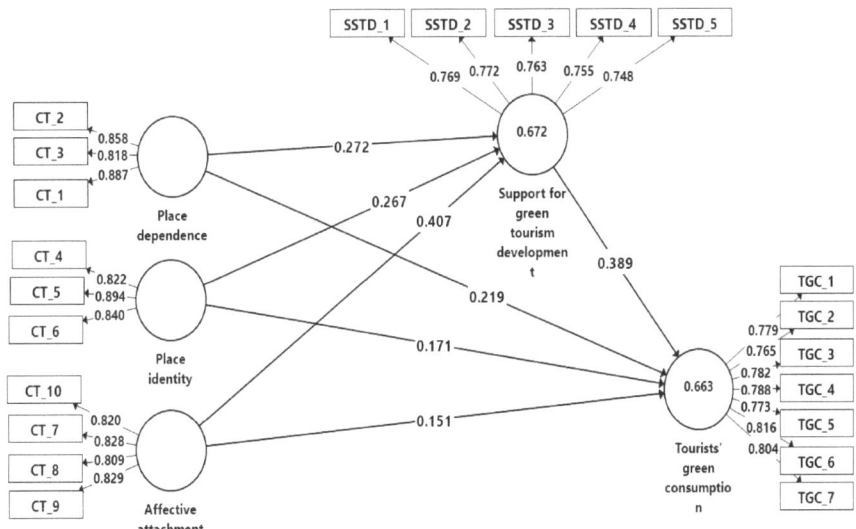

Figure 3. Estimation of the model structure.

6. Discussion and Implications

Green consumption, which has the power to undermine the destination, place, or community environment that tourists target, visit, and stay in, is becoming more interesting as it brings to prominence the link between community attachment and green consumption behavior. Community attachment is a multidimensional construct with various sub-dimensions that might have an effect on different expected outcomes such as individuals' attachments and commitments to their environment and surrounding place. In the current study, community attachment is defined as a multifaceted factor comprising three second-order dimensions including place dependence, place identity, and affective attachment [96]. These factors can independently or collectively shape or change tourists' attitudes, perceptions, and behaviors toward a place, (i.e., green consumption behaviors).

The findings of our study showed a positive significant path from the three dimensions of community attachment to green consumption behaviors of tourists visiting AlUla heritage sites. Each sub-dimension plays a unique role in shaping tourists' environmental behaviors. Place dependence was found to have a significant positive and direct impact on tourists' green consumption behaviors. Tourists who perceive AlUla as a destination that offers eco-friendly facilities, and sustainable green heritage sites are more likely to participate in green consumption behaviors. This result is consistent with Williams and Vaske [97] and Choi and Kim [98], who argued that place-dependent visitors prefer eco-friendly hotels, nature-based attractions, and local services/products that foster environmental sustainability. Likewise, the results of PLS-SEM showed a positive, direct, significant impact of place identity on tourists' green consumption. This finding suggested that visitors to AlUla heritage sites with high place identity had a feeling of attachment, dignity, and physiological connection to the destination with its distinguishing heritage and environmental resources. This result is consistent with Buonincontri et al. [99] and Hernández et al. [100], who found that visitors are motivated to implement green consumption activities as a way to sustain a place's cultural and environmental heritage. Additionally, the PLS-SEM output showed a positive direct path coefficient between affective place attachment and green consumption activities. Tourists who develop emotional connections and constructive feelings toward AlUla heritage sites tend to prefer sustainable practices to ensure the destination's well-being. This argument is consistent with Lewicka [101] and Buckley [102], who stated that affective attachment encourages a feeling of trust and obligation among visitors toward preserving the environment and promoting sustainable practices. In sum, according to our study survey, the community attachment of AlUla was able to motivate tourists to participate in sustainable environmental conservation endeavors, utilize green products, be willing to participate in natural environment conservation, plan to pay extra costs for environmental conservation, utilize public transportation, and stay in green hotels. Community attachment can also prompt tourists to make their colleagues environmentally aware and invite them to visit AlUla through positive word of mouth. Moreover, tourists can play a key role in the achievement and success of green tourism in AlUla city by constructing conscious selections that back sustainability, value local traditions, environments, and cultures, and support responsible travel activities that preserve the local environment. By actively engaging in and fostering green tourism practices, tourists can help safeguard future generations' environmental and cultural sources.

The current study explored the mediating impacts of locals' support for green tourism development in the link from community attachment to tourists' green consumption practices. Residents who dynamically support and contribute to green tourism practices, preservation strategies, and sustainable initiatives are considered good examples and role models for visitors. This is consistent with the view of Leslie [103], who argued that resident support and sponsorship of environmental practices generate a backing environment that fosters visitors to be involved in green consumption activities.

This study has several implications for green consumption behavior and sustainable tourism management in AlUla heritage sites. Involving local citizens as defenders of sustainability and engaging them in the decision-making processes can promote a feeling of proprietorship and empowerment toward fostering green tourism strategies. Additionally, the partnership among residents, AlUla authorities, and private businesses is vital for applying effective sustainability practices that address the diverse factors of community attachment and foster green consumption activities. Educating residents and visitors about the eco-friendly advantages of green consumption can motivate environmental behaviors that can create a culture of sustainability. Finally, implementing regulations and policies that advocate green tourism development, prioritize sustainable programs, and foster eco-certification initiatives can create a supportive environment for sustainable tourism strategies.

7. Conclusions

The results of analyzing the data collected from 328 local residents of AlUla by PLS-SEM demonstrate that the three sub-dimensions of community attachment, i.e., place dependence, place identity, and affective attachment, significantly and positively affect tourists' green consumption. The findings also indicated that support for the green tourism development variables successfully mediates the connection between the three sub-dimensions of community attachment and tourists' green consumption. These results emphasize the significance of community place attachment in stimulating tourists' green consumption practices and its role in improving community support for green tourism.

8. Limitations and Future Avenues

This study, similar to other studies, has a number of limitations and can be addressed in further research and generate research prospects. One limitation is a generalizability issue, which can be expected in our study since the findings are derived from the context of

AlUla only. Community attachment and local advocacy for green tourism development could differ radically in other diverse geographical and cultural settings. Future studies should use this study's model in different heritage sites to evaluate and compare the robustness of the tested inter-relationships. This study used a cross-sectional approach, which lacks the ability to develop causal inferences among the examined relationships. A longitudinal research design or experimental approach could offer a solid perception of the tested paths' causal interactions and dynamics. Comparative research papers on different heritage sites can better explain the tested model's variations. Finally, other mediating factors such as environmental awareness, social norms, and peer effects can be employed in future studies to examine the relationship between community attachment and tourist green consumption behavior.

Author Contributions: Conceptualization, I.A.E. and S.F.; methodology, I.A.E., S.F., M.A. and A.M.S.A.; software, I.A.E. and S.F.; validation, I.A.E., A.M.S.A. and S.F.; formal analysis, I.A.E. and A.M.S.A.; investigation, I.A.E., S.F. and A.M.S.A.; resources, I.A.E.; data curation, I.A.E.; writing—original draft preparation, S.F., I.A.E. and A.M.S.A.; writing—review and editing, I.A.E., S.F., M.A. and A.M.S.A.; visualization, I.A.E.; supervision, I.A.E.; project administration, I.A.E., S.F. and A.M.S.A.; funding acquisition, I.A.E. and A.M.S.A. All authors have read and agreed to the published version of the manuscript.

Funding: This work was supported by the Deanship of Scientific Research, Vice Presidency for Graduate Studies and Scientific Research, King Faisal University, Saudi Arabia [Grant A259].

Data Availability Statement: Data is available upon request from researchers who meet the eligibility criteria. Kindly contact the first author privately through e-mail.

Conflicts of Interest: The authors declare no conflict of interest.

Appendix A

The study questionnaire

Community attachment
Place dependence.

"The settings and facilities of AlUla are the best."
"Prefer living in AlUla to other communities."
"Enjoy living in AlUla more than other communities."

Place identity.

"I identify the living in AlUla."
"I feel that AlUla is a part of me."
"Living in AlUla says a lot about who I am."

Affective attachment.

"Living in AlUla means a lot to me."
"I am very attached to AlUla."
"I feel a strong sense of belonging to AlUla."
"Many of my friends/family prefer living in AlUla to other communities."

Support for sustainable tourism development.

"I support the development of community-based sustainable tourism initiatives."
"I participate in sustainable tourism-related plans and development."
"I participate in cultural exchanges between local residents and visitors."
"I cooperate with tourism planning and development initiatives."
"I participate in the promotion of environmental education and conservation."

Tourists' green consumption.

"Tourists use public transport when possible."
"Tourists walk and/or cycle when possible."

"Tourists buy local eco-products."
"Tourists stay at green hotels when travelling."
"Tourists consider saving energy at hotels."
"Tourists try to dispose of garbage properly."
"Tourists pick up the rubbish that they see during their trip."

References

1. Alsahafi, R.; Alzahrani, A.; Mehmood, R. Smarter Sustainable Tourism: Data-Driven Multi-Perspective Parameter Discovery for Autonomous Design and Operations. *Sustainability* **2023**, *15*, 4166. [CrossRef]
2. Waheeb, S.A.; Zerouali, B.; Elbeltagi, A.; Alwetaishi, M.; Wong, Y.J.; Bailek, N.; AlSaggaf, A.A.; Abd Elrahman, S.I.M.; Santos, C.A.G.; Majrashi, A.A. Enhancing Sustainable Urban Planning through GIS and Multiple-Criteria Decision Analysis: A Case Study of Green Space Infrastructure in Taif Province, Saudi Arabia. *Water* **2023**, *15*, 3031. [CrossRef]
3. Al Fahmawee, E.A.; Jawabreh, O. Sustainability of green tourism by international tourists and its impact on green environmental achievement: Petra heritage, Jordan. *Geo J. Tour. Geosites* **2023**, *46*, 27–36. [CrossRef]
4. Dodds, R.; Joppe, M. Promoting Urban Green Tourism: The Development of the Other Map of Toronto. *J. Vacat. Mark.* **2001**, *7*, 261–267. [CrossRef]
5. Mazzetto, S. Heritage Conservation and Reuses to Promote Sustainable Growth. *Mater. Today Proc.* **2023**, *85*, 100–103. [CrossRef]
6. Erianda, A.; Alanda, A.; Hidayat, R. Systematic Literature Review: Digitalization of Rural Tourism Towards Sustainable Tourism. *Int. J. Adv. Sci. Comput. Eng.* **2023**, *5*, 247–256. [CrossRef]
7. RCU. Initiatives. Available online: https://www.rcu.gov.sa/en/initiatives/ (accessed on 18 May 2024).
8. Goeldner, C.R.; Ritchie, J.R.B. *Tourism Principles, Practices, Philosophies*; John Wiley & Sons: Hoboken, NJ, USA, 2007; ISBN 8126513438.
9. Kim, G.; Duffy, L.N.; Moore, D. Importance of Residents' Perception of Tourists in Establishing a Reciprocal Resident-Tourist Relationship: An Application of Tourist Attractiveness. *Tour. Manag.* **2023**, *94*, 104632. [CrossRef]
10. Choi, H.C.; Murray, I. Resident Attitudes toward Sustainable Community Tourism. *J. Sustain. Tour.* **2010**, *18*, 575–594. [CrossRef]
11. Šegota, T.; Mihalič, T.; Perdue, R.R. Resident Perceptions and Responses to Tourism: Individual vs Community Level Impacts. *J. Sustain. Tour.* **2024**, *32*, 340–363. [CrossRef]
12. Woo, E.; Kim, H.; Uysal, M. Life Satisfaction and Support for Tourism Development. *Ann. Tour. Res.* **2015**, *50*, 84–97. [CrossRef]
13. Hu, J.; Xiong, L.; Lv, X.; Pu, B. Sustainable Rural Tourism: Linking Residents' Environmentally Responsible Behaviour to Tourists' Green Consumption. *Asia Pac. J. Tour. Res.* **2021**, *26*, 879–893. [CrossRef]
14. Sheldon, P.J.; Var, T.; Var, T. Resident Attitudes to Tourism in North Wales. *Tour. Manag.* **1984**, *5*, 40–47. [CrossRef]
15. Wang, Y.; Pfister, R.E. Residents' Attitudes Toward Tourism and Perceived Personal Benefits in a Rural Community. *J. Travel. Res.* **2008**, *47*, 84–93. [CrossRef]
16. Pai, C.K.; Chen, H.; Lee, T.J.; Hyun, S.S.; Liu, Y.; Zheng, Y. The Impacts of Under-Tourism and Place Attachment on Residents' Life Satisfaction. *J. Vacat. Mark.* **2023**, 1. [CrossRef]
17. Strzelecka, M.; Boley, B.B.; Strzelecka, C. Empowerment and Resident Support for Tourism in Rural Central and Eastern Europe (CEE): The Case of Pomerania, Poland. *J. Sustain. Tour.* **2017**, *25*, 554–572. [CrossRef]
18. Kaján, E. Community Perceptions to Place Attachment and Tourism Development in Finnish Lapland. *Tour. Geogr.* **2014**, *16*, 490–511. [CrossRef]
19. Boley, B.B.; McGehee, N.G.; Perdue, R.R.; Long, P. Empowerment and Resident Attitudes toward Tourism: Strengthening the Theoretical Foundation through a Weberian Lens. *Ann. Tour. Res.* **2014**, *49*, 33–50. [CrossRef]
20. Stylidis, D. Place Attachment, Perception of Place and Residents' Support for Tourism Development. *Tour. Plan. Dev.* **2018**, *15*, 188–210. [CrossRef]
21. Stedman, R.C. Toward a Social Psychology of Place. *Environ. Behav.* **2002**, *34*, 561–581. [CrossRef]
22. Gustafson, P. Meanings of place: Everyday experience and theoretical conceptualizations. *J. Environ. Psychol.* **2001**, *21*, 5–16. [CrossRef]
23. Aleshinloye, K.D.; Woosnam, K.M.; Joo, D. The Influence of Place Attachment and Emotional Solidarity on Residents' Involvement in Tourism: Perspectives from Orlando, Florida. *J. Hosp. Tour. Insights* **2024**, *7*, 914–931. [CrossRef]
24. Aytekin, A.; Keles, H.; Uslu, F.; Keles, A.; Yayla, O.; Tarinc, A.; Ergun, G.S. The Effect of Responsible Tourism Perception on Place Attachment and Support for Sustainable Tourism Development: The Moderator Role of Environmental Awareness. *Sustainability* **2023**, *15*, 5865. [CrossRef]
25. Manzo, L.C.; Perkins, D.D. Finding Common Ground: The Importance of Place Attachment to Community Participation and Planning. *J. Plan. Lit.* **2006**, *20*, 335–350. [CrossRef]
26. Simpson, M.C. Community Benefit Tourism Initiatives—A Conceptual Oxymoron? *Tour. Manag.* **2008**, *29*, 1–18. [CrossRef]
27. Manyara, G.; Jones, E. Community-Based Tourism Enterprises Development in Kenya: An Exploration of Their Potential as Avenues of Poverty Reduction. *J. Sustain. Tour.* **2007**, *15*, 628–644. [CrossRef]
28. Nyaupane, G.P.; Morais, D.B.; Dowler, L. The Role of Community Involvement and Number/Type of Visitors on Tourism Impacts: A Controlled Comparison of Annapurna, Nepal and Northwest Yunnan, China. *Tour. Manag.* **2006**, *27*, 1373–1385. [CrossRef]

29. Lee, C.-K.; Kang, S.K.; Long, P.; Reisinger, Y. Residents' Perceptions of Casino Impacts: A Comparative Study. *Tour. Manag.* **2010**, *31*, 189–201. [CrossRef]
30. Bull, C.; Lovell, J. The Impact of Hosting Major Sporting Events on Local Residents: An Analysis of the Views and Perceptions of Canterbury Residents in Relation to the *Tour de France* 2007. *J. Sport Tour.* **2007**, *12*, 229–248. [CrossRef]
31. Yoon, Y.; Gursoy, D.; Chen, J.S. Validating a Tourism Development Theory with Structural Equation Modeling. *Tour. Manag.* **2001**, *22*, 363–372. [CrossRef]
32. Dyer, P.; Gursoy, D.; Sharma, B.; Carter, J. Structural Modeling of Resident Perceptions of Tourism and Associated Development on the Sunshine Coast, Australia. *Tour. Manag.* **2007**, *28*, 409–422. [CrossRef]
33. Nicholas, L.N.; Thapa, B.; Ko, Y.J. Residents' perspectives of a world heritage site. *Ann. Tour. Res.* **2009**, *36*, 390–412. [CrossRef]
34. Gursoy, D.; Rutherford, D.G. Host Attitudes toward Tourism. *Ann. Tour. Res.* **2004**, *31*, 495–516. [CrossRef]
35. Tang, C.; Han, Y.; Ng, P. Green Consumption Intention and Behavior of Tourists in Urban and Rural Destinations. *J. Environ. Plan. Manag.* **2023**, *66*, 2126–2150. [CrossRef]
36. Kyle, G.T.; Mowen, A.J.; Tarrant, M. Linking Place Preferences with Place Meaning: An Examination of the Relationship between Place Motivation and Place Attachment. *J. Environ. Psychol.* **2004**, *24*, 439–454. [CrossRef]
37. Raymond, C.M.; Brown, G.; Robinson, G.M. The Influence of Place Attachment, and Moral and Normative Concerns on the Conservation of Native Vegetation: A Test of Two Behavioural Models. *J. Environ. Psychol.* **2011**, *31*, 323–335. [CrossRef]
38. Nasr, E.; Emeagwali, O.L.; Aljuhmani, H.Y.; Al-Geitany, S. Destination Social Responsibility and Residents' Environmentally Responsible Behavior: Assessing the Mediating Role of Community Attachment and Involvement. *Sustainability* **2022**, *14*, 14153. [CrossRef]
39. Devine-Wright, P.; Howes, Y. Disruption to Place Attachment and the Protection of Restorative Environments: A Wind Energy Case Study. *J. Environ. Psychol.* **2010**, *30*, 271–280. [CrossRef]
40. Son, J.Y.; Yang, J.-J.; Choi, S.; Lee, Y.-K. Impacts of Residential Environment on Residents' Place Attachment, Satisfaction, WOM, and pro-Environmental Behavior: Evidence from the Korean Housing Industry. *Front. Psychol.* **2023**, *14*, 1217877. [CrossRef]
41. Sivek, D.J.; Hungerford, H. Predictors of Responsible Behavior in Members of Three Wisconsin Conservation Organizations. *J. Environ. Educ.* **1990**, *21*, 35–40. [CrossRef]
42. Sharma, T.; Chen, J.S.; Ramos, W.D.; Sharma, A. Visitors' Eco-Innovation Adoption and Green Consumption Behavior: The Case of Green Hotels. *Int. J. Contemp. Hosp. Manag.* **2024**, *36*, 1005–1024. [CrossRef]
43. Scannell, L.; Gifford, R. The Relations between Natural and Civic Place Attachment and Pro-Environmental Behavior. *J. Environ. Psychol.* **2010**, *30*, 289–297. [CrossRef]
44. Vaske, J.J.; Kobrin, K.C. Place Attachment and Environmentally Responsible Behavior. *J. Environ. Educ.* **2001**, *32*, 16–21. [CrossRef]
45. Elshaer, I.A.; Augustyn, M.M. Testing the Dimensionality of the Quality Management Construct. *Total Qual. Manag. Bus. Excell.* **2016**, *27*, 353–367. [CrossRef]
46. Halpenny, E.A. Pro-Environmental Behaviours and Park Visitors: The Effect of Place Attachment. *J. Environ. Psychol.* **2010**, *30*, 409–421. [CrossRef]
47. Ramkissoon, H.; Graham Smith, L.D.; Weiler, B. Testing the Dimensionality of Place Attachment and Its Relationships with Place Satisfaction and Pro-Environmental Behaviours: A Structural Equation Modelling Approach. *Tour. Manag.* **2013**, *36*, 552–566. [CrossRef]
48. Ramkissoon, H.; Smith, L.D.G.; Weiler, B. Relationships between Place Attachment, Place Satisfaction and pro-Environmental Behaviour in an Australian National Park. *J. Sustain. Tour.* **2013**, *21*, 434–457. [CrossRef]
49. Al-Tokhais, A.; Thapa, B. Management Issues and Challenges of UNESCO World Heritage Sites in Saudi Arabia. *J. Herit. Tour.* **2020**, *15*, 103–110. [CrossRef]
50. Bay, M.A.; Koziol, C. *Adobe Fabric and the Future of Heritage Tourism: A Case Study Analysis of the Old Historical City of Alula, Saudi Arabia*; University of Colorado at Denver: Denver, CO, USA, 2014.
51. Pavan, A. A Conceptual Investigation of the Transformation of AlUla into a Global Tourism Destination: Saudi Arabia Rediscovers Its Pre-Islamic Heritage and Bets on Cultural Diplomacy. *J. Tour. Manag. Res.* **2023**, *8*, 1152–1168.
52. Filippi, L.D.; Mazzetto, S. Comparing AlUla and The Red Sea Saudi Arabia's Giga Projects on Tourism towards a Sustainable Change in Destination Development. *Sustainability* **2024**, *16*, 2117. [CrossRef]
53. RCU Is Saudi Arabia the Next Big Heritage Tourism Destination? Available online: https://www.rcu.gov.sa/en/media-gallery/articles/is-saudi-arabia-the-next-big-heritage-tourism-destination/ (accessed on 1 May 2024).
54. Gallego, J.I.; Margottini, C.; Perissé Valero, I.; Spizzichino, D.; Beni, T.; Boldini, D.; Bonometti, F.; Casagli, N.; Castellanza, R.; Crosta, G.B.; et al. Rock Slope Instabilities Affecting the AlUla Archaeological Sites (KSA). In *Progress in Landslide Research and Technology*; Springer Nature: Cham, Switzerland, 2023; pp. 413–429.
55. Saudi Vision, 2030 Discover the Extraordinary Heritage of AlUla—A Living Museum of Sandstone Outcrops, Historic Developments, and Preserved Tombs. Available online: https://www.vision2030.gov.sa/en/projects/alula (accessed on 1 May 2024).
56. Bowlby, J. Attachment Theory, Separation Anxiety, and Mourning. *Am. Handb. Psychiatry* **1975**, *6*, 292–309.
57. Yuksel, A.; Yuksel, F.; Bilim, Y. Destination Attachment: Effects on Customer Satisfaction and Cognitive, Affective and Conative Loyalty. *Tour. Manag.* **2010**, *31*, 274–284. [CrossRef]
58. Lee, T.H.; Shen, Y.L. The Influence of Leisure Involvement and Place Attachment on Destination Loyalty: Evidence from Recreationists Walking Their Dogs in Urban Parks. *J. Environ. Psychol.* **2013**, *33*, 76–85. [CrossRef]

59. Hinds, J.; Sparks, P. Engaging with the Natural Environment: The Role of Affective Connection and Identity. *J. Environ. Psychol.* **2008**, *28*, 109–120. [CrossRef]
60. White, D.D.; Virden, R.J.; van Riper, C.J. Effects of Place Identity, Place Dependence, and Experience-Use History on Perceptions of Recreation Impacts in a Natural Setting. *Environ. Manag.* **2008**, *42*, 647–657. [CrossRef]
61. Stokols, D. People in Places: A Transactional View of Settings. 1981. Available online: https://escholarship.org/content/qt48v387g7/qt48v387g7.pdf (accessed on 25 April 2024).
62. Strzelecka, M.; Boley, B.B.; Woosnam, K.M. Place Attachment and Empowerment: Do Residents Need to Be Attached to Be Empowered? *Ann. Tour. Res.* **2017**, *66*, 61–73. [CrossRef]
63. Williams, D.R.; Patterson, M.E.; Roggenbuck, J.W.; Watson, A.E. Beyond the Commodity Metaphor: Examining Emotional and Symbolic Attachment to Place. *Leis. Sci.* **1992**, *14*, 29–46. [CrossRef]
64. Moore, R.L.; Graefe, A.R. Attachments to Recreation Settings: The Case of Rail-trail Users. *Leis. Sci.* **1994**, *16*, 17–31. [CrossRef]
65. Chen, N.; Dwyer, L. Residents' Place Satisfaction and Place Attachment on Destination Brand-Building Behaviors: Conceptual and Empirical Differentiation. *J. Travel. Res.* **2018**, *57*, 1026–1041. [CrossRef]
66. Wang, S.; Wang, J.; Li, J.; Yang, F. Do Motivations Contribute to Local Residents' Engagement in pro-Environmental Behaviors? Resident-Destination Relationship and pro-Environmental Climate Perspective. *J. Sustain. Tour.* **2020**, *28*, 834–852. [CrossRef]
67. Elshaer, I.A.; Azazz, A.M.S.; Fayyad, S. Residents' Environmentally Responsible Behavior and Tourists' Sustainable Use of Cultural Heritage: Mediation of Destination Identification and Self-Congruity as a Moderator. *Heritage* **2024**, *7*, 1174–1187. [CrossRef]
68. Han, H.; Hyun, S.S. College Youth Travelers' Eco-Purchase Behavior and Recycling Activity While Traveling: An Examination of Gender Difference. *J. Travel. Tour. Mark.* **2018**, *35*, 740–754. [CrossRef]
69. Hernández, B.; Martín, A.M.; Ruiz, C.; Hidalgo, M. del C. The Role of Place Identity and Place Attachment in Breaking Environmental Protection Laws. *J. Environ. Psychol.* **2010**, *30*, 281–288. [CrossRef]
70. Uzzell, D.; Pol, E.; Badenas, D. Place Identification, Social Cohesion, and Enviornmental Sustainability. *Environ. Behav.* **2002**, *34*, 26–53. [CrossRef]
71. Nunkoo, R.; Gursoy, D.; Juwaheer, T.D. Island Residents' Identities and Their Support for Tourism: An Integration of Two Theories. *J. Sustain. Tour.* **2010**, *18*, 675–693. [CrossRef]
72. Kellert, S.R. Experiencing Nature: Affective, Cognitive, and Evaluative Development in Children. In *Children and Nature: PSYCHOLOGICAL, Sociocultural, and Evolutionary Investigations*; MIT Press: Cambridge, MA, USA, 2002; Volume 117151.
73. Su, L.; Swanson, S.R. The Effect of Destination Social Responsibility on Tourist Environmentally Responsible Behavior: Compared Analysis of First-Time and Repeat Tourists. *Tour. Manag.* **2017**, *60*, 308–321. [CrossRef]
74. Su, L.; Swanson, S.R.; Chen, X. Reputation, Subjective Well-Being, and Environmental Responsibility: The Role of Satisfaction and Identification. *J. Sustain. Tour.* **2018**, *26*, 1344–1361. [CrossRef]
75. Jorgensen, B.S.; Stedman, R.C. Sense of place as an attitude: Lakeshore owners attitudes toward their properties. *J. Environ. Psychol.* **2001**, *21*, 233–248. [CrossRef]
76. Confente, I.; Scarpi, D. Achieving Environmentally Responsible Behavior for Tourists and Residents: A Norm Activation Theory Perspective. *J. Travel. Res.* **2021**, *60*, 1196–1212. [CrossRef]
77. Matarrita-Cascante, D.; Stedman, R.; Luloff, A.E. Permanent and Seasonal Residents' Community Attachment in Natural Amenity-Rich Areas. *Environ. Behav.* **2010**, *42*, 197–220. [CrossRef]
78. Adongo, R.; Choe, J.Y.; Han, H. Tourism in Hoi An, Vietnam: Impacts, Perceived Benefits, Community Attachment and Support for Tourism Development. *Int. J. Tour. Sci.* **2017**, *17*, 86–106. [CrossRef]
79. Lee, T.H. Influence Analysis of Community Resident Support for Sustainable Tourism Development. *Tour. Manag.* **2013**, *34*, 37–46. [CrossRef]
80. Gursoy, D.; Chi, C.G.; Dyer, P. Locals' Attitudes toward Mass and Alternative Tourism: The Case of Sunshine Coast, Australia. *J. Travel. Res.* **2010**, *49*, 381–394. [CrossRef]
81. Huong, P.M.; Lee, J.-H. Finding Important Factors Affecting Local Residents' Support for Tourism Development in Ba Be National Park, Vietnam. *For. Sci. Technol.* **2017**, *13*, 126–132. [CrossRef]
82. Wang, S.; Xu, H. Influence of Place-Based Senses of Distinctiveness, Continuity, Self-Esteem and Self-Efficacy on Residents' Attitudes toward Tourism. *Tour. Manag.* **2015**, *47*, 241–250. [CrossRef]
83. Kitnuntaviwat, V.; Tang, J.C.S. Residents' Attitudes, Perception and Support for Sustainable Tourism Development. *Tour. Hosp. Plan. Dev.* **2008**, *5*, 45–60. [CrossRef]
84. Chang, M.-X.; Choong, Y.-O.; Ng, L.-P. Local Residents' Support for Sport Tourism Development: The Moderating Effect of Tourism Dependency. *J. Sport Tour.* **2020**, *24*, 215–234. [CrossRef]
85. Wang, W.; Wu, J.; Wu, M.-Y.; Pearce, P.L. Shaping Tourists' Green Behavior: The Hosts' Efforts at Rural Chinese B&Bs. *J. Destin. Mark. Manag.* **2018**, *9*, 194–203. [CrossRef]
86. Wang, C.; Zhang, J.; Cao, J.; Duan, X.; Hu, Q. The Impact of Behavioral Reference on Tourists' Responsible Environmental Behaviors. *Sci. Total Environ.* **2019**, *694*, 133698. [CrossRef]
87. López, M.F.B.; Virto, N.R.; Manzano, J.A.; Miranda, J.G.-M. Residents' Attitude as Determinant of Tourism Sustainability: The Case of Trujillo. *J. Hosp. Tour. Manag.* **2018**, *35*, 36–45. [CrossRef]

88. Podsakoff, P.M.; Organ, D.W. Self-Reports in Organizational Research: Problems and Prospects. *J. Manag.* **1986**, *12*, 531–544. [CrossRef]
89. Bendakir, M. *At-Turaif District in Ad-Dir'iyah (Saudi Arabia) No 1329*; ICOMOS: Charenton-le-Pont, France, 2010.
90. Hair, J.F.; Hult, G.T.M.; Ringle, C.M.; Sarstedt, M. *A Primer on Partial Least Squares Structural Equation Modeling (PLS-SEM)*, 2nd ed.; SAGE Publications, Inc.: Thousand Oaks, CA, USA, 2017.
91. Leguina, A. A Primer on Partial Least Squares Structural Equation Modeling (PLS-SEM). *Int. J. Res. Method Educ.* **2015**, *38*, 220–221. [CrossRef]
92. Hair, J.F., Jr.; Hult, G.T.M.; Ringle, C.; Sarstedt, M. *A Primer on Partial Least Squares Structural Equation Modeling (PLS-SEM)*; Sage Publications: Thousand Oaks, CA, USA, 2016; ISBN 1483377431.
93. Hair, J.F.; Risher, J.J.; Sarstedt, M.; Ringle, C.M. When to Use and How to Report the Results of PLS-SEM. *Eur. Bus. Rev.* **2019**, *31*, 2–24. [CrossRef]
94. Fornell, C.; Larcker, D.F. Evaluating Structural Equation Models with Unobservable Variables and Measurement Error. *J. Mark. Res.* **1981**, *18*, 39–50. [CrossRef]
95. Tenenhaus, M.; Vinzi, V.E.; Chatelin, Y.-M.; Lauro, C. PLS Path Modeling. *Comput. Stat. Data Anal.* **2005**, *48*, 159–205. [CrossRef]
96. Kim, D.J.; Salvacion, M.; Salehan, M.; Kim, D.W. An Empirical Study of Community Cohesiveness, Community Attachment, and Their Roles in Virtual Community Participation. *Eur. J. Inf. Syst.* **2023**, *32*, 573–600. [CrossRef]
97. Williams, D.R.; Vaske, J.J. The Measurement of Place Attachment: Validity and Generalizability of a Psychometric Approach. *For. Sci.* **2003**, *49*, 830–840. [CrossRef]
98. Choi, S.; Kim, I. Sustainability of Nature Walking Trails: Predicting Walking Tourists' Engagement in pro-Environmental Behaviors. *Asia Pac. J. Tour. Res.* **2021**, *26*, 748–767. [CrossRef]
99. Buonincontri, P.; Marasco, A.; Ramkissoon, H. Visitors' Experience, Place Attachment and Sustainable Behaviour at Cultural Heritage Sites: A Conceptual Framework. *Sustainability* **2017**, *9*, 1112. [CrossRef]
100. Hernández, B.; Carmen Hidalgo, M.; Salazar-Laplace, M.E.; Hess, S. Place Attachment and Place Identity in Natives and Non-Natives. *J. Environ. Psychol.* **2007**, *27*, 310–319. [CrossRef]
101. Lewicka, M. Place Attachment: How Far Have We Come in the Last 40 Years? *J. Environ. Psychol.* **2011**, *31*, 207–230. [CrossRef]
102. Buckley, R. Understanding and Managing Tourism Impacts: An Integrated Approach. *J. Ecotourism* **2011**, *10*, 177–178. [CrossRef]
103. Leslie, D. *Responsible Tourism: Concepts, Theory and Practice*; CABI: Wallingford, UK, 2012; ISBN 1845939883.

Disclaimer/Publisher's Note: The statements, opinions and data contained in all publications are solely those of the individual author(s) and contributor(s) and not of MDPI and/or the editor(s). MDPI and/or the editor(s) disclaim responsibility for any injury to people or property resulting from any ideas, methods, instructions or products referred to in the content.

Article

The Influence of Heritage on the Revealed Comparative Advantage of Tourism—A Worldwide Analysis from 2011 to 2022

Zsuzsanna Bacsi

Institute of Agricultural and Food Economics, Hungarian University of Agriculture and Life Sciences, 2100 Gödöllő, Hungary; bacsi.zsuzsanna@uni-mate.hu

Abstract: A country's development is crucially determined by its cultural and natural heritage, and it is reflected in its industrial structure and its success in the global marketplace. The present paper looks at the global performance of tourism, comparing its performance measured by the Normalised Revealed Comparative Advantage (NRCA) index to the components of natural and cultural heritage, analysing 117 countries of the world. Natural and cultural heritage indicators were derived from the tourism competitiveness reports of the World Economic Forum for the years 2011–2013–2015–2017–2019–2022. Panel regression analysis was applied, with NRCA as the dependent variable and eight indicators of natural and cultural heritage as independent variables, comparing regions of the world. The main findings show considerably differing patterns between regions; Europe and Eurasia being similar to the Americas, with decreasing competitive advantage associated with more focus on endangered species and observance of environmental treaties, while the Middle East and North Africa show a strongly opposite pattern. Cultural heritage has a positive impact only in Sub-Saharan Africa, while Asia and the Pacific benefit mainly from the increase of protected areas and abundance of species. These differences shed light on differences in tourism competitiveness in the global market and may guide policymakers towards utilising heritage items for improving tourism performance.

Keywords: normalised revealed comparative advantage; natural heritage; cultural heritage; tourism infrastructure; panel regression

Citation: Bacsi, Z. The Influence of Heritage on the Revealed Comparative Advantage of Tourism—A Worldwide Analysis from 2011 to 2022. *Heritage* **2024**, *7*, 5232–5250. https://doi.org/10.3390/heritage7090246

Academic Editors: Fátima Matos Silva and Isabel Vaz de Freitas

Received: 1 August 2024
Revised: 13 September 2024
Accepted: 16 September 2024
Published: 18 September 2024

Copyright: © 2024 by the author. Licensee MDPI, Basel, Switzerland. This article is an open access article distributed under the terms and conditions of the Creative Commons Attribution (CC BY) license (https://creativecommons.org/licenses/by/4.0/).

1. Introduction

A country's development is crucially determined by its heritage, including its natural and cultural resources, history, and traditions, often through investment and job creation [1], the recognition and development of local identity [2,3], or by driving learning and innovation [4,5] and stimulating economic diversification [5,6]. Natural heritage activities contribute to rural development, especially through environmental management, primary production, and providing a background to other business activities [7]. In the face of climate change, promoting the process of a circular economy, it has become more and more obvious that the acknowledgement of past traditions and the protection of natural and cultural resources are main components of sustainable development, as a demonstration of the unique potential of heritage [8].

The Sustainable Development Goals (SDGs) also underline the fundamental importance of heritage, as SDG11.4 focuses on the protection and safeguarding of the world's cultural and natural heritage [9]. Similarly, the 2015 UNESCO Policy on World Heritage and Sustainable Development acknowledges the fundamental importance of protecting heritage, with the need of strengthening the three dimensions of sustainable development, environmental sustainability, inclusive social development, and inclusive economic development [9] in the global marketplace, reflecting its capabilities of producing a commodity that is important for the rest of the world. The most successful industries of a country are usually those for which the natural, physical, and human resources are most suited [10–12]. Thus, the existence and appreciation of traditions and the protection of natural and cultural heritage are important factors of socioeconomic development.

It is obvious that various industries depend on heritage factors to a varying extent. The tourism industry is one example in which success in the global market crucially depends on natural and cultural heritage. This is reflected by the Travel and Tourism Competitiveness Reports [13,14] and the recent modifications of these in the Travel and Tourism Development reports of the World Economic Forum [15,16]. These reports list the most important heritage items that are relevant for the tourism industry and provide quantitative indicators, defining the complex Travel and Tourism Competitiveness Index (TTCI) for most countries of the world.

The present paper looks at the global market success of the travel and tourism industry in contrast to the natural and cultural heritage of 117 countries of the world. Indicators of various items of natural and cultural heritage were derived from the World Economic Forum (WEF) databases for the years 2011–2013–2015–2017–2019–2022 [17–24]. The market success of the tourism industry is measured by a modified version of the Revealed Comparative Advantage Index, RCA [25], which was developed in 2009 under the name of Normalised Revealed Comparative Advantage, NRCA [26], as an improvement eliminating the known weaknesses of the original RCA index [26–28].

The components of the TTCI—including cultural and natural heritage indicators—are applied to assess the resources and endowments of countries as facilitators of tourism development, e.g., and their impact on international tourism arrivals and receipts [29–32].

The RCA and the NRCA indices have already been used for several regions of the world to assess the market success of the tourism industry globally and for specific geographic regions, e.g., Eastern Africa [33], Sub-Saharan Africa [34], and the continents of the world [31].

The novel aspect of our research, however, is to pair the two approaches, i.e., revealed comparative advantage as a measure of the market performance of the tourism industry and heritage indicators as major resources. This approach, though applied for specific regions of the world, has not been pursued on this large a scale, comparing 117 countries covering the main geographical regions, nor on this timescale, from 2011 to 2022, using the latest available data on heritage values. Thus, the analysis can give a good global insight into the contribution of heritage to the export competitiveness of tourism, both reflecting temporal changes and regional differences of the world.

2. Materials and Methods

2.1. Data

The analysis covers 117 countries of the world, in the following structure, with Africa having 25 countries in the analysis, the Americas 21, Asia and the Pacific 26, Europa 37, and the Middle East eight. The choice of the countries was determined by the availability of data regarding tourism performance, total exports, and data on cultural and natural heritages available for the 2011–2022 time period. See Table 1 for the data sources.

Table 1. Variables and data sources.

Variable Notation	Variable Name	Meaning	Data Source
WHSC_no	Number of world heritage cultural sites	Number of world heritage cultural sites in country	TTCI/TTDI datasets [17–21] and reports [15,16,22–24] by WEF
WHSN_no	Number of world heritage natural sites	Number of world heritage natural sites in country	
KnownSpecies	Total known species	Total known species in the country	
EnvTreaty	Environmental treaty ratification	Number of environmental treaties ratified by country	
TServInf	Tourist service infrastructure	Scale of 1 (very bad) to 7 (very good)	

Table 1. *Cont.*

Variable Notation	Variable Name	Meaning	Data Source
ProtAreaPct	Protected area percent	Total protected area, as % of country area	[35]
ForestPct	Forest area percentage	Total forest area, as % of country area	[36]
RedListInd	Red List Index	Aggregate extinction risk (0 to 1); with 0: all species extinct; 1: no species to be extinct in near future	[37]
RCA	Revealed Comparative Advantage Index for tourism	Computed by Equation (1) for tourism	International tourism receipts: [38,39]; Total export values: [40]
NRCA	Normalised Revealed Comparative Advantage Index for tourism	Computed by Equation (2) for tourism	

The list of the countries is presented in Appendix A. The analysed variables are the NRCA index, computed by the author based on the international tourism revenues and the total export revenues of the countries, compared to the world total and the availability of various heritage items—components of cultural and natural heritage—as listed by the Travel and Tourism Competitiveness Reports and Travel and Tourism Development Reports by the WEF [14–16,22–25]. The heritage indicators are accompanied by the quality of the tourism service infrastructure, as this feature certainly influences the accessibility of heritage for tourism purposes. The analysis purposefully avoids the inclusion of general economic variables, such as GDP or price levels, as these might mask the influence of heritage on export competitiveness. The complete list of the variables, with their precise meanings and sources of data are listed in Table 1. Some variables were not covered by the WEF reports for all countries or all years, and, in these cases, the research used the original full data sources suggested by the WEF reports. The timescale of the analysis is also limited by the coverage of the WEF reports (6 years: 2011, 2013, 2015, 2017, 2019, and 2022).

2.2. The Computation of the NRCA Index

The index of Revealed Comparative Advantage (RCA) defined by Balassa in 1965 [25] is a tool to assess the export competitiveness of countries. It is computed by the following formula, with values between 0 and 1 meaning comparative disadvantage, while values above 1 reflect comparative advantage (see Equation (1)):

$$RCA_{c,i} = [X_{c,i}/X_c]/[X_{w,i}/X_w] = [X_{c,i}/X_{w,i}]/[X_c/X_w], \quad (1)$$

where X: export value, c: 1...m country, i: 1...n industry, w: world, so $X_c = \sum_{(i=1...n)} X_{c,i}$ and, similarly, $X_{w,i} = \sum_{(c=1...m)} X_{c,i}$ and $X_w = \sum_{(c=1...m)} X_c$.

The Revealed Comparative Advantage index (RCA) compares the share of a country's total exports of the analysed commodity in its total exports to the share of world exports of the same commodity in total world exports. An index value above 1 means that the analysed commodity represents a higher share in the country's total exports than in the world, and this indicates that the country has an advantage in the commodity compared to the world average.

Although RCA is widely used in trade competitiveness analysis, there are some problematic features that may distort its meaning [26–28]. The two most important problems are as follows: (a) RCA is not symmetric because non-competitive values fall between 0 and 1, while competitive values range from 1 to infinity; (b) the so-called small-country bias means that, when X_c is small relative to X_w, then RCA tends to be high, suggesting unreasonably high comparative advantage. To correct these weaknesses, the index of Normalised Revealed Comparative Advantage (NRCA) was developed by Yu et al. [26], as in Equation (2):

$$NRCA_{c,i} = X_{c,i}/X_w - X_c \times X_{w,i}/X_w^2 = [X_{c,i}/X_{w,i} - X_c/X_w] \times X_{w,i}/X_w \quad (2)$$

with the same notations as the ones used for RCA.

The Normalised Revealed Comparative Advantage index (NRCA) calculates the difference between a country's actual exports share of world exports in the analysed commodity and the country's total export share in total world exports. This difference is normalised by the share of the analysed commodity in world exports. If the country's export share of the commodity is the same as its total export share, then the commodity is in a neutral position in the country, producing the same export share as the average commodities of the country. In this case, NRCA equals zero, while positive values represent a successful export commodity, and negative values indicate an unsuccessful one disadvantaged in the world market.

As Yu et al. [26] demonstrated, the NRCA is symmetric, ranging from −1 to +1, with the neutral value being zero, and it is free of small-country bias. In addition, NRCA is additive and transitive, and it correctly assesses temporal changes, i.e., it allows for between-year comparisons.

The application of NRCA for trade analysis is justified by its ability to handle the weaknesses of RCA, as is explained in the Materials and Methods section. As was stated in [27], NRCA has the unique ability to be additive with respect to both countries and commodities, and it is consistent with regard to time. This feature makes NRCA values comparable between countries, commodities, and time periods.

Despite the problems of the original RCA index, it has been applied for the tourism industry, using international tourism revenues as the indicator of tourism exports at the country level [33,34,41]. However, Bogale [34] also used NRCA for tourism, revealing important differences between the assessments by the two methods.

2.3. Analysis of the Relationship between NRCA and Heritage

The NRCA values are compared to the heritage indicators for the 117 countries. The comparison is performed by correlation analysis, as an initial step, followed by a panel regression for the 117 countries and the 6 years, with NRCA being the dependent variables, and heritage indicators and tourism service infrastructure being the independent ones. Panel regression was used, focusing on fixed effects of the independent variables, as the 6 years were not sufficient for doing a proper temporal assessment for the eight independent variables.

The panel regression estimation was performed by the Linear Mixed Model method (LMM) available in the "R" Statistical Software [42]. The assumptions for the LMM require that the explanatory variables are related linearly to the response variable, and the errors have constant variance, are independent, and are normally distributed. These diagnostics are presented as a part of the analysis. However, as Schielzeth et al. [43] proved, the LMM and other mixed-effect procedures are considerably robust, i.e., minor violations of the assumptions do not make the model estimates unreliable.

The general equation of a panel model is [44]

$$y(i,j) = a + c(i) + \sum_{k=1\ldots K} [b(k) + d(i,k)] \times (i,j,k) + e(i,j)$$

where:

- y(i,j): is the observed dependent variable, with two factors, i = 1...n and j: 1...M (e.g., i being the time and j being the subject or vice versa);
- x(i,j,k): is the value of the observed independent variable k (k = 1...K), for i,j;
- a: is the fixed intercept, not depending on i and j;
- b(k): is the fixed slope of the independent variable k;
- c(i): is the random intercept, varying by factor i;
- d(i,k): is the random slope, varying by variable k and factor i;
- e(i,j): is the random error (of normal distribution).

Panel model estimation requires the estimations of a, c(i), b(k), d(i,k), and e(i,j). Various simplifications exist with regard to assumptions of the parameters, and the estimation methods differ according to these simplifications. Several statistical packages provide estimation procedures. The present research uses the LME4 package [45] of „R" statistical software [42]; therefore, the LME4 formulas are also presented with the equations. In these equations Y and X1,...XK refer to the various variable names. Four model versions are described:

1. Fixed intercept and slopes, with random intercepts and slopes: $y(i,j) = a + c(i) + \sum_{k=1...K} [b(k) + d(i,k)] x(i,j,k) + e(i,j)$, i.e., the full model. The relevant LME4 formula is $Y \sim 1 + X1 + X2 +...XK + (1 + Xk | subject)$ or $Y \sim 1 + X1 +...+ XK + (1 + Xk) | time)$ depending on which factor and which Xk independent variable are used for random effects.
2. Fixed intercept and slope with random intercept: $y(i,j) = a + c(i) + \sum_{k=1...K} [b(k)] x(i,j,k) + e(i,j)$. The LME4 formula: $Y \sim 1 + X1 +...+ XK + (1 | subject)$ or $Y \sim 1 + X1 +...+ XK + (1 | time)$.
3. Fixed intercept with random slopes: $y(i,j) = a + \sum_{k=1...K} d(i,k) x(i,j,k) + e(i,j)$. The LME4 formula: $Y \sim 1 + X1 + X2 +...XK + (0 + Xk | subject)$ or $Y \sim 1 + X1 +...+ XK + (0 + Xk | time)$.
4. Fixed slopes and random intercept: $y(i,j) = c(i) + \sum_{k=1...K} b(k) x(i,j,k) + e(i,j)$. The LME4 formula is $Y \sim 0 + X1 + X2 +...XK + (1 | subject)$, or $Y \sim 0 + X1 +...+ XK + (1 | time)$.

From the viewpoint of interpretation, Model 4 seems to be the simplest, as it assumes that the average impact of each independent variable on the dependent variable is the same, regardless of subjects or time, but there are variations around this average according to subjects or times. The following analysis applies this relatively simple approach. If the model had turned out to be unreliable, a more complex structure would have to be explored.

3. Results

Table 2 presents the descriptive statistics of the analysed variables in our database of 2011–2013–2015–2017–2019–2022, for 117 countries, together with regional mean values. The normality testing of the variables rejected the assumption of normality; therefore, we used nonparametric statistical methods that are not sensitive to lack of normalities. The variables were compared for significant differences between regions and years, applying the Kruskal–Wallis nonparametric test. Table 2 shows that significant differences were found between regions for all variables, while, regarding the years, only the ProtAreaPct, KnownSpecies, TServInf, and EnvTreaty variables differed significantly. Figure 1 presents the temporal patterns of each variable by region.

The pairwise correlations (Table 3) indicate that there are moderate but significant correlations between the NRCA and heritage components. The largest absolute values are with the Red List Index and the Tourism Service Infrastructure variables (Spearman's correlation). As the data are not normally distributed, the Pearson and Spearman correlations considerably differ. The small absolute values indicate that the pooled data do not show clear patterns, and the various indicators reflect significant relationships between the heritage components and export competitiveness of tourism.

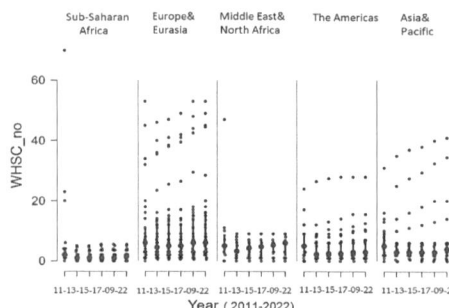
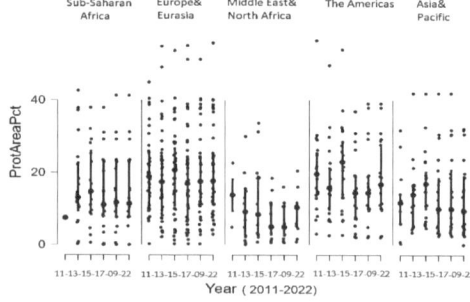

Figure 1. *Cont.*

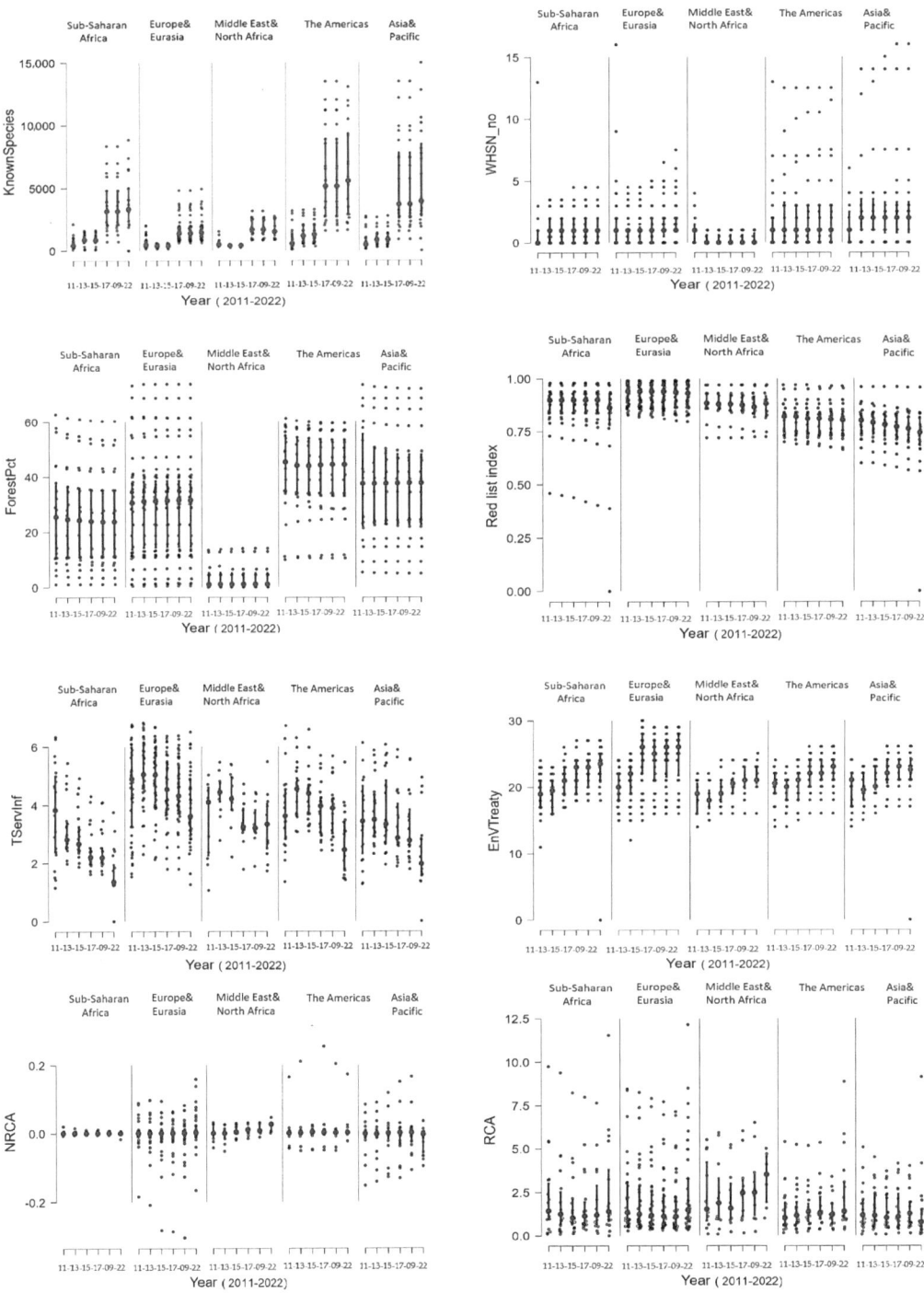

Figure 1. Heritage components and comparative advantages by year and region. Source, author's own construction applied [46].

Table 2. Descriptive statistics of the analysed variables.

	N	Min	Max	Mean	St.Dev.	Region K-W St	p	Year K-W St	p
RCA	595	0.014	12.142	1.968	1.917	13.55820	0.00885	1.53053	0.90952
NRCA	557	−0.305	0.318	0.000	0.049	11.41499	0.02228	2.36870	0.79613
WHSC_no	691	0.000	70	6.943	9.861	105.59840	6.315×10^{-22}	4.88886	0.42959
ProtAreaPct	653	0.000	56.5	16.492	11.527	53.76257	5.901×10^{-11}	12.12895	0.03306
KnownSpecies	692	0.000	15,878.0	2167.63	2663.90	108.31665	1.663×10^{-22}	354.57962	1.807×10^{-74}
WHSN_no	688	0.000	16	1.773	2.669	61.31917	1.532×10^{-12}	4.31315	0.50527
ForestPct	696	0.000	73.74	29.746	19.688	206.34547	1.623×10^{-43}	0.01236	1.00000
Red list index	695	0.000	0.992	0.856	0.115	284.35238	2.568×10^{-60}	2.30220	0.80594
TServInf	692	0.000	7.00	3.820	1.410	160.33738	1.238×10^{-33}	70.42547	8.358×10^{-14}
EnvTreaty	686	0.000	30.00	21.613	3.900	64.62533	3.086×10^{-13}	99.37025	7.174×10^{-20}
RCA	595	0.014	12.142	1.968	1.917	13.55820	0.00885	1.53053	0.90952
NRCA	557	−0.305	0.318	0.000	0.049	11.41499	0.02228	2.36870	0.79613

Means by region	Sub-Saharan Africa	Europe and Eurasia	Middle East and North Africa	The Americas	Asia and Pacific
WHSC_no	2.65	10.02	5.09	5.52	7.38
ProtAreaPct	15.15	19.22	9.10	18.09	13.87
KnownSpecies	2253.98	1077.09	1145.73	3833.75	3266.27
WHSN_no	1.18	1.43	0.42	2.66	3.01
ForestPct	25.15	30.32	3.85	41.53	36.87
Red list index	0.85	0.93	0.87	0.80	0.76
TServInf	2.7554	4.5878	3.7276	3.8123	3.3432
EnvTreaty	20.675	23.098	19.786	21.032	21.061
RCA	2.14	2.12	2.52	1.65	1.53
NRCA	0.00137	−0.00390	0.00664	0.01098	−0.00978

Note: K-W = Kruskal-Wallis test for significant differences between regions and years.

Table 3. Pairwise correlations of RCA, NRCA, and heritage-related indicators.

		WHSC _no	Prot AreaPct	Known Species	WHSN _no	Forest Pct	RedList Ind	TServ Inf	Env Treaty	RCA
WHSC_no	P	—								
	S	—								
Prot Area Pct	P	0.148 ***	—							
	S	0.156 ***	—							
Known Species	P	0.135 ***	−0.016	—						
	S	0.139 ***	−0.02	—						
WHSN_no	P	0.494 ***	0.087 *	0.454 ***	—					
	S	0.471 ***	0.199 ***	0.390 ***	—					
ForestPct	P	0.061	0.294 ***	0.240 ***	0.095 *	—				
	S	0.069	0.311***	0.229 ***	0.206 ***	—				
RedListInd	P	0.017	0.181 ***	−0.284 ***	−0.111 **	−0.119 **	—			
	S	0.035	0.129 **	−0.336 ***	−0.168 ***	−0.165 ***	—			
TServInf	P	0.293 ***	0.160 ***	−0.237 ***	0.164 ***	0.080 *	0.178 ***	—		
	S	0.329 ***	0.167 ***	−0.361 ***	0.106 **	0.076 *	0.135 ***	—		
EnvTreaty	P	0.363 ***	0.260 ***	0.184 ***	0.168 ***	0.188 ***	0.231 ***	0.314 ***	—	
	S	0.423 ***	0.258 ***	0.228 ***	0.308 ***	0.177 ***	0.103 **	0.281 ***	—	
RCA	P	−0.124 **	−0.170 ***	−0.087 *	−0.113 **	−0.065	−0.130 **	0.075	−0.104 *	—
	S	−0.065	−0.119 **	−0.034	−0.04	−0.037	−0.122 **	0.122 **	−0.104 *	—
NRCA	P	−0.091 **	−0.091 *	0.138 ***	0.236 ***	−0.058	−0.060	0.114 **	−0.184 ***	0.280 ***
	S	0.044	−0.046	0.099 *	0.059	0.009	−0.186 ***	0.174 ***	−0.096 *	0.816 ***

Significance of correlations: *: $p < 0.05$, **: $p < 0.01$, ***: $p < 0.001$, P: Pearson's, S: Spearman's correlation.

The heritage-related independent variables were tested for collinearity, and, as the TOL and VIF factors indicate (Table 4), no issues of multicollinearity were detected.

As correlations did not reveal any clear patterns, the analysis continues with a panel regression based on the LME4 package of „R", as described in the Methodology section. The first step was using all the countries and years for a panel regression with NRCA as a dependent variable, heritage components as independent ones, and year as a random factor. The results are presented in Table 5.

Table 4. Multicollinearity test for heritage-related variables.

	Tolerance (TOL)	Variance Inflation Factor (VIF)
WHSC_no	0.706	1.416
ProtAreaPct	0.784	1.276
KnownSpecies	0.428	2.338
WHSN_no	0.584	1.712
ForestPct	0.804	1.244
Red list index	0.793	1.261
TServInf	0.649	1.54
EnvTreaty	0.591	1.692
Year	0.512	1.955

Table 5. Fixed- and random-effect estimates for all countries.

Variable	Fixed-Effects Estimate	Std. Err	t Value	Pr(>|t|)		Random Intercept for Years
WHSC_no	−0.001205	0.0003	−4.413	1.25×10^{-5}	***	2011: −0.0013872238
ProtAreaPct	−0.000217	0.0002	−1.064	0.28775		2013: −0.0064993186
KnownSpecies	0.000003	0.0000	2.447	0.01538	*	2015: 0.0006474683
WHSN_no	0.005909	0.0011	5.462	7.46×10^{-8}	***	2017: −0.0010744230
ForestPct	−0.000065	0.0001	−0.487	0.62641		2019: 0.0003912855
RedListInd	−0.022490	0.0216	−1.044	0.2971		2022: 0.0079222115
TServInf	0.009548	0.0020	4.879	1.62×10^{-6}	***	
EnvTreaty	−0.003565	0.0007	−5.379	1.31×10^{-7}	***	
Region Sub-Sah Afr	0.061850	0.0211	2.928	0.00358	**	
Region Europe & Eurasia	0.062270	0.0235	2.653	0.00825	**	
Region Middle East & NAfr EEastEsast&NENA	0.064940	0.0224	2.901	0.0039	**	
Region The Amers	0.054300	0.0213	2.545	0.01127	*	
Region Asia & Pacific	0.040820	0.0206	1.981	0.04823	*	
	$R^2 = 0.223408$, N = 508					

Number of obs: 508, groups: Year, 6; Significance: *: $p < 0.05$, **: $p < 0.01$, ***: $p < 0.001$.

Panel data analysis and linear mixed models are widely used in social science, economics, business analytics, and in many other fields of science. The fixed-effect and random-effect estimations of these models reveal important features of the relationship between the dependent and the independent variables. Fixed effects account for variables or factors that remain constant across observations, in our case across countries.

Random effects are used to account for variability and differences between different entities, which can be countries or time periods. In our model, the random effects were estimated for the various years.

The model estimates show the influence of a unit changes in the independent variables on the NRCA value, on average, and the random intercept values modify this estimated value by adding the relevant constant for each year. This means, using the results in Table 5, that the following equation can estimate NRCA for the Asia and Pacific region in 2017 (keeping only the significant estimates in the equation):

$$NRCA = -0.001205\ WHSC\_no + 0.000003\ KnownSpecies + 0.005909\ WHSN\_no + 0.009548\ TServInf - 0.003565\ EnvTreaty + 0.040820 - 0.0010744230$$

This type of modelling is particularly suitable in panel analyses, when we work with a large group of countries (or individuals, subjects) and have multiple-year observations for them. The fixed-effects estimations in these models can estimate the impacts of time-invariant factors, and the random effect estimations account for variations in outcomes that cannot be explained by observed variables alone. Thus, the fixed effect of a variable can be considered as the average estimate of its impact, while the random effect adds the individual variations to this. This type of modelling technique is very popular in cross-country economic comparisons of several years, and is also favoured in sociology and psychology when repeated observations are taken about a large group of individuals.

Model diagnostics require the testing of the residuals for normality, and it was approved by visual analysis as well as by the Kolmogorov–Smirnov test (see the first panel in Appendix B). As the p-value of the statistical test is higher than 0.05 ($p = 0.0768$), the normality of the residual series cannot be rejected.

The model suggests that NRCA is negatively influenced by WHSC and EnvTreaty values and positively influenced by TServIng, KnownSpecies, WHSN, and each of the various regions, though to different extents. The random effects of the years are also visible, though their magnitude is less than 10% of the regional fixed effects. The positive impacts of the factors are reasonable, but the negative effects are difficult to explain, as cultural heritage should reasonably be considered as a tourist attraction, and environmental treaties should also enhance a country's image as devoted to the protection of natural heritage. The correspondence between observed and predicted NRCA values is not too strong (R = 0.474771 and R^2 = 0.223408), and this further supports the need for more refined analysis. As the regions show significant impacts, this suggests that the analysis could be performed by region.

To analyse regions separately, the Region variable was removed from the model structure, and the fixed effects of the years were added to see if there were considerable differences between years. Each region has several—but varying—significant fixed slope effects, but the random intercepts are at the 10^{-19} to 10^{-16} range, so their impact is negligible.

Table 6 presents the results by region.

Table 6. Regionwide analysis by panel regression.

Fixed Effects: N = 75, R^2 = 0.61468	Sub-Saharan Africa Estimate	Std. Error	t Value	Pr(>|t|)		Random Intercept (Year)
WHSC_no	0.000672	0.000316	2.125000	0.037660	*	2011: 8.590075×10^{-19}
ProtAreaPct	−0.000034	0.000045	−0.749000	0.456930		2013: 1.040383×10^{-17}
KnownSpecies	0.000000	0.000000	0.973000	0.334540		2015: 3.467942×10^{-18}
WHSN_no	0.000761	0.000477	1.595000	0.115950		2017: 8.538866×10^{-18}
ForestPct	0.000010	0.000030	0.327000	0.745070		2019: $-9.517754 \times 10^{-18}$
RedListInd	−0.007073	0.002521	−2.805000	0.006730	**	2022: $-3.155572 \times 10^{-18}$
TServInf	0.001364	0.000508	2.685000	0.009320	**	
EnvTreaty	−0.000138	0.000125	−1.102000	0.274850		
factor(Year)2011	0.022120	0.004432	4.990000	0.000005	***	
factor(Year)2013	0.003733	0.003357	1.112000	0.270550		
factor(Year)2015	0.004210	0.003350	1.257000	0.213700		
factor(Year)2017	0.003442	0.003174	1.085000	0.282370		
factor(Year)2019	0.003335	0.003197	1.043000	0.300990		
factor(Year)2022	0.002173	0.002870	0.757000	0.451830		
Fixed effects: N = 198 R^2 = 0.36054	Europe and Eurasia Estimate	Std. Error	t value	Pr(>|t|)		Random intercept (Year)
WHSC_no	−0.001283	0.000401	−3.196	0.001638	**	2011: 2.783644×10^{-17}
ProtAreaPct	−0.000247	0.000317	−0.78	0.436233		2013: $-4.839659 \times 10^{-17}$
KnownSpecies	0.000031	0.000007	4.326	2.49×10^{-5}	***	2015: 1.156400×10^{-16}
WHSN_no	−0.000544	0.002359	−0.231	0.817764		2017: $-1.007347 \times 10^{-18}$
ForestPct	0.000500	0.000223	2.241	0.026229	*	2019: $-3.525714 \times 10^{-17}$
RedListInd	−0.277600	0.071520	−3.882	0.000144	***	2022: $-3.449263 \times 10^{-17}$
TServInf	0.006441	0.003723	1.73	0.085327	+	
EnvTreaty	−0.004024	0.001256	−3.204	0.001596	**	

Table 6. Cont.

	Estimate	Std. Error	t value	Pr(>|t|)		Random intercept (Year)
factor(Year)2011	0.297100	0.071990	4.127	0.999989		
factor(Year)2013	0.291200	0.071390	4.079	0.999989		
factor(Year)2015	0.303000	0.071080	4.263	0.99999		
factor(Year)2017	0.262500	0.074210	3.537	0.999988		
factor(Year)2019	0.263500	0.073840	3.569	0.999988		
factor(Year)2022	0.292200	0.073770	3.962	0.999988		

Fixed effects: N = 47, R^2 = 56136	Middle East and North Africa					Random intercept (Year)
	Estimate	Std. Error	t value	Pr(>|t|)		
WHSC_no	−0.00169	0.00119	−1.41700	0.16580		2011: 4.250664×10^{-19}
ProtAreaPct	−0.00058	0.00033	−1.76800	0.08630	+	2013: $-8.032700 \times 10^{-19}$
KnownSpecies	0.00001	0.00001	1.61100	0.11660		2015: 9.908962×10^{-19}
WHSN_no	0.01544	0.00984	1.56800	0.12630		2017: 1.213061×10^{-18}
ForestPct	0.00187	0.00083	2.24200	0.03180	*	2019: 8.087071×10^{-19}
RedListInd	−0.08722	0.05476	−1.59300	0.12070		2022: 1.455326×10^{-18}
TServInf	0.00274	0.00349	0.78600	0.43760		
EnvTreaty	0.00390	0.00172	2.26500	0.03020	*	
factor(Year)2011	−0.01478	0.03980	−0.37100	0.99950		
factor(Year)2013	−0.01176	0.03729	−0.31500	0.99960		
factor(Year)2015	−0.00489	0.03742	−0.13100	0.99970		
factor(Year)2017	−0.01893	0.03707	−0.51100	0.99960		
factor(Year)2019	−0.01691	0.03714	−0.45500	0.99960		
factor(Year)2022	−0.01592	0.03737	−0.42600	0.99960		

Fixed effects: N = 96, R^2 = 0.74999	The Americas					Random intercept (Year)
	Estimate	Std. Error	t value	Pr(>|t|)		
WHSC_no	−0.00390	0.00073	−5.375	7.04×10^{-7}	***	2011: $-9.632035 \times 10^{-18}$
ProtAreaPct	0.00059	0.00042	1.41	0.16245		2013: $-6.476665 \times 10^{-18}$
KnownSpecies	0.00000	0.00000	0.164	0.87038		2015: $-1.050447 \times 10^{-17}$
WHSN_no	0.01059	0.00185	5.731	1.60×10^{-7}	***	2017: $-8.136377 \times 10^{-18}$
ForestPct	−0.00049	0.00027	−1.821	0.07228	+	2019: $-9.595060 \times 10^{-18}$
RedListInd	−0.14170	0.04569	−3.101	0.00264	**	2022: $-5.729711 \times 10^{-18}$
TServInf	0.00309	0.00448	0.69100	0.49187		
EnvTreaty	−0.00944	0.00143	−6.605	3.70×10^{-9}	***	
factor(Year)2011	0.26150	0.05065	5.163	0.99892		
factor(Year)2013	0.25250	0.05056	4.994	0.99893		
factor(Year)2015	0.27020	0.05099	5.298	0.99889		
factor(Year)2017	0.28570	0.05286	5.405	0.99872		
factor(Year)2019	0.28340	0.05304	5.344	0.99871		
factor(Year)2022	0.29800	0.05159	5.776	0.99882		

Fixed effects: N = 92, R^2 = 0.43755	Asia and Pacific					Random intercept (Year)
	Estimate	Std. Error	t value	Pr(>|t|)		
WHSC_no	−0.00213	0.00096	−2.21300	0.02986	*	2011: 2.915842×10^{-19}
ProtAreaPct	0.00187	0.00067	2.76600	0.00708	**	2013: $-6.460514 \times 10^{-19}$
KnownSpecies	0.00001	0.00000	2.60000	0.01114	*	2015: $-1.761062 \times 10^{-19}$
WHSN_no	0.00189	0.00241	0.78300	0.43619		2017: $-4.110190 \times 10^{-19}$
ForestPct	−0.00124	0.00030	−4.11700	0.00009	***	2019: $-1.370761 \times 10^{-19}$
RedListInd	−0.12640	0.06485	−1.94900	0.05491	+	2022: $-3.720638 \times 10^{-19}$
TServInf	0.00309	0.00448	0.69100	0.49187		
EnvTreaty	−0.00169	0.00274	−0.61600	0.53948		
factor(Year)2011	0.15490	0.08204	1.88800	0.95579		
factor(Year)2013	0.13190	0.08173	1.61400	0.95824		
factor(Year)2015	0.13110	0.08326	1.57500	0.95592		
factor(Year)2017	0.11630	0.08667	1.34200	0.95212		
factor(Year)2019	0.12580	0.08719	1.44200	0.95006		
factor(Year)2022	0.09867	0.08714	1.13200	0.95392		

Significance: +: $p < 0.1$, *: $p < 0.05$, **: $p < 0.01$, ***: $p < 0.001$.

The diagnostic testing of the residuals justifies their reasonable normality, though occasionally this required the deletion of a few outliers (see Appendix B). As the diagnostics and the R^2 values indicate, all regional models provide reliable estimates. Table 7 sums up the main findings, presenting only the significant fixed effects of the heritage components. It is quite clear that the regions considerably differ in the following aspects:

- The Year factor did not show any significant effects except for the first year (2011) in Sub-Saharan Africa.
- The tourism service infrastructure (TServInf) had a positive effect in Europe and Eurasia, the Americas, and Sub-Saharan Africa, but no impact was found in the Middle East and North Africa or in the Asia and Pacific region.
- The number of cultural world heritage sites (WHSC_no) had a negative significant effect except in Sub-Saharan Africa, where a positive impact was noted, and in the Middle East and North Africa, where no effect was found.
- Regarding the natural heritage components:
    - The Red List Index had a significant negative impact everywhere except the Middle East and North Africa.
    - The forest area proportion (ForestPct) was positive in Europe and Eurasia and in the Middle East and North Africa and negative in the Americas and in the Asia and Pacific region.
    - The number of ratified environmental treaties (Envreaty) showed significantly positive impacts in the Middle East and North Africa, and negative impacts in Europe and Eurasia and in the Americas.
    - The proportion of protected areas (ProtAreaPct) was significantly positive in the Asia and Pacific region and significantly negative in the Middle East and North Africa region.
    - The number of known species had significant positive impacts in Europe and Eurasia and in the Asia and Pacific region.
    - The number of natural world heritage sites (WHSN_no) had a positive impact only in the Americas.

Table 7. Summary of significant estimates of fixed effects by regions.

	Europe and Eurasia	The Americas	Asia and Pacific	Sub-Saharan Africa	Middle East and North Africa
TServInf	0.00644	0.01357		0.00136	
WHSC_no	−0.00128	−0.00390	−0.00213	0.00067	
ProtAreaPct			0.00187		−0.00058
KnownSpecies	0.00003		0.00001		
WHSN_no		0.01059			
ForestPct	0.00050	−0.00049	−0.00124		0.00187
RedListInd	−0.27760	−0.14170	−0.12640	−0.00707	
EnvTreaty	−0.00402	−0.00944			0.00390
factor(Year)2011				0.02212	

As various heritage components have different magnitudes, the coefficients were compared, multiplying them by the relevant regional mean values; thus, their values are made comparable (Figure 2, top panel). The bottom panel of Figure 2 presents these impacts in percentages, with the full impact scaled to 100 percent in each region.

The impact of the Red List Index seems to be the largest, representing approximately 58% of the impacts in Europe and Eurasia, 52% in Sub-Saharan Africa, 47% in Asia and Pacific, and 26% in the Americas, and these impacts are all negative. The single largest positive impact is seen in Middle East and North Africa, where the number of environmental treaties take up 86% of all impacts (with a positive sign), while this factor has a considerable importance in the Americas (−46%) and in Europe (−21%). The tourism infrastructure is rather influential in Sub-Saharan Africa (+ 33%), while its influence is much less in the

Americas (+12%) and in Europe and Eurasia (+7%). The forest areas have a 23% influence, and the protected areas a 13% influence in the Asia and Pacific region, while the number of cultural world heritage sites show a 15% positive influence in Sub-Saharan Africa. The rest of the factors—natural world heritage sites, known species, and protected areas—play a much smaller role, with only 0% to 13% importance.

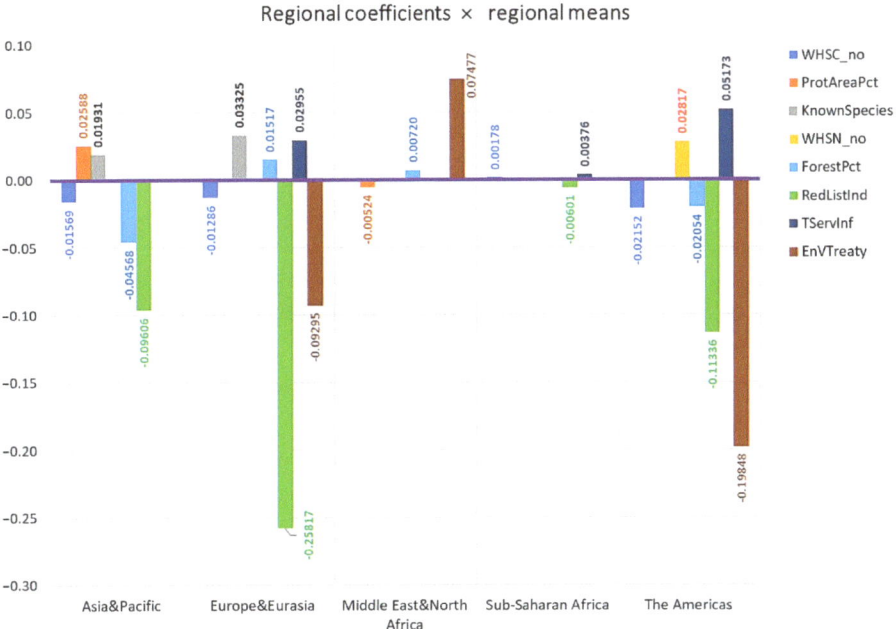

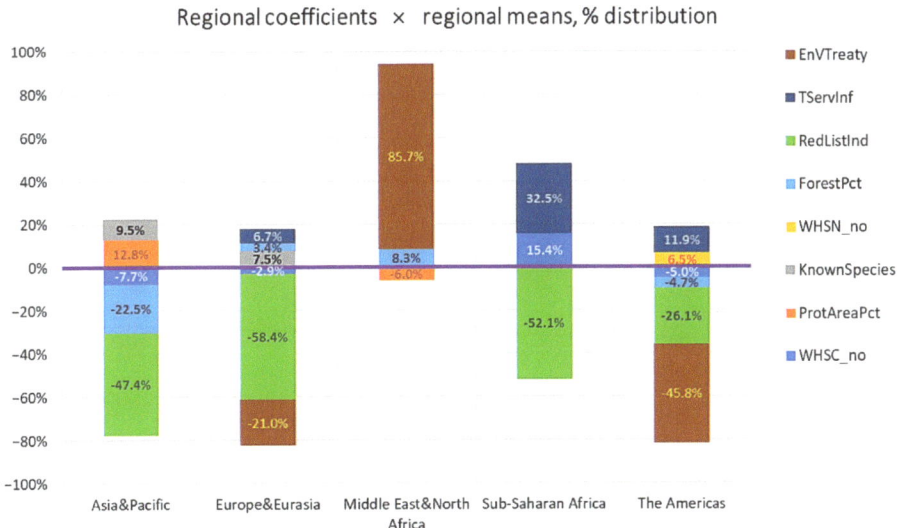

Figure 2. Coefficients multiplied by regional means of the various heritage components (as actual values and as % shares).

In this respect, the Americas and Europe and Eurasia seem to have much in common, while the Asia and Pacific region slightly differs from them. The Middle East and North Africa region and Sub-Saharan Africa seem to have rather unique impact structures towards NRCA, which is not surprising considering their unique natural conditions, development levels, and cultural endowments.

4. Discussion

The revealed comparative advantage of the tourism industry has been analysed widely for many countries, using various forms of the RCA index [33,34,47–49], while the NRCA index has been applied less frequently [34]. The present study applies NRCA, which is a new approach for a worldwide tourism analysis. The differences shown in Table 7 and Figure 2 are reflections of the strengths and weaknesses of various regions and the manifestations of these in international tourism export competitiveness [50].

The most influential factor on NRCA seems to be the Red List Index, and this has a considerable negative impact on all regions except the Middle East and North Africa, which is not very rich in species compared to the other regions. The negative impact of this on all regions means that the higher the index (i.e., the fewer species that are threatened by extinction), the less comparative advantage is shown by NRCA. This seems to suggest that tourists seek the sensational and are attracted to endangered sights; therefore, tourism exports increase where such unique living systems are found. This is true for all regions, except the Middle East and North Africa, where the mean number of known species is much lower than elsewhere; thus, wildlife is not a determinant factor in tourism. Increased tourist interest in regions with endangered wildlife was found in [51] about the wildlife of the Swedish Arctic and in [52] about ecotourism in general. The concept of wildlife tourism utilises this interest in observing and interacting with animals that may be endangered, threatened, or rare in worldwide destinations [53–55].

Other aspects of natural heritage also show diverse influences on NRCA. The positive impact of the number of known species (Europe and Eurasia, Asia and Pacific), the proportion of protected areas (Asia and Pacific), the proportion of forest areas (the Americas) and the number of ratified environmental treaties (Middle East and North Africa) are in line with our logical expectations, and their positive impact on tourism competitiveness is supported by previous research [31,56,57]. The more surprising findings in this respect are the negative impacts. The negative impacts of protected areas in the Middle East and North Africa may be due to their outstanding low level, which may make them less accessible to tourists. The negative effects of forest areas in the Americas and the Asia and Pacific region may be explained by their exceptionally high proportion, which may be a barrier to transport and travel. The negative impact of the number of ratified environmental treaties in Europe and Eurasia, the Americas, and the Asia and Pacific region may easily lead to increased costs related to environmentally threatened tourist attractions. Further research may be needed to clarify these impacts.

Tourism infrastructure is expected to have a significant impact on tourism market success, as better services increase tourism revenues [33,56,57], and, in certain destinations, tourism services infrastructure was found to be a more influential factor than natural or cultural endowments [48]. In our findings, the surprising fact is that tourism services infrastructure proved to be non-influential in the Asia and Pacific region and in the Middle East and North Africa. According to the last two WEF reports [15,16], these two regions performed the worst regarding the tourism services and infrastructure indicators, both in 2019 and 2022, and this may be the explanation why this factor could not contribute to the export competitiveness of the industry.

Cultural world heritage sites (WHSCs) are also influential factors in four of the five regions, but their impact is more controversial. They contribute to the increase of tourism export competitiveness only in Sub-Saharan Africa. Their impact is negative in Europe and Eurasia, in the Americas, and in the Asia and Pacific region, while in the Middle East and North Africa region, their impact is not significant. This can be explained by the fact that these sites are rather rare in Sub-Saharan Africa, so their existence is a comparative advantage for the countries possessing them, while, in the rest of the world, their prevalence is far higher. In the regions with negative impact, their presence may lead to overtourism features [50,58,59], leading to congestion, and may even lead to the deterioration of these sites or enhanced conservation costs that decrease their net positive impacts. Prior research indicated negative impacts of cultural heritage sites on RCA index values in Eastern Africa [33], but, when suitable infrastructure was available, this impact turned positive. Positive impacts of cultural world heritage sites were demonstrated on tourism arrivals and receipts in Europe and America [29,31], but these cannot be equated to tourism export competitiveness, and this points to the need to assess how tourism export competitiveness relates to the export competitiveness of other industries.

5. Conclusions

Based on the above results, the following conclusions may be drawn. First, the application of the NRCA index can reveal important features of the export competitiveness of the tourism industry and its regional differences. While RCA-based tourism analysis is frequent in the literature, the more reliable NRCA index is less applied for tourism analysis; therefore, the present research is an important contribution to the state of the art in this respect. Second, the analysis revealed that the destination competitiveness approach, based on the tourism endowments of destinations, usually shows a very different picture from the actual revealed comparative advantages, i.e., the market-based analysis of tourism performance. The panel regression pointed out that the main factors that are used in the evaluation of destination endowments do not always contribute positively to the export competitiveness revealed by the NRCA index. The regions considerably differ as to which endowments are successfully translated to market performance and which ones are more a drawback than a benefit. This points to the factors most valued by international tourists either for their abundance or for their actual uniqueness. There is a general tendency that what is most sought after tends to be overused, and, therefore, there is a danger that formerly appreciated resources may disappear and become no longer a positive factor in tourism performance.

Economic theory appreciates the economic importance of cultural heritage, as it provides positive externalities, enhancing employment, and improving human and social capital, while following the principles of sustainability. Cultural heritage provides a unique perspective on history and tradition and is an important component of cultural capital. As such, it is invaluable in educating younger generations, enhancing a sense of identity and belonging, critical thinking, and empathy towards diverse cultures. However, the phenomenon of overtourism poses a significant threat to heritage sites; therefore, risk-mitigation strategies should be developed to prevent damage to heritage. Such strategies could start with the careful analysis of the carrying capacity of heritage sites, followed by limiting rules on the access times, traffic restrictions favouring on-foot access, tax regulations, and higher prices of access and accommodation in the neighbourhood. The idea of tourism demarketing may serve the protection of such sites, though at the expense of lower economic benefit.

Based on the above, policy implications may be outlined, both for regions of developed tourism sectors and for those still underdeveloped in this respect. One such implication is that cultural heritage sites should be protected against overtourism in the most developed tourism regions, while they provide untapped opportunities in less developed areas. The same is true about environmentally threatened areas and the protection measures applied, which may become negative factors in tourism market success, especially when the natural (and cultural) resources are affected by overtourism. In spite of their negative impacts, these steps should be strengthened in the development of sustainable tourism, and new ways of maintaining and presenting these resources should be invented, utilising tourism innovations, such as artificial intelligence, virtual tours, and limited access to endangered resources.

The limitations of the present research include data availability, as only 117 countries could be covered for a 10-year-long period. More detailed analysis within and between region comparisons could also be performed, not stopping at the regional level but looking at sub-regions of continents or at climatic zones regardless of continental breakup. The same methodology can be applied to more localised areas, small, socio-economically similar groups of nations, to identify which direction tourism exports should take in the coming decades. The analysis could be performed on a continent-driven basis instead of a region-driven approach, to see the impact of regional differences within continents and regional similarities across continents. The influence of variables, such as a safety index, a digitalisation index, or access to online information, could also be used for grouping the countries, as they may also have an impact on tourism performance.

Tourism competitiveness can also be assessed by other indicators besides NRCA, and comparisons between indicators could also reveal interesting and unexpected patterns in export performance of the industry. These issues are areas of future research.

Funding: This research received no external funding.

Data Availability Statement: Data were downloaded from publicly available databases, as is indicated in Table 1 of the current research. The actual websites are listed in the References Section in detail.

Conflicts of Interest: The author declares no conflicts of interest.

Appendix A. Country List with Continent and Region

Continent	Region	Country (Code)
Africa	Middle East and North Africa	Egypt (EGY), Morocco (MAR), Tunisia (TUN), Yemen (YEM)
	Sub-Saharan Africa	Angola (AGO), Benin (BEN), Botswana (BWA), Cameroon (CMR), Cape Verde (CPV), Chad (TCD), Côte d'Ivoire (CIV), Ghana (GHA), Kenya (KEN), Lesotho (LSO), Malawi (MWI), Mali (MLI), Mauritius (MUS), Namibia (NAM), Nigeria (NGA), Rwanda (RWA), Senegal (SEN), Sierra Leone (SLE), South Africa (ZAF), Tanzania (TZA), Zambia (ZMB)
America	The Americas	Argentina (ARG), Bolivia (BOL,) Brazil (BRA), Canada (CAN), Chile (CHL), Colombia (COL), Costa Rica (CRI), Dominican Republic (DOM), Ecuador (ECU), El Salvador (SLV), Guatemala (GTM), Honduras (HND), Mexico (MEX), Nicaragua (NIC), Panama (PAN), Paraguay (PRY), Peru (PER), Trinidad and Tobago (TTO), United States (USA), Uruguay (URY), Venezuela (VEN)
Asia	Asia-Pacific	Australia (AUS), Bangladesh (BGD), Cambodia (KHM), China (CHN), Hong Kong SAR (HKG), India (IND), Indonesia (IDN), Japan (JPN),Korea, Rep. (KOR), Lao PDR (LAO), Malaysia (MYS), Mongolia (MNG), Nepal (NPL), New Zealand (NZL), Pakistan (PAK), Philippines (PHL), Singapore (SGP), Sri Lanka (LKA), Thailand (THA), Vietnam (VNM)
	Europe and Eurasia	Armenia (ARM), Azerbaijan (AZE), Kazakhstan (KAZ), Kyrgyz Republic (KGZ), Tajikistan (TJK), Georgia (GEO)
Europe	Europe and Eurasia	Albania (ALB), Austria (AUT), Belgium (BEL),Bosnia-Herzegovina (BIH), Bulgaria (BGR), Croatia (HRV), Cyprus (CYP), Czech Republic (CZE), Denmark (DNK), Estonia (EST), Finland (FIN), France (FRA), Germany (DEU), Greece (GRC), Hungary (HUN), Iceland (ISL), Ireland (IRL), Italy (ITA), Latvia (LVA), Lithuania (LTU), Luxembourg (LUX), Macedonia FYR (MKD), Malta (MLT), Moldova (MDA), Montenegro (MNE), Netherlands (NLD) Poland (POL), Portugal (PRT), Romania (ROU), Serbia (SRB), Slovak Republic (SVK), Slovenia (SVN), Spain (ESP), Sweden (SWE), Switzerland (CHE), Turkey (TUR), United Kingdom (GBR)
Middle East	Middle East and North Africa	Bahrain (BHR), Israel (ISR), Jordan (JOR), Kuwait (KWT), Lebanon (LBN), Qatar (QAT), Saudi Arabia (SAU), United Arab Em (ARE)

Appendix B. Residual Histograms and Kolmogorov–Smirnov Statistics for Residual Normality, for the Global Model, and for the Regions Separately

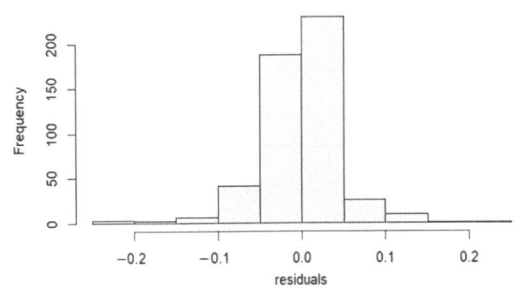

All countries, global model
K-S statistics: D = 0.06461; K = 1.2678; P = 0.07687

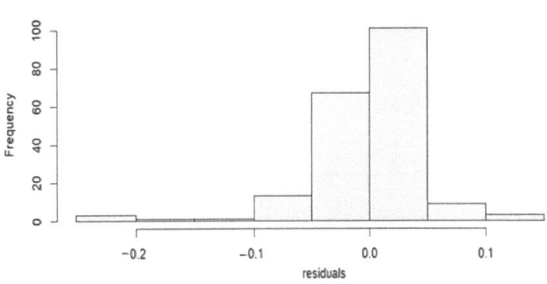

Europe and Eurasia
K-S statistics: D = 0.06187; K = 0.8484; P = 0.4498

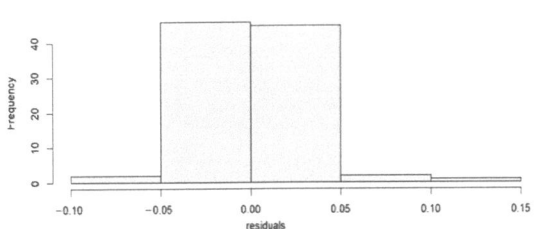

The Americas
K-S statistics: D = 0.06984; K = 0.6735; P = 0.7280

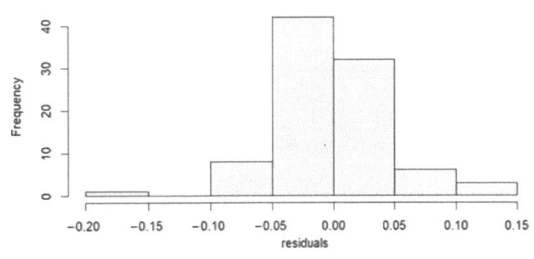

Asia and Pacific
K-S statistics: D = 0.1213; K = 1.1634; P = 0.1227

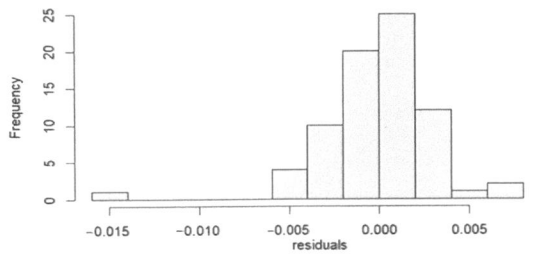

Sub-Saharan Africa
K-S statistics: D = 0.06475; K = 0.5494; P = 0.9042

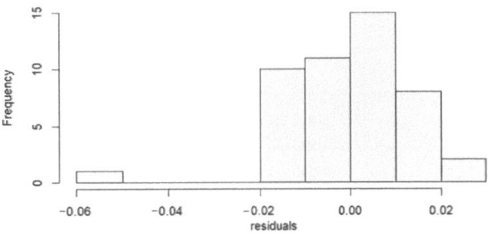

Middle East and North Africa
K-S statistics: D = 0.07916; K = 0.5427; P = 0.9075

References

1. Guzmán, P.C.; Roders, A.R.P.; Colenbrander, B.J.F. Measuring links between cultural heritage management and sustainable urban development: An overview of global monitoring tools. *Cities* **2017**, *60*, 192–201. [CrossRef]
2. Ashworth, G.J. Heritage and local development: A reluctant relationship. In *Handbook on the Economics of Cultural Heritage*; Rizzo, I., Mignola, A., Eds.; Edward Elgar: Cheltenham, UK, 2013; pp. 824–855. [CrossRef]
3. VanBlarcom, B.L.; Kayahan, C. Assessing the economic impact of a UNESCO World Heritage designation. *J. Herit. Tour.* **2011**, *6*, 143–164. [CrossRef]
4. Camagni, R.; Capello, R.; Cerisola, S.; Panzera, E. The cultural heritage—Territorial capital nexus: Theory and empirics/Il nesso tra Patrimonio Culturale e Capitale Territoriale: Teoria ed evidenza empirica. *Il Capitale Cult. Stud. Value Cult. Herit.* **2020**, *11*, 33–59. [CrossRef]
5. Muštra, V.; Škrabić Perić, B.; Pivčević, S. Cultural heritage sites, tourism and regional economic resilience. *Pap. Reg. Sci.* **2023**, *102*, 465–482. [CrossRef]
6. Cellini, R.; Cuccia, T. Do behaviours in cultural markets affect economic resilience? An analysis of Italian regions. *Eur. Plan. Stud.* **2019**, *27*, 784–801. [CrossRef]
7. Courtney, P.; Hill, G.; Roberts, D. The role of natural heritage in rural development: An analysis of economic linkages in Scotland. *J. Rural Stud.* **2006**, *22*, 469–484. [CrossRef]
8. Barrientos, F.; Martin, J.; De Luca, C.; Tondelli, S.; Gómez-García-Bermejo, J.; Casanova, E.Z. Computational methods and rural cultural & natural heritage: A review. *J. Cult. Herit.* **2021**, *49*, 250–259. [CrossRef]
9. Labadi, S. *Rethinking Heritage for Sustainable Development*; UCL Press: London, UK, 2022. [CrossRef]
10. Gupta, S.D. Comparative Advantage and Competitive Advantage: An Economics Perspective and a Synthesis. *Athens J. Bus. Econ.* **2015**, *1*, 9–22. [CrossRef]
11. Bansal, S.; Sharma, G.D.; Rahman, M.M.; Yadav, A.; Garg, I. Nexus between environmental, social and economic development in South Asia: Evidence from econometric models. *Heliyon* **2021**, *7*, e05965. [CrossRef]
12. Chopra, R.; Magazzino, C.; Shah, M.I.; Sharma, G.D.; Rao, A.; Shahzad, U. The role of renewable energy and natural resources for sustainable agriculture in ASEAN countries: Do carbon emissions and deforestation affect agriculture productivity? *Res. Policy* **2022**, *76*, 102578. [CrossRef]
13. Dupeyras, A.; MacCallum, N. *Indicators for Measuring Competitiveness in Tourism: A Guidance Document*; OECD Tourism Papers; OECD Publishing: Paris, France, 2013; Volume 2.
14. World Economic Forum. *The Travel & Tourism Competitiveness Report 2017: Paving the Way for a More Sustainable and Inclusive Future*; World Economic Forum: Geneva, Switzerland, 2017. Available online: https://www3.weforum.org/docs/WEF_TTCR_2017_web_0401.pdf (accessed on 10 May 2024).
15. World Economic Forum. *The Travel & Tourism Development Index 2021: Rebuilding for a Sustainable and Resilient Future*; World Economic Forum: Geneva, Switzerland, 2022. Available online: https://www3.weforum.org/docs/WEF_Travel_Tourism_Development_2021.pdf. (accessed on 10 May 2024).
16. World Economic Forum. *Travel & Tourism Development Index 2024—Insight Report*; World Economic Forum: Geneva, Switzerland, 2024. Available online: https://www3.weforum.org/docs/WEF_Travel_and_Tourism_Development_Index_2024.pdf (accessed on 10 May 2024).
17. World Economic Forum. TTCI Dataset 2024. Available online: https://www3.weforum.org/docs/WEF_TTDI_2024_edition_data.xlsx (accessed on 10 May 2024).
18. World Economic Forum. TTDI Dataset 2021. Available online: https://www3.weforum.org/docs/WEF_TTDI_2021_data_for_download.xlsx (accessed on 10 May 2024).
19. World Economic Forum. TTCI Dataset 2019. Available online: http://www3.weforum.org/docs/WEF_TTCR19_data_for_download.xlsx (accessed on 10 May 2024).
20. World Economic Forum. TTCI Dataset 2017. Available online: http://www3.weforum.org/docs/WEF_TTCR17_data_for_download.xlsx (accessed on 10 May 2024).
21. World Economic Forum. TTCI Dataset 2015. Available online: http://www3.weforum.org/docs/TT15/WEF_TTCR_Dataset_2015.xlsx (accessed on 10 May 2024).
22. World Economic Forum. TTCR 2019 Report. Available online: https://www3.weforum.org/docs/WEF_TTCR_2019.pdf (accessed on 10 May 2024).
23. World Economic Forum. TTCR 2015 Report. Available online: https://www3.weforum.org/docs/TT15/WEF_Global_Travel&Tourism_Report_2015.pdf (accessed on 10 May 2024).
24. World Economic Forum. TTCR 2013 Report. Available online: https://www3.weforum.org/docs/WEF_TT_Competitiveness_Report_2013.pdf (accessed on 10 May 2024).
25. Balassa, B. Trade liberalization and revealed comparative advantage. *Manch. Sch. Econ. Soc. Stud.* **1965**, *33*, 92–123.
26. Yu, R.; Cai, J.; Leung, P. The normalized revealed comparative advantage index. *Ann. Reg. Sci.* **2009**, *43*, 267–282. [CrossRef]
27. Stellian, R.; Danna-Buitrago, J.P. Which revealed comparative advantage index to choose? Theoretical and empirical considerations. *CEPAL Rev.* **2022**, *138*, 45–66. [CrossRef]

28. Stellian, R.; Danna-Buitrago, J.P. Revealed Comparative Advantage and Contribution-to-the-Trade-Balance indexes. *Int. Econ.* **2022**, *170*, 129–155. [CrossRef]
29. Roh, T.S.; Bak, S.; Min, C. Do UNESCO Heritages Attract More Tourists? *World J. Manag.* **2015**, *6*, 193–200. [CrossRef]
30. Din, H.B.; Habibullah, M.S.; Tan, S.H. The Effects of World Heritage Sites and Governance On-Tourist Arrivals: Worldwide Evidence. *Int. J. Econ. Manag.* **2017**, *11*, 437–448.
31. Bacsi, Z.; Tóth, É. Word Heritage Sites as soft tourism destinations—Their impacts on international arrivals and tourism receipts. *Bull. Geography. Socio-Econ. Ser.* **2019**, *45*, 25–44. [CrossRef]
32. González-Rodríguez, M.R.; Díaz-Fernández, M.C.; Pulido-Pavón, N. Tourist destination competitiveness: An international approach through the travel and tourism competitiveness index. *Tour. Manag. Perspect.* **2023**, *47*, 101127. [CrossRef]
33. Bacsi, Z.; Yasin, A.S.; Bánhegyi, G. Tourism Competitiveness in Eastern Africa: RCA and TTCI. *Heritage* **2023**, *6*, 5997–6015. [CrossRef]
34. Bogale, M.; Ayalew, M.; Mengesha, W. The Competitiveness of Travel and Tourism Industry of Sub-Saharan African Countries in the World Market. *Afr. J. Hosp. Tour. Leis.* **2021**, *10*, 131–144. [CrossRef]
35. World Population Review Website. Available online: https://worldpopulationreview.com/country-rankings/protected-land-by-country (accessed on 10 May 2024).
36. FAOSTAT SDG Indicators. Available online: https://www.fao.org/faostat/en/#data/SDGB (accessed on 10 May 2024).
37. Red List Index Website. Available online: https://ourworldindata.org/grapher/red-list-index?tab=table (accessed on 10 May 2024).
38. UNWTO Tourism Statistics Database. World Tourism Organization Madrid. 2023. Available online: https://www.unwto.org/tourism-statistics/tourism-statistics-database (accessed on 10 May 2024).
39. UNWTO Tourism Data Dashboard. World Tourism Organization Madrid. 2024. Available online: https://www.unwto.org/tourism-data/global-and-regional-tourism-performance (accessed on 10 May 2024).
40. World Development Indicators Database. World Bank. 2024. Available online: https://databank.worldbank.org/source/world-development-indicators# (accessed on 10 May 2024).
41. Jackman, M.; Lorde, T.; Lowe, S.; Alleyne, A. Evaluating tourism competitiveness of small island developing states: A revealed comparative advantage approach. *Anatolia* **2011**, *22*, 350–360. [CrossRef]
42. R Core Team. *R: A Language and Environment for Statistical Computing*; R Foundation for Statistical Computing: Vienna, Austria, 2021; Available online: https://www.R-project.org/ (accessed on 5 January 2023).
43. Schielzeth, H.; Dingemanse, N.J.; Nakagawa, S.; Westneat, D.F.; Allegue, H.; Teplitsky, C.; Réale, D.; Dochtermann, N.A.; Garamszegi, L.Z.; Araya-Ajoy, Y.G. Robustness of linear mixed-effects models to violations of distributional assumptions. *Methods Ecol. Evol.* **2020**, *11*, 1141–1152. [CrossRef]
44. Barr, D.J. Learning Statistical Models through Simulation in R: An Interactive Textbook. Version 1.0.0. 2021. Available online: https://psyteachr.github.io/stat-models-v1 (accessed on 5 January 2023).
45. Bates, D.; Mächler, M.; Bolker, B.; Walker, S. Fitting Linear Mixed-Effects Models Using lme4. *J. Stat. Softw.* **2015**, *67*, 1–48. [CrossRef]
46. JASP Team. JASP (Version 0.19.0)—Computer Software. 2024. Available online: https://jasp-stats.org/ (accessed on 10 May 2024).
47. Majidli, F. International Comparative and Competitive Advantage of Post-Soviet Countries in Tourism. *Res. World Econ.* **2020**, *11*, 369–379. [CrossRef]
48. Le, N.H. International trade in travel of selected ASEAN nations from comparative advantage theory and value-added trade approach. *Proc. Next Gener. Glob. Workshop* **2020**, *13*, 1–16. [CrossRef]
49. Labanauskaité, D.; Gedvilas, E. Lithuanian tourism competitiveness in the context of Baltic States. *Reg. Form. Dev. Stud.* **2013**, *2*, 111–122.
50. Gómez-Vega, M.; Picazo-Tadeo, A.J. Ranking world tourist destinations with a composite indicator of competitiveness: To weigh or not to weigh? *Tour. Manag.* **2019**, *72*, 281–291. [CrossRef]
51. Larm, M.; Elmhagen, B.; Granquist, S.M.; Brundin, E.; Angerbjörn, A. The role of wildlife tourism in conservation of endangered species: Implications of safari tourism for conservation of the Arctic fox in Sweden. *Hum. Dimens. Wildl.* **2018**, *23*, 252–272. [CrossRef]
52. Weaver, D.B. The evolving concept of ecotourism and its potential impacts. *Int. J. Sustain. Dev.* **2002**, *5*, 251–264. [CrossRef]
53. Cousins, J.A. The role of UK-based conservation tourism operators. *Tour. Manag.* **2007**, *28*, 1020–1030. [CrossRef]
54. Orams, M.B. Feeding wildlife as a tourism attraction: A review of issues and impacts. *Tour. Manag.* **2002**, *23*, 281–293. [CrossRef]
55. Woods, B.; Moscardo, G. Enhancing wildlife education through 19 mindfulness. *Aust. J. Environ. Educ.* **2003**, *19*, 97–108. [CrossRef]
56. Reisinger, Y.; Michael, N.; Hayes, J.P. Destination competitiveness from a tourist perspective: A case of the United Arab Emirates. *Int. J. Tour. Res.* **2019**, *21*, 259–279. [CrossRef]
57. Arumugam, A.; Nakkeeran, S.; Subramaniam, R. Exploring the Factors Influencing Heritage Tourism Development: A Model Development. *Sustainability* **2023**, *15*, 11986. [CrossRef]

58. Ladki, S.; Abimanyu, A.; Kesserwan, L. The Rise of a New Tourism Dawn in the Middle East. *J. Serv. Sci. Manag.* **2020**, *13*, 637–648. [CrossRef]
59. Dodds, R.; Butler, R. The phenomena of overtourism: A review. *Int. J. Tour. Cities* **2019**, *5*, 519–528. [CrossRef]

Disclaimer/Publisher's Note: The statements, opinions and data contained in all publications are solely those of the individual author(s) and contributor(s) and not of MDPI and/or the editor(s). MDPI and/or the editor(s) disclaim responsibility for any injury to people or property resulting from any ideas, methods, instructions or products referred to in the content.

Article

In Search of New Dimensions for Religious Tourism: The Case of the Ancient City of Nessebar

Sonia Mileva [1,*] and Milena Krachanova [2]

[1] Faculty of Economics and Business Administration, Sofia University "St. Kliment Ohridski", 1504 Sofia, Bulgaria
[2] Faculty of Philosophy, Sofia University "St. Kliment Ohridski", 1504 Sofia, Bulgaria; m.krachanova@phls.uni-sofia.bg
* Correspondence: smileva@feb.uni-sofia.bg

Abstract: Religious tourism is a growing phenomenon that is increasingly intertwined with cultural tourism, particularly in countries like Bulgaria, which possesses a wealthy heritage of religious interest. The Ancient City of Nessebar, a UNESCO World Heritage site with a rich religious history, provides a unique case study for examining this dynamic. This study employed a mixed-methods approach, including documentary analysis and case study methodology, to investigate the current state of religious tourism in Nessebar. The research revealed the underutilization of Nessebar's religious sites for tourism purposes, despite their potential value. A complex interplay between mass tourism, cultural preservation, and the decline of traditional religious practices was identified.

Keywords: religious tourism; Nessebar; UNESCO World Heritage

1. Introduction

The intersection of religion, culture, and tourism has become an increasingly complex and dynamic phenomenon in the contemporary world. This paper explores heritage tourism in Nessebar, focusing on the adaptive reuse of religious spaces and the complexities of managing them as World Heritage (WH) sites, including their overlap with other mainstream forms of tourism.

Religious tourism, also known as faith tourism or spiritual tourism, is defined as travel where religious beliefs or a desire for spiritual experiences drive the tourist's motivation [1]. While religious tourism has been a growing market segment, the adaptation of sacred spaces to accommodate tourist demands raises important questions about the preservation of cultural heritage and the authenticity of religious experiences.

According to UNESCO [2], approximately 20% of WH Sites have religious significance, encompassing a broad spectrum of cultures and faiths. Primarily revered for their spiritual or religious importance, these sites have also accrued substantial cultural and historical value.

The Ancient City of Nessebar was chosen as a case study due to its unique religious heritage adaptively reused for cultural purposes following a significant waned religious practice after the early 19th century. Ongoing debates about its future, including the potential inscription of Nessebar on the List of World Heritage in Danger, the controversial impacts of tourism, and its under-explored potential for religious tourism, make it a compelling subject of study.

UNESCO recognition interconnects global, national, and local governance, balancing the preservation of Nessebar's rich religious and cultural heritage with sustainable tourism development, while addressing the challenges posed by the negative impacts of increasing mass tourism from the nearby Sunny Beach Resort.

The topic is extremely interesting for Bulgaria, keeping in mind that the country has inscribed ten cultural and natural properties on the WH list, six of which are of religious

Citation: Mileva, S.; Krachanova, M. In Search of New Dimensions for Religious Tourism: The Case of the Ancient City of Nessebar. *Heritage* 2024, 7, 5373–5389. https://doi.org/10.3390/heritage7100253

Academic Editors: Fátima Matos Silva and Isabel Vaz de Freitas

Received: 16 August 2024
Revised: 24 September 2024
Accepted: 26 September 2024
Published: 28 September 2024

Copyright: © 2024 by the authors. Licensee MDPI, Basel, Switzerland. This article is an open access article distributed under the terms and conditions of the Creative Commons Attribution (CC BY) license (https://creativecommons.org/licenses/by/4.0/).

interest. Traditional religious practices have been abandoned in most places, except Rila Monastery, which continues to serve its original purpose. Nevertheless, the churches of Nessebar still dominate the city's skyline and testify to the centuries when the city was a religious center.

As religion is an essential element of culture, religious sites, and their remains, are important components of cultural tourism [3]. Theoretical insights from Critical Heritage Studies (CHS) enable us to consider heritage not as static, preserved relics, but as dynamic processes shaped by socio-cultural, political, and economic factors [4]. This perspective emphasizes the processual nature of heritage, acknowledging that religious heritage sites evolve and acquire new meanings as they adapt to tourism, globalization, and cultural shifts [5].

Religious tourism is approached as a part of broader cultural and heritage tourism, referring to destinations, recognized as WH cities, integrating visitation to diverse types of religious sites, some of which have lost their initial purpose. This research delves into this dynamic by examining the interplay between tourism and the transformation of sacred spaces and churches, which have lost their original spirituality and purpose. This case study evaluates the adaptation strategies employed by the national and local management authorities in response to the evolution of religious tourism. To be successfully approached and managed, it is essential to understand how these ancient spiritual places are navigating the pressures of tourism while maintaining their significance as cultural and religious symbols. By examining the case of Nessebar, we aim to contribute to the broader discourse on the management and valorization of religious heritage in the context of tourism.

2. Materials and Methods

The main research objectives are to analyze the potential of religious tourism in Nessebar within the broader context of cultural heritage tourism.

A mixed-methods approach provides holistic understanding on how religious tourism in WH cities is influenced by tourism and heritage management in broader cultural and sustainable aspects. The documentary analysis aims to review national, regional, and local tourism policies and strategies related to religious tourism. The research design combines qualitative and quantitative methods, including documentary analysis and case study methodology for the WH Site Nessebar.

Documentary analysis was conducted to examine the national framework for tourism, regional tourist zoning, and the specific focus on religious tourism at the municipal level. This includes reviewing relevant documents such as the Bulgarian National Strategy for Sustainable Development of Tourism (2014–2030) [6], Concept of National Tourist Zoning [7], Cultural Tourism Policy analysis [8], and Cultural Heritage Strategy of Nessebar Municipality (2023–2032) [9]. The scope of analysis was to identify the extent to which religious tourism is integrated into broader tourism strategies and how these policies influence the development and management of religious heritage in the city. Most relevant national and regional tourism policy documents, UNESCO monitoring reports [10,11], and municipal strategic plans were used as primary data sources.

The Case Study Analysis is applied for the "Faith in Nessebar" project, including its objectives, implementation, outcomes, and challenges. The impact and outcomes of the project are evaluated as experience to develop religious tourism in Nessebar. The case study approach is particularly suitable for this research as it allows for an in-depth exploration of a complex phenomenon within its real-world context. The main objective was critical analysis of past initiatives and measures, their implementation strategies, the challenges they encountered, and most importantly the impacts on religious tourism development in Nessebar. Data were collected from project reports, official documentation, and strategic plans that are publicly available.

A comprehensive analysis of Bulgarian religious sites inscribed on the World Heritage List is conducted, examining their current function and accessibility. The Ancient City of Nessebar serves as a compelling case study for this research for several reasons, including its status as a heritage city and richness of religious places and churches, most of which have lost their purpose. Moreover, among many other factors, it is perceived as problematic for religious tourism development and faces tourism pressure from the neighboring Sunny Beach Resort.

To systematically assess the current state of religious sites, and churches in particular, within the Ancient City of Nessebar, a data collection, documentary analysis, and categorization process were undertaken. A statistical analysis of visitor data for churches repurposed as museums was conducted to identify trends and patterns.

By observing the religious tourism in Nessebar to assess the current state of offerings and visitor experiences. Information about churches was compiled from the official website of the Archaeological Museum in Nessebar and Cultural Heritage Strategy of Nessebar Municipality for the period 2023–2032 [9]. The data are presented in the table and figures. The categorization aims to present, in a comprehensive form, the factors relevant to their current functionality (active church, museum, ruins), accessibility (entrance fees), and their state of preservation from the point of view of integrity and conservation.

This study acknowledges several limitations. The reliance on secondary data sources restricts the depth of analysis and prevents direct observation of the churches' current conditions. Additionally, focusing solely on Nessebar limits the generalizability of findings to other heritage sites. The position and opinion of different stakeholders is not directly addressed, but rather considered based on secondary available data and previous publications on the topic.

3. Literature Review

The CHS framework enables us to analyze heritage as an ongoing process, shaping sites like the Ancient City of Nessebar into entities continually influenced by social, political, and economic factors. Tourism, globalization, and local communities play significant roles in this dynamic process [5,12].

The influence of political power [13] and the interplay between global and local governance [14] frame the major challenges faced by Nessebar. The designation of Nessebar as a UNESCO World Heritage site in 1983 brought with it global recognition and strict regulations. These rules, aimed at preserving the "authenticity and integrity" of the Ancient City of Nessebar, clashed with the livelihood strategies of the residents who depend on tourism for economic survival [15]. The locals were put in a position to navigate between adhering to UNESCO regulations and leveraging their cultural heritage for economic gain and tourism development.

The representation of religious heritage in Nessebar as part of collective memory and identity [16] is not sharply controversial as part of the Eastern Orthodox religion, despite the significant influences of Byzantine and Greek cultures (education, language) over time.

Often religious tourism is discussed as a form of tourism that is not confined to one religion, involving subjective travel for spiritual betterment and self-discovery [17]. The concept of religious tourism is evolving, covering a diverse range of studies, and is concentrated in a few leading social science journals [18–23]. However, the "the ties between tourism and pilgrimage are currently unclear and poorly classified" [24]. In addition to demand, destination development, and marketing [25], the studies shift to de-differentiation, the experience economy [26], gender, tourist perceptions, technology [27], sustainable development, and the interaction between religious sites, cultural contexts, and other tourism sectors [28,29].

While pilgrimages to sacred places remain important, beyond them, there is a big shift in visits and the selection of destinations. Tourist motivations are expanding, and visitors are increasingly drawn to destinations with deeper personal significance, reflecting their individual religious affiliations and overall religiosity [30]. Religious tourism has prompted many countries, destinations, and tourism businesses to capitalize on this niche market by investing in religious and spiritual experiences [31].

Religious tourism and route-based pilgrimage is considered as a form of slow, responsible, and sustainable tourism with positive effects on the development of the regional cultures, strengthening the local traditions and cultures [32]. Religious tourism should be studied in a broader approach related to the historical buildings, sacred places, churches, etc., particularly those with religious significance. The preservation of their cultural value is crucial and should align with a more sustainable future.

Globalization, with its emphasis on consumerism and individual choice, has led to a decline in traditional religious practices and a rise in personalized spirituality [33]. According to McKercher, religious travel is under the category "Personal Quest", which refers to travel for more personal reasons associated with self-development, improvement, and/or learning. The tourist motivation about selection of destinations and places of interest expands and evolve [34]. People are defining their own religious beliefs and experiences, creating a more diverse marketplace [35].

This new emphasis on individual experiences opens new possibilities for tourism and new religious and spiritual movements, with potential to be packaged and offered as a tourist experience. This niche has become highly explored as a new customer segment. Consequently, the religious tourism offerings have become more diverse, putting behind their core purpose and encompassing a variety of events, including the organization of weddings, themed vacations, volunteer opportunities, and visits to religious sites [36].

Drawing from Waterton and Watson's [4] emphasis on performativity in heritage, the religious spaces in Nessebar are approached not just static relics, but as sites of performative interaction. This means exploring how tourists, locals, and religious actors engage with these spaces, turning them into venues for spiritual, cultural, and even economic performances. From a tourist perspective, this also means the creation of new memories, reshaping their understanding of spiritual and heritage significance. The process of recollection through photographs, social media posts, and personal reflections—further contributes to the evolving narrative of the site, blurring the boundaries between the past and the present [37]. A different relationship between the tourist and the heritage site is emerging and the tourist's presence influences the cultural life of the location.

The enrichment of pilgrimage through so called "rejuvenation" has led to the loss of its religious element and unique identity that differentiates them from other types of travel [26]. Our starting point is the thesis that tourism transcends the traditional divide between the sacred and secular. Instead, it fosters a hybrid experience where pilgrimage destinations become spaces that are neither strictly modern nor traditional, but rather a blend of both [21]. As noted by Pouls et al. [3], p. 326 "Tourism, pilgrimage, religion, and spirituality mix in an ever-changing manner, at both the individual and societal level".

While acknowledging the ongoing debate around defining religious tourism, with two primary interpretations, one focused on religious motivations and destinations and the other on tourist activities [38], this research seeks to move beyond these established approaches.

Traditional pilgrimages in Nessebar, referred as a "miniature Constantinople" began in the Early Christian period and continued through the Middle Ages, but ceased during the Ottoman rule. Currently most of the religious celebrations are following the traditional religious calendar and August 15 Assumption of the Mother of God, converted into official municipality holiday and major cultural event.

Contemporary religious tourists are driven by a shift in motivation, emphasizing the significance of the destination itself rather than solely the sacred place. As a result, religious tourism is seen as going beyond simply visiting a religious site; it involves connecting with the history, culture, and spiritual essence of the place [1,30].

The adaptive reuse of the sites goes beyond just preservation and conservation, rather than their integration into sustainable development, including tourism considering environmental, economic, and social aspects. Most successful reuse plans consider the impact on local communities, inclusive economic growth, and social justice [39]. The process of "recycling" by giving new life and by respecting their history and architecture offers a sustainable solution for outdated buildings [40]. This is achieved by engagement and active participation of local communities and stakeholders in the decision-making process at local level.

Local community involvement is crucial for the sustainable management of heritage sites, especially in the context of Nessebar, where the residents have a long-standing connection to their town's cultural and spiritual legacy [15]. Activating local stakeholders through a community-based management approach could alleviate tensions between heritage preservation and economic interests, allowing for a more inclusive process that considers local needs alongside UNESCO's global conservation standards. Additionally, this fosters a sense of ownership and ensures that chosen solutions are effective, beneficial, and acceptable for all stakeholders. A core issue is the achievement of a balance of interests between preserving the past and adapting these structures to meet the needs of the present and future [41]. Engaging with local stakeholders allows for a more sustainable and authentic development of religious tourism, aligning with critical heritage studies' focus on inclusivity and participatory governance.

Religious tourism offers a multifaceted experience that goes beyond mere devotion, fostering cultural exploration and enriching tourists' understanding of the local heritage. In Nessebar, the increased mass tourism puts a strain on sacred sites and their surrounding communities, raising questions about impacts (both positive and negative) on sustainable development.

According to Waterton and Watson [4], the so-called "in heritage" theories provide tools to evaluate tourism and management effectiveness, focusing on economic aspects such as visitor numbers and profits. In contrast, "of heritage" theories analyze heritage as a social and cultural phenomenon [4]. In this context, tourist engagement with heritage sites is not only an act of consumption but also a deeply embedded process of memory-making. Heritage spaces serve for both individual and collective identity formation [16,37].

4. Religious Tourism and Heritage Preservation: A Case Study of Nessebar, Bulgaria

The Bulgarian National Strategy for Sustainable Development of Tourism (2014–2030) identifies religious tourism as an integral part of the broader cultural and sightseeing tourism sector [6]. The strategy highlights several advantages that make Bulgaria a favorable destination for religious tourism, including the absence of religious restrictions, religious freedom, and dedicated support from government, society, and religious institutions. These factors create an optimal environment for incorporating religious elements into cultural and sightseeing experiences, thereby enhancing the country's overall tourism potential. However, despite the abundance of religious sites, the strategy points out a significant shortcoming: religious tourism remains underdeveloped across Bulgaria's tourist regions.

In the 2018 Product Analysis of Cultural Tourism [8], religious tourism in Bulgaria is traced back to the ancient practice of pilgrimage, a tradition spanning various religions. The primary motivations for religious tourists include affirming, deepening, or re-evaluating their personal faith. The key forms of religious tourism encompass visits to religious sites, pilgrimages, monastery stays, religious camps, participation in religious events, and forming connections with fellow believers. The analysis underscores the significant potential of religious tourism for both domestic and international markets, driven by a resurgence of interest in religion and religious monuments. Moreover, this type of tourism is recognized for fostering community and intercultural connections as part of the broader cultural tourism framework. In 2016, the Bulgarian Patriarch Neofit established the Pilgrimage and Educational Center of the Sofia Metropolis. Collaborating with a working group that included representatives from the Bulgarian Orthodox Church, the tourism industry, and the cultural sector, the center developed ten pilgrimage routes that highlight Bulgaria's Orthodox heritage [42]. The developed pilgrimage routes connect Bulgaria's World Heritage sites—Rock-Hewn Churches of Ivanovo, Nessebar, Boyana Church, and Rila Monastery—with internationally renowned religious centers in Athos (Greece), Jerusalem and the Holy Land (Israel), and Rome (Italy).

In 2016, The Bulgarian Patriarch Neofit established the Pilgrimage and Educational Center of the Sofia Metropolis. Collaborating with a working group that included representatives from the Bulgarian Orthodox Church, the tourism industry, and the cultural sector, the center developed pilgrimage routes that highlight Bulgaria's Orthodox heritage [42].

The implementation of pilgrimage routes in Bulgaria focuses on WH properties and several international religious centers recognized by UNESCO. Table 1 outlines the pilgrimages and visits undertaken or planned by the Pilgrimage and Educational Center of the Sofia Metropolis in 2024, including both WH sites of religious significance in Bulgaria and other major internationally recognized religious centers. Although these pilgrimage routes were established in 2021, the data indicates that religious tourism at WH in Bulgaria remains underdeveloped. In 2024, visits by Bulgarian pilgrims to internationally recognized religious centers (both undertaken and planned) account for 81% of all visits to religious sites in Bulgaria [43]. The data suggests that religious sites that have preserved their original purpose attract higher visitor numbers. While Bulgarian tourists currently show a preference for well-established international religious centers, there is a significant opportunity to promote national sites of religious interest. This highlights the need for targeted strategies to increase the appeal and visibility of these national sites, along with collaboration of different stakeholders (government, tourism, religious institutions, etc.)

Table 1. Pilgrimages and visits to sites of religious interest/international religious centers carried out in 2024 by The Pilgrimage and Educational Centre of the Holy Metropolis, Sofia, Bulgaria.

WH Sites	WH Sites/Centers	Preserved Original Function	Pilgrimages/Visits per 2024
WH Sites of religious interest in Bulgaria	Rock-Hewn Churches of Ivanovo	no	1
	Boyana Church	no	2
	Rila Monastery	yes	2
	Ancient City of Nessebar	no	1
Internationally recognized religious centres (WH Sites)	Rome (Italy)	yes	13
	Jerusalem and Holy land (Israel)	yes	3
	Athos (Greece)	yes	1

Source: own elaboration based on data from Pilgrimage and Educational Centre of the Holy Metropolis, Sofia, Bulgaria [43].

Table 2 provides an overview of Bulgaria's WH Sites of religious interest, examining their religious affiliation, site type, original and current function, and their association with pilgrimage routes. Bulgaria is home to seven cultural and three natural UNESCO WH Sites, with six of the cultural sites holding religious significance. The developed pilgrimage routes exclusively feature sites associated with Orthodox Christian practices, both historical and contemporary. Meanwhile, sites connected to Thracian beliefs and burial customs are integrated into broader cultural routes that highlight general heritage or specific WH elements.

Bulgaria's national tourism zoning, outlined in the Concept of Tourism Zoning [7], identifies religious tourism as a component within six of the country's nine regions. One of them, the Burgas Black Sea Tourism Region, focuses particularly on the Ancient City of Nessebar. As main asset is pointed that the city has been a "remarkable spiritual center of Christianity for a thousand years" and "today it is a developing and vibrant urban organism" [10]. Special attention is given to its medieval churches as the most significant attribute conveying the outstanding universal value (OUV) of the WH property.

Table 2. Bulgarian properties of religious interest on the WH list.

	WH Property	Inscription	Religious Affiliation	Site Type	Original Function	Current Function	National Pilgrimage Route
1.	Rock-Hewn Churches of Ivanovo	1979	Christian	monastery	religious/ pilgrimage	cultural	Rock Monasteries in Bulgaria[1]
2.	Boyana Church	1979	Christian	church	religious/ pilgrimage	cultural	Sofia's Sacred Mountains
3.	Thracian Tomb of Kazanluk	1979	Thracian beliefs	tomb	burial	cultural	-
4.	Rila Monastery	1983	Christian	monastery	religious/ pilgrimage	religious/ pilgrimage/ cultural	St. John of Rila—the heavenly Patron Saint of Bulgaria
5.	Ancient City of Nessebar	1983	Christian	religious center	religious/ pilgrimage	cultural	Nessebar—"encyclopedia" of Christian church construction[2]
6.	Thracian Tomb of Sveshtari	1985	Thracian beliefs		burial	cultural	-

Source: own elaboration.

Nessebar boasts a rich history spanning over 3000 years. Successive civilizations, from Thracians to Byzantines, have left their mark on the city [44]. Once home to over 40 churches, the peninsula now preserves just 13 of them belonging to the Antiquity (3), Middle Ages (7), and the Renaissance period (1). Renowned for its architectural and artistic heritage, Nessebar is particularly celebrated for its medieval churches, exemplifying a unique blend of Byzantine and local styles craft practices—Northeast or the Turnovo-Nessebar school [45]. These churches, architectural masterpieces adorned with intricate details, are integral to Nessebar's Outstanding Universal Value (OUV). However, their deteriorating condition necessitates urgent preservation efforts.

The Ancient City of Nessebar's history is vibrant in terms of demographic shifts, cultural exchanges, and political governance. This turbulent past has left an indelible mark on its social and cultural development, particularly on its rich religious heritage. As it is stated in the Retrospective Statement of the Outstanding Universal value of the WH property, the Ancient City of Nessebar has been "a remarkable spiritual canter of Christianity for a thousand years" [46]. Ruins from a temple and churches from the Hellenistic period and Late Antiquity present part of the religious heritage of the city. Ivanchev [45] notes that the spread of Christianity in the early Middle Ages is marked by the beginning of serious construction of many Christian churches and monasteries on the western shore of the Black Sea. The medieval Nessebar's churches are one of the most impressive symbols for the religious past of the city. They are valuable as architectural and artistic heritage with their impressive decoration of the facades representing the early Byzantine architecture in the "Tarnovo-Nessebar" style. They have a significant role as a factor in the historic urban environment and cultural landscape, being significant benchmarks in the urban space for centuries. The change in street structure during the Middle Ages because of the churches and the formation of the "mosaic route", typical of the period, serve as evidence of their significant role in medieval cultural and spiritual life.

In 1829, due to heavy bombardment from the sea by the Russian army, Nessebar's churches suffered severe damage ([45], p. 16) followed by the loss of the original functions and their integrity. Only the church "Assumption of the Mother of God" built in the 19th century has kept its original purposes even currently.

After the Liberation in 1878, migrations and demographic shifts led to a partial loss of cultural context, marking the beginning of Nessebar's decline. The city transitioned from a district center to a town center within the newly established Eastern Rumelia. This once vibrant town experienced a downturn, characterized by abandoned houses and a dwindling population, consisting of Greeks with a small number of Bulgarians and Turks. Despite Bulgaria's liberation and subsequent political changes, these churches never returned to their authentic functions. In 1927, they were designated as cultural monuments and following Nessebar's declaration as an architectural reserve in 1956, some of them evolved into exhibition halls, presenting the frescoes, icons, and other objects from the movable cultural heritage of ancient Nessebar. The process of transformation of the city into a museum and leading tourist destination started at that point. Aiming to exploit the ancient city's cultural potential and develop international tourism, politically supported, a new resort complex "Sunny Beach" was built nearby (1959), sharply shaping the tourism development of Nessebar [15].

A key turning point for Nessebar, with lasting consequences for its sustainability, was its designation as a UNESCO WH Site. On the one hand, this status helped preserve the authenticity of the churches. On the other hand, an official national policy of promoting sea, sand, and sun tourism development has begun, along with investments in the nearby Sunny Beach Resort, which triggered a surge in seasonal mass tourism, bringing controversial results. This tourism boom led to extensive changes across the city, straining infrastructure, impacting local livelihoods, city's townscape by transforming the traditional functions of Renaissance houses.

Chronologically, the effort at municipal level regarding religious tourism start back in 2011 with the realization of "Faith in Nessebar" project aimed to enhance religious tourism by restoring, socializing, and improving accessibility to three churches: St. Paraskeva, St. John the Baptist, and St. Spas. A "Spiritual Path", also as tourist route, linking these churches with "St. Archangels Michael and Gabriel" was also created.

The proposed route comprised three distinct itineraries: a primary route featuring prominent, well-preserved churches; a supplementary route encompassing less accessible or renowned sites; and a virtual route commemorating the former church's location. To enhance the visitor experience, the project introduced thematic zones based on various criteria: chronological, artistic, and architectural features; the history of lost churches; educational and interpretive elements; the churches' role in urban development; and the integration of contemporary arts.

The realization of the Project "Faith in Nessebar" itself suffered several challenges since the focus was on the restauration and socialization. The route's structure, which was overall complex, was not entirely understandable for both tourists and local stakeholders. The route, as a tourist product, relied on the museum as the sole promotional channel and was coupled with a dearth of collaboration with other stakeholders (e.g., guides, travel agencies, etc.), which hindered the project's outreach.

The failure of previous projects like "Faith in Nessebar" highlights the need for a more systematic approach to heritage management. It demonstrates that while "in heritage" theories may be useful for analytical project justification, they do not necessarily lead to the development of a reassembled product with tangible exchange value [4,12]. A critical approach to heritage acknowledges that it is not possible to fully understand heritage by focusing solely on material culture or discursive representation, as demonstrated in the discussed project. Instead, attention must be given to the embodied, extra-discursive, and pre-cognitive responses that visitors experience [4,37,47].

Additionally, the official strategic document [48] was not accepted and implemented. Recognizing the significant potential of its rich ecclesiastical heritage, the Nessebar Municipality's 2013–2017 Tourism Development Program prioritized religious tourism and outlined plans to improve the proposed religious route. While the project successfully restored and displayed several churches, the broader strategy encountered severe challenges. Consequently, the Nessebar Municipality's Sustainable Tourism Development Program (2018–2024) relegated religious tourism to a secondary position, prioritizing emerging sectors such as wedding and festival tourism, while acknowledging the significance of cultural heritage tourism.

A SWOT analysis (Table 3) summarizes how the religious and cultural heritage of Nessebar can be best utilized, considering the affective and performative dimensions of tourism and the community's role in preserving the site's spiritual and historical significance.

Table 3. SWOT Analysis.

Strengths	Weaknesses
- WH Status adding prestige and global recognition, attracting a diverse range of tourists interested in both religious and cultural heritage. - Rich religious history and unique cultural heritage - Developed tourism infrastructure and facilities - The local community's long-standing connection to the religious celebrations	- Over-commercialization and danger of loss of authenticity - Lack of community-based tourism (CBT) engagement and local stakeholders - Negative impacts related to mass, seasonal tourism growth and challenges for site management
Opportunities	Threats
- Heritage as soft power could be used as a tool for "heritage diplomacy" [49], strengthening both local identity and international ties through religious tourism. - Growing demand for tourism based on spiritual experience in cultural context, driven by increased interest in authentic travel. - Increasing recognition of community-based approaches as essential for achieving sustainable development and enriching cultural production of place, reflecting a broader shift toward more inclusive practices.	- Threats to loss of WH status if listed as heritage in danger - Loss of authenticity and commercialization in favor of tourism revenue

Source: own elaboration.

5. Results

The state of the churches and their attractiveness as spiritual and sacred places for religious tourism is included in the UNESCO Reactive Monitoring Report (2018) [11], suggesting the decline of Nessebar as a spiritual center. Legends suggest the peninsula once hosted around forty churches and monasteries. Presently, only thirteen churches remain, dated from the 5th to 19th centuries, some of them preserved as ruins. The Middle Ages marked a zenith of religious and commercial activity, a period during which Nessebar's distinctive architectural style emerged, influenced by Byzantine traditions yet displaying local crafts. These churches, adorned with intricate facades and vibrant frescoes, are quintessential to Nessebar's OUV [46].

However, historical evidence indicates that the city's medieval churches ceased serving their original religious purpose well before the late 19th century, when Nessebar's architectural allure began captivating travelers and scholars alike.

Nessebar's strategic location on the Black Sea coast facilitated cultural exchange between the Byzantine Empire, the Bulgarian Empire, and other regions. This is reflected in the church's architecture and artistic style, which exhibit a blend of Byzantine, local, and potentially Western influences [50].

Its religious rise started in the Early Christian Period, being recognized as an Episcopal Center, resembling "miniature Constantinople", and as a part of Haemimont Province subordinated to an Adrianoupolis diocese [44]. The Middle Ages were pivotal for the religious and spiritual prominence of the city with its monasteries and basilicas. Despite scholarly debates on the historical accuracy of past events [51], certain texts reference the discovery and subsequent transfer of the relics of Saint Theodore Stratilatus, the hand and jaws of the Apostle Andrew the First-Called, the skull of Saint Sist, and the hand of the Apostle Bartholomew from Mesemvria to Venice in 1257 [50,52,53]. These accounts highlight the city's longstanding importance as a religious and pilgrimage center, justifying its continued inclusion on official pilgrimage routes. Between the 13th and 14th centuries, Nessebar become significant center for the commercial expansion of Italian cities in the Black Sea region. Following the fall of Constantinople in 1453, it attracted and become a host to Byzantine aristocrats drawn to its spirituality, culture, and openness. According to Giuzelev [52] "in the Balkan possessions of Byzantium after Constantinople the second most important church-religious center became Nessebar".

The Ottoman conquest of the Balkans in the 14th century presented new challenges and opportunities for the Orthodox Churches being forced to adapt under Islamic religion. Despite its ancient and medieval prominence, by the 16th century, Nessebar was a provincial Ottoman town that was small and showed a more profound cultural and spiritual heritage beyond its commercial status. The "high" stylistics of Eastern Orthodox culture interconnects with the "Greek", understood as transfer and combination of the language, education, and the Ecumenical Patriarchate, originating from nominally Greek territories [50]. As early as the mid-19th century, General Helmut Moltke observed the peninsula's picturesque charm and the remnants of its once-grand Byzantine churches, highlighting the city's architectural evolution over time [45,54,55].

The city's religious power underwent significant transformation by the 19th century. The medieval churches, having lost their original liturgical functions, were repurposed for secular uses. Today, most have been adapted into cultural heritage sites, preserving their architectural and artistic legacy. Notably, only one church remains active, the 19th-century "Assumption of the Mother of God. The current state of conservation and functionality of Nessebar's churches are systematized in Table 4.

Most of the churches (six out of thirteen) currently are reused as cultural spaces hosting exhibitions related to Nessebar's cultural heritage. Some of the churches (three out of thirteen) are in ruins, surrounded by open public spaces. Only one of the churches serves its original purposes. This indicates that they have a limited ability to be purely religious or spiritual places of interest and are important for cultural appreciation and tourism. While some churches are in a good state of conservation (eight out of thirteen), the others require conservation activities. This highlights the need for ongoing efforts to safeguard these historic structures as cultural heritage. Seven of the thirteen churches are accessible to the public without an entry fee, primarily those in a state of ruin and located in public areas.

Table 5 presents the number of visits to churches re-used as exhibition halls and venues for cultural events between 2019 and 2022. From all churches transformed into exhibition halls, the dominant one is "St. Stefan" as most popular attraction, contributing significantly to the overall visitor numbers. The pandemic years 2020 and 2021 exhibit a sharp decline in visitor numbers due to the global pandemic, reflecting a broader trend in the tourism industry. The data for 2022 indicates a strong recovery, with visitor numbers surpassing pre-pandemic levels in some cases. While the churches contribute significantly to overall museum visitation, they represent a portion of the total museum visits, suggesting that the city offers a diverse range of cultural attractions.

Table 4. Current functionality and state of conservation of the churches in Nessebar.

N	Churches	Construction Period/Century	Current Function	Access	Entrance Fee	Integrity of the Building	State of Conservation (SOC)
1	Old Metropolis	end of 5th–beginning of 6th century	-	accessible, located in open public spaces	No	preserved—in ruin form	Good SOC
2	"St. Virgin of Eleusis"	5th century	-	accessible, located in open public spaces	No	preserved—in ruin form	Fair—need for conservation activities
3	"St. Kliment"	Before 9th century	Cultural	restricted access	No	incorporated in a building	Good SOC
4	"St. Vlaherna"	12th century	-	accessible, located in open public spaces	No	preserved—in ruin form (only the apse)	Fair—need for conservation activities
5	"St. Joan Aliturgetos"	14th century	-	restricted access	No	preserved in ruin form	Good SOC
6	"St. Archangels Michael and Gabriel"	13th century	-	restricted access	No	preserved in ruin form	Fair—need for conservation activities
7	"St. John the Baptist"	9th–10th century	Cultural	accessible	Yes	preserved	Good SOC
8	"St. Paraskeva"	13th century	Cultural	accessible	Yes	preserved	Good SOC
9	"St. Stefan"	10th–beginning of 11th century	Cultural	accessible	Yes	preserved	Good SOC
10	"Christ Pantocrator"	13th century	Cultural	closed for restoration	Yes	preserved	Fair—need for conservation activities
11	"St. Todor"	13th century	-	restricted access	Yes	preserved	Fair—need for conservation activities
12	"St. Spas"	no data	Cultural	restricted access	Yes	preserved	Good SOC
13	"Assumption of the Mother of God"	19th century	Active church	accessible	No	preserved	Good SOC

Source: own elaboration.

Table 5. Visits to the churches that reused exhibition halls and other cultural events.

Reused Churches	Visitors/per Year			
	2019	2020	2021	2022[3]
"St. John the Baptist"	7164	2953	6898	11,391
"St. Paraskeva"	6199	2780	5501	7739
"St. Stefan"	53,289	8454	19,152	22,990
"Christ Pantocrator" *	18,671	–	–	–
"St. Todor" **	3168	–	–	–
"St. Spas"	11,014	2803	5321	8492
Total visits in churches	99,505	16,990	36,872	50,612
Total visits to all museums	137,883	24,945	56,312	79,740

Source: Nessebar Museum. * closed for restoration, ** restricted access

6. Discussion

Regarding political power [37], it should be noted that it was by a political decision back in 1956 that the government declared Nessebar and its coastline a "museum, tourist, and resort complex of national and international importance". This strategic decision aimed to use the ancient town as a cultural resource for the planned nearby Sunny Beach Resort, linking its heritage protection and exhibition with regional tourism development. This initiative represented the government's intention and authority to involve all relevant institutions in this large-scale project. The tourism growth changed Nessebar's development, and the latest UNESCO monitoring reports [10] highlighted several issues that should be addressed at both national and local levels in tension with economic interests and tourism development. Although the official national position is clear and in line with global governance, concrete responsibilities are deferred or postponed, often left for future administrations to handle (both on national and local level).

With respect to collective memory and identity [16], tensions could be found back in time even in the religious field (Eastern Orthodox). The migration waves after the Convention for the Exchange of Minorities between Bulgaria and Greece (1919), the Greek-Turkish War (1921–1922), and the Inter-Government Decision of 1926 for the exchange of refugees between Bulgaria and Greece (Aegean Thrace) changed the profile of the local population. This migration waves reflected on the religious field, when in 1926, some of the saints' relics were taken by the leaving Greek population from Nessebar (Mesemvria) to the twin town of Nea Mesemvria in Greece, as noted by Mutafov [46].

The historical and architectural significance of Nessebar's churches appeals to visitors interested in the cultural heritage of religious sites. However, the potential for religious (pilgrimage) tourism is quite limited due to the significant decline in religious practices after the early 19th century. Instead, there is an opportunity to develop spiritual tourism by attracting visitors seeking reflection and a deeper connection with historical spirituality. The tourists engage with religious sites not just as passive observers but as active participants in the creation of new memories, reshaping their understanding of the spiritual and historical significance of these spaces. This process of recollection—through photographs, social media posts, and personal reflections—further contributes to the evolving narrative of the site, blurring the boundaries between the past and the present.

From a governance perspective, the focus is on the adaptive reuse and conservation of these religious sites, highlighting them as integral elements of the broader cultural heritage narrative.

The analysis of strategic documents reveals a clear shift in recent years from a narrow definition of religious tourism. Bulgaria, with its rich religious heritage and supportive policy environment, holds significant potential for religious tourism. However, pilgrimage, particularly to Orthodox Christian sites, remains the dominant form of religious tourism. Unfortunately, many religious sites, especially those outside the Orthodox tradition, suffer from poor accessibility and limited promotion. Additionally, a considerable number of Bulgarian religious tourists prefer visiting internationally recognized religious centers.

Efforts to revitalize and develop religious tourism in Nessebar, such as the "Faith in Nessebar" project, have not met with success. The project mistakenly assumed that restoring churches, improving accessibility, and creating a thematic route would automatically attract year-round visitors, positioning Nessebar as a cultural and pilgrimage destination. This approach overlooked the fact that many of these churches had ceased to function as active religious sites and were viewed more as tourist attractions or cultural landmarks. Despite careful efforts to present the churches in ways that respected their sacred nature and historical significance, these initiatives failed to stimulate religious tourism. Insufficient marketing and a lack of collaboration with local stakeholders further hindered the project's impact.

Traditional religious tourism and pilgrimage, as narrowly defined, have limited potential for revitalization. The focus on religious and spiritual aspects, reliance on predefined routes, and static signage resulted in a passive tourist experience, and, crucially, failed to make the route recognizable among tourists. Moreover, Nessebar's image as a destination offering diverse experiences contrasts with a heavy emphasis on religious tourism, which appeals to a limited visitor segment. This is especially true given the seasonal, sun–sea–sand tourist profile centered around the nearby Sunny Beach Resort.

Data shows that most of Nessebar's churches have been repurposed as venues for cultural events and exhibitions, emphasizing their cultural rather than purely religious significance. Only one church has retained its original function, indicating a limited capacity to serve as primary destinations for religious tourism. While this transformation helps preserve the city's architectural heritage, it also underscores the need to diversify visitor experiences beyond religious themes.

The varying states of conservation among the churches highlight the ongoing challenges in preserving these historical structures. Despite these challenges, offering free access to most churches reflects a commitment to cultural accessibility and tourism. While these 5th–14th century churches hold immense historical and religious significance, shifting demographics, evolving religious practices, and the need for sustainable preservation necessitate a re-evaluation of their role in contemporary Nessebar. Reusing and integrating these structures into a broader cultural tourism framework offers an opportunity to expand their reach, attract a wider audience, and ensure their long-term preservation [40].

In Nessebar, adapting churches for events such as concerts and weddings represents a departure from their original purpose but has successfully integrated them into contemporary urban life. This trend is reflected in the rising number of weddings held in the city, increasing from 107 in 2017 to 151 in 2023. These historic sites have also become popular photography backdrops, supported by the development of tourist infrastructure and promotional effort at the municipal level.

7. Conclusions

Bulgaria has a substantial foundation for developing religious tourism, characterized by a rich religious heritage and supportive policy environment. However, the sector remains untapped, with a particular focus on Orthodox Christianity and limited exploration of other religious traditions.

To fully capitalize on the potential of religious tourism, concerted efforts are needed to develop infrastructure, promote diverse religious sites, and create engaging pilgrimage experiences. Addressing the imbalance between domestic and international religious tourism is also crucial for the sector's growth.

By strategically developing religious tourism, Bulgaria can enhance its tourism offerings, foster intercultural exchange, and preserve its valuable religious heritage.

Our findings suggest that places of religious interest can impact and bring religious tourism into a new dimensions and perspectives. The overwhelmed focus on architectural aspects of the churches, as in Nessebar, should be diversified by different events and thematic tours to add more value and spiritually enrich visitor experiences. Nessebar is not an exception, as religious heritage-making often relies heavily on infrastructural dynamics [56].

Nessebar keeps a strong connection with religion. While religious tourism, narrowly defined, faces challenges in resurgence, the deep-rooted Christian influence in Nessebar is evident in the transformation of traditional religious celebrations, such as the August 15 Assumption God, into official municipality holidays. There are many legends related to the Holy Virgin of Magara church, known today as "The Holy Assumption", the "miracles of the Holy Virgin", and wonder-working icons, performing miracles and fulfilling requests [44].

Nessebar's image as a destination is far away from any religious or pilgrimage destination. Nessebar lacks "refreshment", which is what happened in the nearly located Sozopol, where the relics of St John the Baptist were found in 2010. The attempts to develop specialized "spiritual paths" and tours around churches did not manage to enhance religious tourism. In contrast to other pilgrimage routes (e.g., Camino de Santiago in Spain, via Francigena in Italy, Shikoku Pilgrimage in Japan), the Nessebar route is of extremely limited duration and of intensive concentration in small space of area, overwhelmed by distinct cultural layers of heritage and history. The focus on the long way lost the initial religious purpose, and the architecture does not contribute to the re-emergence of religious tourism.

Post-secular tourism transcends the traditional divide between the sacred and secular. Instead, it fosters a hybrid experience where pilgrimage destinations become spaces that are neither strictly modern nor traditional, but rather a blend of both. According to Hill et al. [57], spirituality can be found in anything that fosters a sense of transcendence, connection, or deep personal meaning through associated rituals. This concept extends to travel, where the "sacred" for the modern traveler (the "post-tourist") lies in the intent behind the journey and the lasting impact anticipated upon completion [58]. The power of the journey and tourist experience does not solely reside in the destination itself, but in the personal transformation sought by the traveler.

Nessebar, as a former religious center, has shifted from being a pilgrimage destination to a site where tourists seek both cultural and spiritual experiences. This shift reflects broader global trends in tourism, where visitors engage with religious heritage not only for its historical value but also for personal spiritual enrichment [24,38].

As noted by Benussi [56], the way religion, infrastructure, society, and technology interact in the present day creates a complex and dynamic environment. This environment brings together various aspects of life, leading to both positive and negative consequences for religious tourism.

Religious tourism has the potential to attract a different demographic, focusing on individuals and smaller groups who seek spiritual enrichment and cultural and heritage appreciation. From a governance perspective, religious tourism can be strategically developed as an off-peak or shoulder season attraction. This shift towards quality over quantity can alleviate the pressure of over tourism. The adaptive reuse of religious sites offers an opportunity to create unique experiences that blend religious heritage with cultural and event activities. This can add value and enrich visitors' experience. Developing religious tourism with an emphasis on sustainability involves engaging local communities and stakeholders in the process.

Developing religious tourism in Nessebar necessitates a collaborative, sustainable approach involving public, private, and community stakeholders. Religious sites and churches, often publicly owned, offer a unique opportunity to invest in infrastructure while mitigating local concerns about authenticity. By adaptively reusing these spaces, Nessebar can stimulate economic growth, enhance residents' quality of life, and preserve its OUV without compromising its heritage.

Moreover, religious tourism should not be a "battle" for one city or municipality, but rather be integrated at the national level through targeted marketing and international promotion, together with Orthodox Church participation and involvement.

In conclusion, this paper reframes the study of Nessebar's heritage tourism through a CHS lens, emphasizing the importance of authenticity, identity, community engagement, and the development of religious heritage, which are continuously shaped by social, cultural, and political forces.

Author Contributions: Conceptualization, S.M.; methodology, S.M.; formal analysis, S.M.; investigation, M.K.; resources, M.K.; writing—original draft preparation, S.M.; writing—review and editing, S.M. and M.K.; supervision, S.M. All authors have read and agreed to the published version of the manuscript.

Funding: This research was funded by the European Union-NextGenerationEU, through the National Recovery and Resilience Plan of the Republic of Bulgaria, project N° BG-RRP-2.004-0008-C01 and by the project CultUrEn (Culture Urban Environment)—Cultural Heritage as a Factor for Achieving a Sustainable Urban Environment, funded by Bulgarian National Science Fund—MES/KP-06-N45/6.

Data Availability Statement: The data in Table 1 is sourced from Poklonnik.bg, reference number [43]. Table 2 combines data from Table 1 with information available on the UNESCO website (https://whc.unesco.org/en/statesparties/bg). The data in Tables 4 and 5 is derived from the Cultural Heritage Strategy of Nessebar Municipality (2023–2032), available in Bulgarian on the official website of the Municipality of Nessebar (https://nesebar.bg/programi/StrategyCH_Nessebar%20BG-edited.pdf), reference number [9].

Conflicts of Interest: The authors declare no conflicts of interest.

Notes

1. The pilgrimage route "The rock monasteries in Bulgaria" is part of a larger pilgrimage route called "Varna—the city of St. Ap. St. Andrew the First-Called".
2. Pilgrimage route "Nessebar—"encyclopedia" of Christian church construction" is part of a larger route called "The Southern Black Sea Coast—the spirit of the holy apostles".
3. Up to 16 October.

References

1. Olsen, D.; Timothy, D. Tourism and religious journeys. In *Tourism, Religion and Spiritual Journeys*; Routledge: London, UK, 2006; pp. 1–21.
2. UNESCO. World Heritage Convention. Available online: https://whc.unesco.org/en/religious-sacred-heritage/ (accessed on 20 June 2024).
3. Polus, R.; Carr, N.; Walters, T. Conceptualizing the Changing Faces of Pilgrimage through Contemporary Tourism. *Int. J. Sociol Leis* **2022**, *5*, 321–335. [CrossRef]
4. Waterton, E.; Watson, S. Framing theory: Towards a critical imagination in heritage studies. *Int. J. Herit. Stud.* **2013**, *19*, 546–561. [CrossRef]
5. Harrison, R. Heritage and Globalization. In *The Palgrave Handbook of Contemporary Heritage Research*; Palgrave Macmillan: London, UK, 2015; pp. 297–312.
6. National Strategy. Ministry of Tourism. 2018. Available online: https://www.tourism.government.bg/sites/tourism.government.bg/files/documents/2018-01/nsurtb_2014-2030.pdf (accessed on 10 July 2024).
7. Concept of Toursm Zoning. Ministry of Tourism. 2015. Available online: https://www.tourism.government.bg/sites/tourism.government.bg/files/uploads/raionirane/koncepcia.pdf (accessed on 10 July 2024).
8. Product analysis of Cultural Tourism, Ministry of Tourism. 2018. Available online: https://www.tourism.government.bg/sites/tourism.government.bg/files/uploads/2019_gg/produktov_analiz_-_kulturen_turizam.pdf (accessed on 10 July 2024).
9. Centre for Cultural Heritage and Architecture. Cultural Heritage Strategy of Nessebar Municipality for the period 2023–2032. 2023. Available online: https://nesebar.bg/programi/StrategyCH_Nessebar%20BG-edited.pdf (accessed on 22 May 2024).
10. UNESCO. State of Conservation Ancient City of Nessebar. 2023. Available online: https://whc.unesco.org/en/soc/4439 (accessed on 20 December 2023).
11. UNESCO. Report of the joint UNESCO World Heritage Centre/ICOMOS Reactive Monitoring mission to the Ancient City of Nessebar Bulgaria, 22 to 26 October 2018. 2018. Available online: https://whc.unesco.org/en/documents/175150/ (accessed on 20 December 2023).
12. Smith, L. *Uses of Heritage*; Routledge: London, UK, 2006.
13. Harrison, R. The politics of heritage. In *Understanding the Politics of Heritage*; Manchester University Press: Manchester, UK, 2010; Volume 5, pp. 154–196.

14. Askew, M. The magic list of global status: UNESCO, World Heritage and the agendas of states. In *Heritage and globalization*; Routledge: London, UK, 2010; pp. 33–58.
15. Luleva, A. The Ancient city of Nessebar between the Outstanding universal value and the Tourist Indsutry. An Ethographic study of a conflict. *Anthropology* **2014**, *1*, 28–64.
16. McDowell, S. Heritage, memory and identity. In *The Routledge Research Companion to Heritage and Identity*; Routledge: London, UK, 2016; pp. 37–53.
17. Smith, M. Spiritual Tourism. In *Encyclopedia of Tourism Management and Marketing*; Buhalis, D., Ed.; Edward Elgar Publishing: Cheltenham, UK, 2022; pp. 209–211.
18. Rinschede, G. Forms of Religious Tourism. *Ann. Tour. Res.* **1992**, *19*, 51–67. [CrossRef]
19. Terzidou, M. Re-materialising the religious tourism experience: A post-human perspective. *Ann. Tour. Res.* **2020**, *83*, 102924. [CrossRef]
20. Thouki, A. The role of ontology in religious tourism education—Exploring the application of the postmodern cultural paradigm in European religious sites. *Religions* **2019**, *10*, 649. [CrossRef]
21. Nilsson, M.; Tesfahuney, M. The post-secular tourist: Re-thinking pilgrimage tourism. Tourist Studies. *Tour. Stud.* **2018**, *18*, 159–176. [CrossRef]
22. Nyaupane, G.; Timothy, D.; Poudel, S. Understanding tourists in religious destinations: A social distance perspective. *Tour. Manag.* **2015**, *48*, 343–353. [CrossRef]
23. Durán-Sánchez, A.; Álvarez-García, J.; del Río-Rama, M.D.L.C.; Oliveira, C. Religious tourism and pilgrimage: Bibliometric overview. *Religions* **2018**, *9*, 249. [CrossRef]
24. Collins-Kreiner, N. A review of research into religion and tourism Launching the Annals of Tourism Research Curated Collection on religion and tourism. *Ann. Tour. Res.* **2020**, *82*, 102892. [CrossRef]
25. Heidari, A.; Yazdani, H.R.; Saghafi, F.; Jalilvand, M.R. The perspective of religious and spiritual tourism research: A systematic mapping study. *J. Islam. Mark.* **2018**, *9*, 747–798. [CrossRef]
26. Collins-Kreiner, N. Pilgrimage tourism-past, present and future rejuvenation: A perspective article. *Tour. Rev.* **2019**, *75*, 145–148. [CrossRef]
27. Das, A.; Kondasani, R.K.R.; Deb, R. Religious tourism: A bibliometric and network analysis. *Tour. Rev.* **2024**, *79*, 622–634. [CrossRef]
28. Yala, İ. Bibliometric Analysis of Literature on Religious Tourism in Web of Science. *J. Tour. Gastron. Stud.* **2023**, *11*, 1844–1856. [CrossRef]
29. Rashid, A.G. Religious tourism–a review of the literature. *J. Hosp. Tour. Insights* **2018**, *1*, 150–167. [CrossRef]
30. Kim, B.; Kim, S.; King, B. Religious tourism studies: Evolution, progress, and future prospects. *Tour. Recreat. Res.* **2020**, *45*, 185–203. [CrossRef]
31. Olsen, D.H. Heritage, tourism, and the commodification of religion. *Tour. Recreat. Res.* **2003**, *28*, 99–104. [CrossRef]
32. Iliev, D. The evolution of religious tourism: Concept, segmentation and development of new identities. *J. Hosp. Tour. Manag.* **2020**, *45*, 131–140. [CrossRef]
33. McKercher, B. Towards a taxonomy of tourism products. *Tour. Manag.* **2016**, *54*, 196–208. [CrossRef]
34. Seyer, F.; Müller, D. Religious tourism: Niche or mainstream? In *The Long Tail of Tourism: Holiday Niches and Their Impact on Mainstream Tourism*; Gabler: Wiesbaden, Germany, 2011; pp. 45–56.
35. Gauthier, F.; Martikainen, T.; Woodhead, L. Acknowledging a global shift: A primer for thinking about religion in consumer societies. *Implicit Relig.* **2013**, *16*, 261–276. [CrossRef]
36. Ron, A.S. Towards a typological model of contemporary Christian travel. *J. Herit. Tour.* **2009**, *4*, 287–297. [CrossRef]
37. Harrison, R. Heritage as social action. In *Understanding Heritage in Practice*; Manchester University Press: Manchester, UK, 2010; pp. 240–276.
38. Olsen, D.; Timothy, D. *The Routledge Handbook of Religious and Spiritual Tourism*; Routledge: Abingdon, UK, 2022.
39. Leus, M.; Verhelst, W. Sustainability Assessment of Urban Heritage Sites. *Buildings* **2018**, *8*, 107. [CrossRef]
40. Ijla, A.; Broström, T. The sustainable viability of adaptive reuse of historic buildings: The experiences of two world heritage old cities; Bethlehem in Palestine and Visby in Sweden. *Int. Invent. J. Arts Soc. Sci.* **2015**, *2*, 52–66.
41. Yıldırım, M.; Turan, G. Sustainable development in historic areas: Adaptive re-use challenges in traditional houses in Sanliurfa, Turkey. *Habitat Int.* **2012**, *36*, 493–503. [CrossRef]
42. Mihalev, I. *Piligrimage Sites in Bulgaria*, 1st ed.; The Pilgrimage and Educational Centre of the Holy Metropolis: Sofia, Bulgaria, 2021.
43. Poklonnik.bg. The pilgrimage and educational center of the Sofia Holy Metropolis. Available online: https://www.poklonnik.bg/ (accessed on 22 May 2024).
44. Theoklieva-Stoycheva, E. *The Golden Book of Nessebar*; Informa Print: Burgas, Bulgaria, 2019.
45. Ivanchev, I. *Nesebar i Negovite Kashti (Nessebar and Its Houses) (In Bulgarian)*; Nauka i Izksutvo: Sofia, Bulgaria, 1957.
46. UNESCO. Retrospective Statement of Outstanding Universal Value. 2010. Available online: https://whc.unesco.org/en/list/217 (accessed on 18 September 2023).
47. Harrison, R.; Dias, N.; Kristiansen, K. (Eds.) *Critical Heritage Studies and the Futures of Europe*; UCL Press: London, UK, 2023.
48. Plan for Protection and Management. Project "Faith in Nesebar". 2012. Available online: https://pou-nesebar.org/bg/prilozhni-produkti/proekt-vyara-v-nesebar/ (accessed on 18 May 2024).

49. Winter, T. Heritage diplomacy; an afterword. *Int. J. Cult. Policy* **2023**, *29*, 130–134. [CrossRef]
50. Mutafov, E. *Mitropolitskiyat Hram "Sv. Stefan" v Nesebar i Negoviyat Hudozhestven Krag: Kulturen Kontekst, Intertekstualnost i Intervizualnost*; Prof. Marin Drinov: Sofia, Bulgaria, 2022. (In Bulgarian)
51. Markov, N. Za mnimoto napadenie na venetsianskia flot nad Nesebar prez 1257 g (On the supposed attack of the Vencie fleet on Nessebar in 1257). *Epohi* **2015**, *23*, 359–370. (In Bulgarian)
52. Giuzelev, V. Novi danni za istoriyata na Balgariya i na grad Nesebar prez 1257 g. (*New data on the history of Bulgaria and the city of Nessebar in 1257*). *Vekove* **1972**, *1*, 5–16.
53. Cornaro, F. Venetae antiquis monumentis nunc etiam primum editis illustratae. In *Venetiis* T.2. 1749, pp. 258–259. Available online: https://archive.org/details/ecclesiaeveneta00unkngoog/page/n2/mode/2up (accessed on 20 May 2024).
54. Jirechek, K. Travels in Bulgaria. In *Bulgarian*; Nauka i Izkustvo: Sofia, Bulgaria, 1974.
55. Kanitz, F. *Donau-Bulgarien und der Balkan: Historisch-Geographisch-Ethnographische Reisestudien aus den Jahren 1860–1875*; Hermann Fries: Leipzig, Germany, 1875.
56. Benussi, M. Afterword: Religious Infrastructure, or Doing Religion in the Contemporary Mode. *Relig. State Soc.* **2024**, *52*, 235–249. [CrossRef]
57. Hill, P.; Pargament, K.; Hood, R.; McCullough, M.; Swyers, J.; Larson, D.; Zinnbauer, B. Conceptualizing religion and spirituality: Points of commonality, points of departure. *J. Theory Soc. Behav.* **2000**, *30*, 51–77. [CrossRef]
58. Leite, N.; Graburn, N. Anthropological interventions in tourism studies. In *The Sage Handbook of Tourism Studies*; Jamal, M.R.T., Ed.; Sage: London, UK, 2009; pp. 35–64.

Disclaimer/Publisher's Note: The statements, opinions and data contained in all publications are solely those of the individual author(s) and contributor(s) and not of MDPI and/or the editor(s). MDPI and/or the editor(s) disclaim responsibility for any injury to people or property resulting from any ideas, methods, instructions or products referred to in the content.

 heritage

Article

The Impact of the Sociodemographic Profile on the Tourist Experience of the Fiesta de los Patios of Córdoba: An Analysis of Visitor Satisfaction

Lucía Castaño-Prieto [1], Lucía García-García [2], Minerva Aguilar-Rivero [1] and José E. Ramos-Ruiz [1,*]

[1] Department of Applied Economics, University of Cordoba, 14071 Córdoba, Spain; lcastano@uco.es (L.C.-P.); u52agrim@uco.es (M.A.-R.)
[2] Department of Business Organization, University of Cordoba, 14071 Córdoba, Spain; z12gagal@uco.es
* Correspondence: d22raruj@uco.es

Abstract: The Festival of the Patios of Cordoba, declared an Intangible Cultural Heritage (ICH) by UNESCO in 2012, serves as an emblematic case of how this designation acts as a tourist brand, attracting a greater number of visitors and granting a competitive advantage to the city's tourist market. This research is focused on analyzing the differences and similarities in the satisfaction, lived experience and behavioral intention of tourists according to their sociodemographic profile during the 2022 edition of the Patios Festival. The study's main objective is to understand the sociodemographic profile of the tourist who visits this event and if there are features of this profile that influence the satisfaction and lived experience with the event. Using a quantitative methodological approach, field work was carried out during the Fiesta de los Patios of Cordoba (Spain) in its 2022 edition, which took place between 3 and 15 May 2022, obtaining 383 valid surveys. The results reveal differences in the perception and satisfaction of the experience depending on the sociodemographic profile of the visitors. These findings highlight the need to adapt the tourism offerings to improve the visitor experience and also contribute to the scarcity of studies on ICH to help tourism managers formulate strategies that maximize the cultural and economic benefits of these Word Heritage inscriptions.

Keywords: cultural tourism; Fiesta de los Patios (Cordoba); Intangible Cultural Heritage; satisfaction; sociodemographic profile

1. Introduction

Europe's great cultural and heritage wealth [1] means that knowledge of culture is one of the main motivations for travel, which in turn has led to an increase in cultural tourism in recent years [2]. The Intangible Heritage of Humanity list is one of the three lists published annually by United Nations Educational, Scientific and Cultural Organization (UNESCO), with the objective not only of protection, but also of safeguarding, in the sense of ensuring its continuous transmission and recreation [3]. While this is UNESCO's main objective, the role of such inscriptions as a tourism resource is undeniable [4]. Due to the visibility that this inscription provides, many scientific studies link Intangible Cultural Heritage (ICHs) to increased tourism flows, particularly of international tourists [5,6].

ICHs are an important tourism resource, which in addition to being a force for attracting tourists [7], have another indirect effect, which is that they are used as a tool or starting point for the creation of places and tourist attractions in those places where such inscriptions are located [8]. In other words, the UNESCO declaration of an ICH often means that the place obtains a seal of value that becomes a competitive advantage in the global tourism market [9]. However, this great publicity that comes with inscription raises problems of balance between authenticity and commodification, and the entities in charge of tourism management must take actions that allow for sustainable tourism development in these places [9,10].

The Cordoba Patios Festival was declared an ICH in 2012 as it is considered to "promote the function of the patio as an intercultural meeting place and foster a sustainable collective way of life, based on the establishment of strong social ties and networks of solidarity and exchanges between neighbors, while stimulating the acquisition of knowledge and respect for nature" [11]. As is the case with the other declarations, its declaration leads to an increase in the number of visits due to its greater popularity [12]. Due to the limited number of studies carried out in ICH, this research aims to detect similarities and differences in the tourism profile factors (age, gender, income, educational level and employment status) of the participants of the Fiesta de los Patios in its 2022 edition. In this sense, the research questions that this study aims to answer are the following.

Research Question 1 (RQ1): What is the sociodemographic profile of the tourist who visits La Fiesta de los Patios?

Research Question 2 (RQ2): Are there features of the sociodemographic profile that influence satisfaction, behavioral intention and lived experience?

2. Literature Review

2.1. Cultural Tourism

Culture is to be understood according to the definition given by [13] as the set of distinctive spiritual, material, intellectual and emotional features that are characteristic of and specific to a society or a group within a society. In other words, culture does not only encompass artistic and literary expressions, but refers to a broader set of practices including traditions and beliefs. Culture encompasses everything that enables a person to express and learn more about themself. Linked to culture and the interest in its knowledge is the concept of cultural tourism, which is defined by the United Nations International Tourism Organization as a "type of tourism activity in which the essential motivation of the visitor is to learn about, discover, experience and consume the tangible and intangible cultural attractions/products of a tourism destination" [14]. These tourist attractions refer to all those traditions of the place and expressions of culture that are characteristic of a society and that allow people to learn more about the place, the way of life and the beliefs of its inhabitants, with all these aspects including the decisive ones that make a tourist visit the location [1]. Therefore, cultural tourism involves immersing tourists in the art, cultural heritage, thoughts and institutions of another country, city or region [15].

Knowing this motivation is key for local communities, as they will be able to make use of the cultural elements of the place to attract tourists, and, in addition, this cultural tourism should be used to bring the characteristics of the place closer to tourists by explaining its rich heritage [16]. All of these efforts to attract tourists based on their cultural motivation mean that cultural tourism becomes a mechanism for the continuous improvement of cultural resources and the quality of the service provided to tourists to increase tourism flows in these places [1]; it is also proof of the evident relationship between heritage and tourism [17].

Cultural tourism includes both tangible and intangible elements that the destination offers. This is why many authors consider World Heritage Sites (WHS) tourism as a form of cultural tourism [18]. The research carried out on WHS is very extensive; however, the research that relates tourism and WHSs is limited, with some examples such as the research on the patrimonialization of tango in Buenos Aires or on the elaboration of the toquilla straw hat [2]. Despite this, there is currently a trend in scientific research in which research on intangible cultural tourism is taking on greater relevance than that on tangible heritage [19].

2.2. Destination Attributes

The attributes of a destination are defined as the set of elements presented by a destination, which will influence the perception of tourists [20]. Thus, the attributes displayed by a destination will have a direct impact on the destination's image, as well as on the tourists' intention to visit and recommend it. In other words, attributes affect tourists' satisfaction and future behavior [21,22]. Attributes such as the attractiveness of the place are configured as a key in the creation of the cognitive image of the destination. Therefore, the unique attractions of the destination influence the creation of its cognitive and unique image, so much so that the cognitive image is influenced by the rational decisions that tourists make in relation to aspects such as accommodation, accessibility or attraction [21]. Therefore, destination attributes become a competitive advantage of destinations and in this way, tourism managers, in an effort to increase and improve place perception, make destination attributes a major part of tourism planning and tourist attraction strategies. In this way, destinations try to achieve certain attributes as a means to promote tourism in the destination [23]. However, not all attributes have a positive influence on the tourism experience, and therefore not all of them provide a competitive advantage [24]. Thus, the attributes of a destination should be configured in a way that helps to increase tourist attachment.

Such is the importance of the attributes of a destination that research such as that carried out by [25] uses attributes such as accommodation, climate, products and accessibility (among others) to assess tourists' satisfaction with a destination [26].

Studies such as [27] distinguish between two elements in tourist destinations:

- Those characteristics or attributes that are given by the destination. In this case, reference is made to aspects such as the climate, location, existing heritage, etc. In other words, they are aspects that determine the type of tourist that the destination attracts.
- Those attributes that have been "purposely" designed by tourism operators to attract tourists and to satisfy tourists' needs.

Destination attributes capable of generating a competitive advantage include food and public safety [28,29] or attributes such as accommodation, reasonable prices, public safety and the possibility of a relaxing holiday [30]. Therefore, following [21], destination attributes can be classified into five different dimensions: accommodation, food and beverage, transport, attractions and public safety.

2.3. Satisfaction and Loyalty

Satisfaction can be defined as the comparison or relationship between tourists' expectations and the experience [31], so that for authors such as [32], satisfaction involves both cognitive and emotional aspects. Tourist satisfaction is a key aspect in the consolidation of a destination. Obtaining dissatisfaction from visitors to a place not only affects those tourists who experience that feeling, but that dissatisfaction will have a ripple effect affecting other people, who will change their previous perception of the place [33].

Tourist satisfaction is a key aspect in the management of tourist destinations; thus, studies such as that of [34] analyze how the factors that influence the satisfaction of tourists visiting Montañita (Ecuador), concluding that the most valued factors were the location of hotel services and the quality of food and beverages in restaurant services. Other studies, such as that of [35], through the study of measures of centrality, dispersion and the correlation of variables, study the impact of the authenticity of a cultural tourism destination on tourist loyalty and satisfaction, concluding that there is a positive correlation between these variables. [36] reviews the literature on the different factors that influence tourist satisfaction, in addition to conducting fieldwork that allowed them to conclude that tourist satisfaction is strongly affected by factors such as the cultural characteristics of the trip, sociodemographic characteristics and the previous source of information for visiting the destination.

Due to the importance of tourist satisfaction, several studies are found linking tourist satisfaction with the evaluation of the tourist experience [37]. Likewise, both the motivation to travel and the sociodemographic characteristics of tourists are directly related to tourist satisfaction [38]. The study of tourist satisfaction can be carried out for different reasons. For example, it may be of interest for effective advertising [39], as well as to increase loyalty to a destination, as tourist satisfaction is considered one of the ways to explain destination loyalty [40].

2.4. Behavioral Intetion

Behavioral intention refers to the degree to which a person is predisposed to perform a certain behavior [41]. Some studies, such as the one conducted by [42] to find out tourists' behavioral intention, will use three different items. The first one refers to whether the tourist will recommend the city to others as a holiday destination. The second one refers to whether the tourist will return to the city in the future and the final one refers to whether the tourist will say positive things about the city. It is with these items that the intention to revisit and the intention to recommend can be ascertained.

In the tourism and hospitality marketing literature [43], behavioral intention is considered one of the most important aspects of consumer behavior, being the main construct for predicting future consumer behavior [44]. Furthermore, several studies analyze the relationship between behavioral intention and consumer satisfaction [45]. The destination image that the tourist creates and the affective image that the tourist develops towards the destination will predict the tourist's behavioral intention [46].

2.5. Sociodemographic Profile

Knowing the sociodemographic profile of heritage tourists is very useful, as it allows to know how these visitors rate the different attributes of the place, which enables the efficient management of the destination, adapting tourism resources for the different tourist groups or segments identified [47]. Several research studies have been carried out in this field, mainly considering the variables of age, gender, income level and education.

Most studies attempt to establish the general profile of heritage tourists. The study carried out by [48] identifies the heritage tourist as a young woman with a high level of education who is either a full-time student or employee.

With regard to the age variable, it is, together with gender, one of the variables on which there is a lack of consensus. Studies such as those by [49] establish a wide range, placing the heritage tourist between 26 and 45 years of age, similar to that indicated by [50]. While authors such as [51] indicate an age between 21 and 35 years, profiling the heritage tourist as a young tourist, a statement which is opposed to that of the authors [52] who indicate ages over 45 years.

Regarding gender, there is no consensus in the scientific literature, with studies indicating a preference for these destinations by men [50,51], while others indicate the opposite [49,52], although in these studies, the preference for these destinations of one gender over the other is small.

In terms of educational level, most studies point to the high educational level of tourists visiting destinations with a rich heritage [49,50,53].

Finally, with respect to income level, there is a consensus in the literature, pointing to the medium, medium-high level [51,52].

Therefore, despite the differences found between the different studies, the heritage tourist can be profiled as a young tourist, with higher education, a high level of education and a medium, medium-high income level. This study aims to contribute to profiling the characteristics of tourists in destinations with a great cultural and heritage wealth, and which have a UNESCO inscription.

3. Methodology

3.1. Survey Design and Data Collection

This study is based on data obtained through a questionnaire which was completed by tourists attending the 2022 edition of the Córdoba Patios Festival, which took place between 3 and 15 May 2022, and is one of the main tourist attractions of the city of Córdoba. The total number of valid surveys obtained was 383 surveys. This questionnaire was used to analyze tourists' perceptions of both the Fiesta de los Patios and Cordoba as a tourist destination. The formulation of the items was based on previous research to ensure the validity of the questionnaire [49,54,55]. The items used to create the questionnaire are inspired by various studies related to the study of cultural tourism [54,55]. Likewise, a

7-point Likert scale has been used, since it is considered that the results obtained with it are more precise than with the use of the 5-point Likert scale [56]. The 7-point Likert scale has been successfully used in previous studies on cultural tourism that analyze the satisfaction, perception and intention of future behavior of these cultural tourists [57]. The development of the questionnaire was carried out in three different phases. Firstly, after having selected the relevant scientific literature on the topic to be addressed, the items were evaluated by an expert in heritage tourism in the city of Córdoba. Secondly, local tourism managers reviewed the selected items. Finally, a pilot study was carried out with 25 tourists to make the final adjustments. After these three phases, the questionnaire was refined to avoid questions that generated doubts and to make the questionnaire as clear as possible, avoiding the questionnaire from being excessively long and trying to make the questions as specific as possible to achieve quality results [58].

Convenience sampling was used to collect the sample, which is commonly used in this type of research, in which tourists are available to be surveyed in a specific space and time [59]. Because there were no previous studies to support stratification by gender, age or any other variable, no stratification was conducted. The refusal rate of the questionnaire was low and not significant for any question. Assuming simple random sampling and infinite population, the sampling error would be 5.01%, with a reliability rate of 95%. The level of significance used was a p-value < 0.05, which is shown in the results for each of the items.

The questionnaire was made up of two blocks. The first block contained a group of questions, measured on a 7-point Likert scale, which analyzed the perception and satisfaction of tourists with both the city of Cordoba and their participation in the Fiesta de los Patios. On the other hand, the second block of the questionnaire consisted of a set of questions relating to the sociodemographic profile of the tourists surveyed, such as gender, age, income level, educational level and employment situation.

3.2. Analysis Method

SPSS Statistics v28 was used for the statistical analysis of the data. Subsequently, the Kolmogorov–Smirnov [60,61] and Shapiro–Wilk [62] normality tests were performed on the questionnaire, both confirming that the sample did not correspond to a normally distributed population for all questions (p-value < 0.05). Therefore, it was decided to use non-parametric tests. Since the samples were independent, it was considered appropriate to use the Mann–Whitney (MW) U-test to compare two samples [63], and in cases with more than two samples, the Kruskal–Wallis (KW) test was used [64]. In the latter case, an additional analysis was also performed by applying the U-test (MW) to all possible combinations of pairs of samples. The MW and KW tests are commonly used to identify statistically significant differences between groups of data according to variables.

4. Results

4.1. Research Question 1

In relation to the sociodemographic profile, the different variables (gender, income, age, academic training, daily spend and employment status) are shown in Table 1. In the gender variable, the presence of women is slightly higher than that of men, although it is a variable in which there is relative parity. In relation to the age variable, the age range that stands out most is between 18 and 35 years of age. It is also noteworthy that almost 85% of the participants in this event are under 55 years of age, which indicates that this event has a greater diffusion and popularity among young people. The average household income is medium-high, mainly between EUR 1000 and EUR 2500. On the other hand, most of the participants in the Fiesta de los Patios were university graduates (72.85%) and employed (78.59%). Finally, in relation to the variable of daily expenditure, which is very important to know the economic effect that this event has on the city of Cordoba, this is mainly between EUR 0 and EUR 100 per day, although the range of EUR 200 or more per day also had a high response.

Table 1. Sociodemographic profile.

GENDER (GEN)		INCOME (INC)	
Man	40.73%	Less than EUR 1000	12.01%
Woman	59.27%	From EUR 1000 to EUR 2500	52.22%
Total	n = 383	More than EUR 2500	35.77%
AGE (AGE)		ACADEMIC TRAINING (ATR)	
18–35	47%	Non-university	27.15%
36–55	36.81%	University	72.85%
More than 55	16.19%		
DAILY SPEND (DSP)		EMPLOYEMENT STATUS (EMP)	
0–100	39.69%	Not employed (unemployed, students, retirees, housework)	21.41%
101–200	27.68%	Employee (public, private and businessmen)	78.59%
More than 200	32.64%		

Therefore, it can be concluded that the profile of the tourist participating in the Fiesta de los Patios in the year 2022 was that of a female university graduate, under 35 years of age, who is in employment and with a medium-high income.

4.2. Research Question 2

The questions that formed part of the survey are presented in Table 2, together with the mean and standard deviation of each of these questions. The questionnaire is grouped into four thematic sections: destination, event, perception and finally experience and behavior intention. In general terms, all of the questions have a high score, and it is noteworthy that in no case is the average rating of the tourists participating in the Fiesta de los Patios lower than 5 points, with the sole exception of question E11, regarding the waiting time to start the visit. On the other hand, the highest averages were obtained in the section on the evaluation of the Fiesta de los Patios as an event, specifically the questions relating to the beauty of the patios as a whole and their surroundings, their state of conservation and cleanliness, and the safety felt during the visit (E03, E04, E05 and E10). This highlights the potential of this event, given the image it projects and how it is perceived by tourists.

Table 2. Set of questions.

Code	Question	Mean	Std. Dev
	DESTINATION (D)		
D01	The historic and monumental center	6.49	0.755
D02	The conservation of monumental and artistic heritage	6.26	0.844
D03	The beauty of the city	6.53	0.711
D04	Accessibility to emblematic buildings and monuments	5.74	1.06
D05	Tourist information	5.61	1.327
D06	Attention and quality of tourist accommodation	5.77	1.169
D07	Attention and quality of restaurants and bars	5.96	1.055
D08	Attention and quality of the tour guides	5.94	1.051
D09	Diversity and quality of local gastronomy	6.12	1.062
D10	Opportunity to purchase handicrafts and traditional food items	6.05	1.096
D11	Complementary leisure offer	5.48	1.223
D12	Citizen security	6.14	1.015
D13	Care and cleaning of the city	6.14	0.922
D14	Resident hospitality	6.39	0.934
D15	Public transportation services	5.61	1.14
D16	City value for money	5.7	1.197

Table 2. *Cont.*

Code	Question	Mean	Std. Dev
	FIESTA DE LOS PATIOS EVENT (E)		
E01	Accessibility to patios and surrounding spaces	5.68	1.385
E02	Kindness and hospitality of residents	6.31	0.998
E03	Joint beauty of the patios and their surroundings	6.56	0.742
E04	State of conservation of the patios visited	6.6	0.72
E05	Care and cleaning of the patios	6.67	0.652
E06	Availability of restaurants and bars in the surrounding area	5.83	1.132
E07	Diversity and variety of patios that can be visited	6.14	0.991
E08	Tourist information and signage points	5.48	1.332
E09	Opportunity to make purchases of interest: crafts, etc.	5.81	1.147
E10	Security during the visit	6.37	0.935
E11	Waiting time to start the visit	4.11	1.835
	PERCEPTION		
P01	During the visit to the patios, I felt part of the cultural and heritage of the city of Cordoba	5.71	1.362
P02	The visit to the patios has especially moved me	5.74	1.222
P03	The visit to the patios has contributed to increasing my level of knowledge about the culture and traditions of the city	5.69	1.276
P04	The visit to the patios has helped me relax	5.49	1.626
	EXPERIENCE AND BEHAVIOR INTENTION		
X01	My level of satisfaction with the patios has been very important	5.97	1.166
X02	My choice to visit the patios was correct	6.17	1.12
X03	Cordoba is a quality tourist destination	6.35	1.007
X04	When I talk about the patios, I will say positive things	6.31	1.04
X05	I will encourage my family and friends to visit the patios	6.23	1.131
X06	After my experience, I think I will visit the patios again in future festivals	5.58	1.635
X07	I would recommend a visit to the patios if someone asked me for advice	6.22	1.204

Regarding the block of questions relating to perception, question P02, about the emotion felt during the visit, is noteworthy. There are several studies that highlight how in cultural tourism activities, the various emotions generated give rise to the configuration of existential experiences [65].

Table 3 shows those questions where significant differences by gender were found. To include the questions, the criterion was to select those whose *p*-value was equal to or less than 0.05. Although in general terms all the scores are very high, both for men and women, those questions in which there is the greatest difference between the assessments of both genders are the following. First, the question on the state of conservation of the monumental and historical heritage of the city of Cordoba (D02) showed a greater sensitivity on the part of women towards artistic preservation. Second, the questions relating to the attention and quality of tourist accommodation (D06) and public safety (D12) indicated the greater importance given by women to these aspects during their experience. Third, the question referring to tourist information points and signposting (E08), where the higher rating by women may suggest a greater importance of tourist orientation for them. Finally, differences were also found in the question on the emotion felt during the visit (P02), which shows the greater emotional connection of women with the experience. In all questions, the scores given by women were higher than those given by men.

The significant differences found by age are reflected in Table 4 and in Figure 1. As a consequence of there being more than two age groups, a KW test was performed to check the differences between the groups, and subsequently a pairwise MW analysis was performed. As can be seen, significant differences were found between the 18–35 age group and the 36–55 and over 55 age groups, but in no case between the 36–55 and over 55 age groups. What can be seen from this table is the more critical spirit of the younger visitors with respect to the variables relating to the historic and monumental quarter (D01) and the friendliness and hospitality of the residents (E02). However, visitors between 18 and 35 years of age rate the opportunity to buy traditional crafts and food items (D10) higher than those over 55 years of age.

Table 3. Mann–Whitney's U test of the question compared by gender.

Code	Male (Mean)	Female (Mean)	p-Values	T-Values
D01	6.40	6.55	0.018 *	19,879.000
D02	6.08	6.38	0.001 *	20,831.500
D03	6.41	6.62	0.003 *	20,364.000
D06	5.62	5.87	0.026 *	19,977.000
D12	5.99	6.24	0.046 *	19,681.000
D15	5.48	5.70	0.018 *	20,141.500
E05	6.57	6.74	0.050 *	19,302.000
E08	5.30	5.60	0.020 *	20,107.000
P02	5.54	5.87	0.026 *	19,980.000
X06	5.45	5.67	0.043 *	19,770.000
X07	6.10	6.30	0.030 *	19,756.000

* Statistical significance at $p < 0.05$.

Table 4. Kruskal–Wallis' test and Mann–Whitney's U test pairwise of the question compared by age.

Code	18–35 Years Old (Mean)	36–55 Years Old (Mean)	More Than 55 Years Old (Mean)	Pairwise Comparison	p-Values	T-Values
D01	6.40	6.58	6.52	Only differences between 18 and 35 and 36–55	0.044 *	6.256
D10	6.16	6.03	5.81	Only differences between 18 and 35 and over 55	0.050 *	5.999
E02	6.20	6.42	6.40	Only differences between 18 and 35 and 36–55	0.023 *	7.523

* Statistical significance at $p < 0.05$.

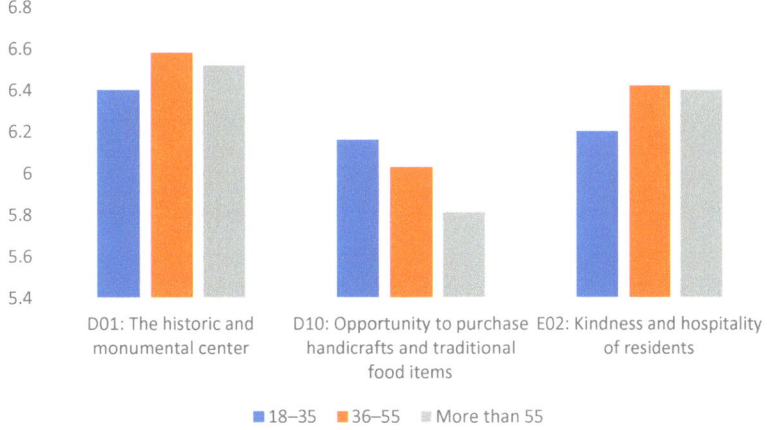

Figure 1. Significant differences by age.

Table 5 and Figure 2 show the questions that show significant differences because of the average daily expenditure; in the same terms as in Table 4, while as there were 3 groups for this variable, a KW test was carried out to see the differences between the groups, as was a subsequent analysis using the MW pairs. The aspects in which significant differences were found for this variable were the perception of the availability of restaurants and bars (E06), safety during the visit (E10), and the intention to visit the event again in future editions (X06). The importance of satisfaction with the gastronomic offer (E06) and perceived safety (E10) by visitors with different levels of expenditure stands out, mainly the latter. In these two areas, people with a higher daily expenditure tended to rate these variables significantly lower than those whose average daily expenditure was in the lower range.

Table 5. Kruskal–Wallis' test and Mann–Whitney's U test pairwise of the question compared by daily spend.

Code	EUR 0–100 (Mean)	EUR 101–200 (Mean)	More Than EUR 200 (Mean)	Pairwise Comparison	p-Value	T-Value
E06	5.99	5.83	5.62	Only differences between 0 and 100 and more than 200	0.041 *	6.405
E10	6.53	6.42	6.14	Only differences between 0 and 100 and more than 200	0.022 *	7.634
X06	5.77	5.48	5.43	Differences between all	0.049 *	6.041

* Statistical significance at $p < 0.05$.

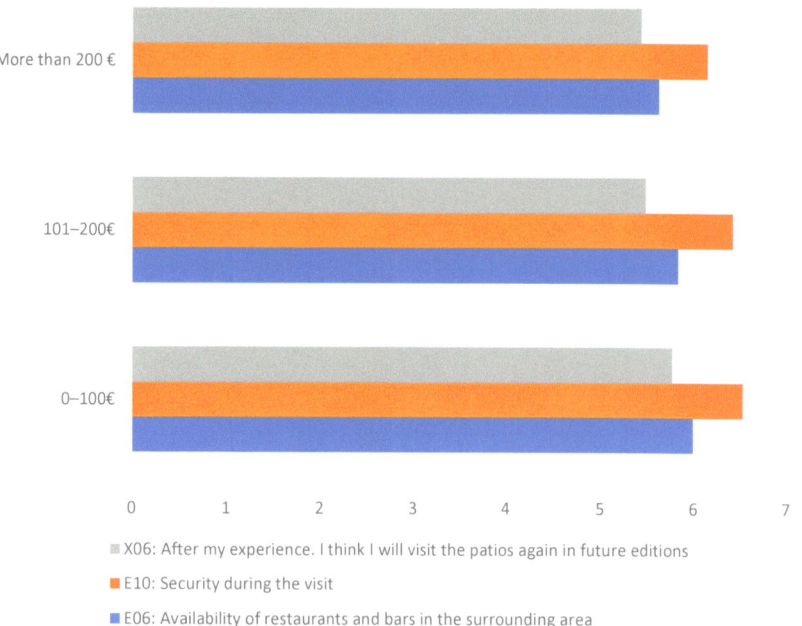

Figure 2. Significant differences by daily spend.

In relation to educational level, the differences found between non-university graduates (primary and secondary education) and university graduates are reflected in Table 6. In general terms, in all the variables studied, non-university graduates tend to give a more positive assessment than those with university degrees, which shows a more critical spirit on the part of the latter group. Specifically, those questions in which the differences are greater between these two groups are those relating to the accessibility of the patios and surrounding areas (E01), the waiting time for the initial visit (E11) and the intention to visit the patios again in future editions (X06). These differences will be relevant in that they should be taken into consideration when studying visitor satisfaction and perception, and it is important to consider the impact of educational level on the assessment and evaluation of the tourism experience. Regarding the monthly household income level variable (INC), no significant differences were found between the different income ranges considered (less than EUR 1000, between EUR 1000 and EUR 2500 and more than EUR 2500).

Table 6. Mann–Whitney's U test of the question compared by educational level.

Code	Non-University (Mean)	University (Mean)	p-Value	T-Value
D05	5.94	5.49	0.002 *	11,556.000
D06	5.99	5.68	0.013 *	12,215.000
D08	6.16	5.86	0.014 *	12,244.500
D09	6.35	6.04	0.006 *	12,048.500
D15	5.82	5.54	0.034 *	12,531.000
E01	6.10	5.53	0.000 *	10,886.500
E02	6.54	6.23	0.014 *	12,388.500
E09	5.99	5.75	0.020 *	12,362.500
E11	4.51	3.97	0.008 *	11,986.000
P01	6.07	5.58	0.008 *	12,053.000
P02	6.01	5.63	0.023 *	12,403.000
P04	5.87	5.35	0.009 *	12,082.500
X01	6.19	5.89	0.042 *	12,655.000
X06	6.00	5.42	0.014 *	12,237.000
X07	6.42	6.15	0.042 *	12,769.000

* Statistical significance at $p < 0.05$.

Regarding the employment status variable, two groups were compared: those in employment (public sector, private sector or self-employed) and those not in employment (unemployed, students, retired and people working in the home). Table 7 and Figure 3 show that in those questions where significant differences were found for this variable, the evaluations of the unemployed were more positive than those of the employed. The biggest differences were found mainly in the experience block, specifically in the intention to revisit in future editions (X06) and the recommendation to third parties (X07), which shows greater loyalty, both in terms of revisiting and recommendation of those not in employment. Also noteworthy is the higher score of non-employees on the questions relating to residents' hospitality and friendliness (D14 and E02).

Table 7. Mann–Whitney's U test of the question compared by employment status.

Code	Unemployed (Mean)	Employed (Mean)	p-Value	T-Value
D14	6.59	6.34	0.011 *	10,363.000
E02	6.54	6.25	0.039 *	10,694.000
E03	6.71	6.52	0.027 *	10,714.500
E05	6.80	6.63	0.038 *	10,930.000
X06	5.94	5.48	0.007 *	10,042.500
X07	6.49	6.15	0.040 *	10,724.000

* Statistical significance at $p < 0.05$.

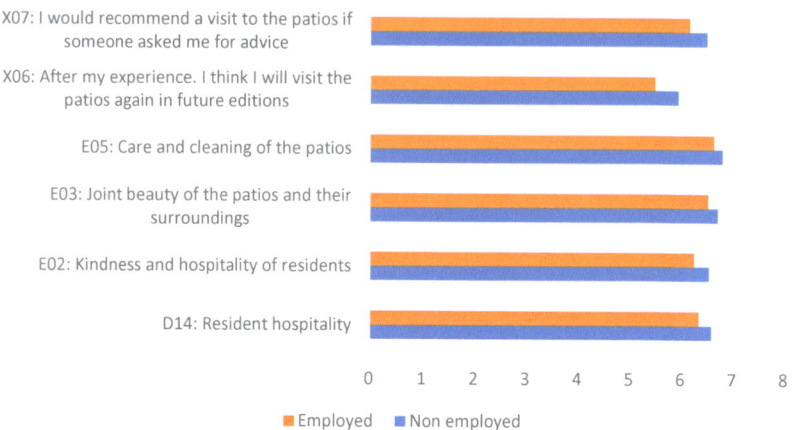

Figure 3. Significant differences by employment status.

In summary, it can be said that unemployed people compared to employed people have a more positive perception in terms of the hospitality and friendliness of the residents, the beauty of the patios and the care and cleanliness, as well as the highest intention to revisit and recommend. These results can be very useful to understand how employment status influences the tourism experience and perception.

Table 8 shows the questions considered relevant in this study, presenting the relationship with the items of the sociodemographic profile in which significant statistical differences were observed between the groups that make up each of the items. One aspect to note is that the determining variable in a greater number of questions is the educational level (ATR), although the gender variable (GEN) is also a determining variable for a large number of questions. In contrast, monthly household income (INC) does not appear in any of the questions, as there are no significant differences in any of the ranges considered. It should also be noted that the daily expenditure variable (DSP), although it does appear as a variable that shows significant differences in some questions, is only relevant in the section on the evaluation of the Fiesta de los Patios event, and in the section on the lived experience, specifically in the intention to revisit.

Table 8. Questions presenting significant statistical differences by sociodemographic profile items.

Code	Question	Mean
DESTINATION		
D01	The historic and monumental center	GEN, AGE
D02	The conservation of monumental and artistic heritage	GEN
D03	The beauty of the city	GEN
D05	Tourist information	ATR
D06	Attention and quality of tourist accommodation	GEN, ATR
D08	Attention and quality of the tour guides	ATR
D09	Diversity and quality of local gastronomy	ATR
D10	Opportunity to purchase handicrafts and traditional food items	AGE
D12	Citizen security	GEN
D14	Resident hospitality	EMP
D15	Public transportation services	GEN, ATR

Table 8. *Cont.*

Code	Question	Mean
FIESTA DE LOS PATIOS EVENT		
E01	Accessibility to patios and surrounding spaces	ATR
E02	Kindness and hospitality of residents	AGE, ATR, EMP
E03	Joint beauty of the patios and their surroundings	EMP
E05	Care and cleaning of the patios	GEN, EMP
E06	Availability of restaurants and bars in the surrounding area	DSP
E08	Tourist information and signage points	GEN
E09	Opportunity to make purchases of interest: crafts, etc.	ATR
E10	Security during the visit	DSP
E11	Waiting time to start the visit	ATR
PERCEPTION		
P01	During the visit to the Patios, I felt part of the cultural and heritage of the city of Cordoba	ATR
P02	The visit to the Patios has especially moved me	GEN, ATR
P04	The visit to the Patios has helped me relax	ATR
EXPERIENCE		
X01	My level of satisfaction with the Patios has been very important	ATR
X06	After my experience, I think I will visit the Patios again in future editions	GEN, DSP, ATR, EMP
X07	I would recommend a visit to the Patios if someone asked me for advice	GEN, ATR, EMP

Note. GEN (gender), INC (income), AGE (age), DSP (daily spend), ATR (academic training), EMP (employment status).

This summary provides an overview of the sections in which significant differences have been found in the perception of the tourist experience according to different sociodemographic characteristics. This will be of great use when developing tourism improvement strategies.

5. Discussion

The results obtained in this study shown in Table 9, in which they are related to the review of the existing literature, showed regarding the sociodemographic profile reinforce the greater presence of women in destinations with great heritage and cultural richness. This aspect is in line with research such as that carried out by [15,49]. However, this is an aspect on which, as detailed in the literature review, there is no clear consensus, with a large group of authors defending the greater presence of men in this type of destination. The profile of young tourists is also in line with the research of [55]. Regarding the level of education, this research corroborates that visitors to places with the WHS or ICH inscription by UNESCO are mainly people with a university degree, which is also affirmed in much of the existing research in this field [50,52].

One of the aspects highlighted in this research is the greater critical nature of younger people and those with university studies, which confirms what has already been stated in other research [66].

In relation to the importance of emotions in shaping the tourist experience, this research is in line with the research carried out by the authors [65]. Among the questions asked to the respondents, the score that stands out is that awarded to the emotion felt during the visit, which reinforces the literature in this field, since different studies point out the crucial importance of the emotions felt during the visit to places with cultural and heritage wealth in the creation of unique experiences [65]. Aspects such as pleasure, disconnection or excitement are feelings that are frequently associated with tourism [67].

Their importance is such that they are often considered indicators of satisfaction during the visit and intention to visit in the future [68].

Table 9. Discussion.

SD Profile	Literature Review		This Study	
	References	Difference	Findings	Match
Age	[49–52,64]	There is no consensus. Wide range: 21–45 years old or over 45, according to the study	Age between 18 and 35 years	Yes, with the Refs. [51,64]
Gender	[49–52]	There is no consensus. Preference for men or women according to the study	Parity in the presence of men and women, with a slightly higher presence of women	Yes
Education	[49,50,53]	Consensus on a high educational level	High educational level	Yes
Income	[51,52]	Consensus on medium or upper-middle income	Medium-high income level	Yes

The causality between emotions and felt satisfaction is latent in this research, since after expressing the felt emotions, they show high satisfaction with the event and loyalty towards it, manifested in the form of a recommendation and intention to visit in the future.

Regarding gender differences, one of the main results obtained has been the greater emotional connection of women with the destination, which is in line with research such as that carried out by [69], who highlights the fact that women are more emotional in tourist destinations. This greater emotional connection of women can be explained with previous research that supports the idea that women show and recognize feelings more easily and that they feel emotionally closer to their environment, a difference from the male gender, which is a more rational analysis [70].

Likewise, in the results, it can be seen how women, in addition to feeling more emotionally connected to the event, show greater loyalty towards it than men, both in terms of revisits and recommendations. These two aspects are linked by various research, which highlights the fact that the emotion transmitted by the place, brand or destination has positive effects on the loyalty created [71].

6. Conclusions
6.1. Theoretical Conclusions

Cultural heritage and specifically the places inscribed by UNESCO as locations of Intangible Heritage of Humanity are a key tourist resource for the destinations that obtain this accolade. Not only does it entail its protection, but this registration implies an increase in tourist flows [15], which is demonstrated in the event of the Fiesta de los Patios (Córdoba).

The research presented highlights the results that should be taken into account, since after having related the different questions to the items of the sociodemographic profile, it can be revealed that the determining variable in a great number of questions is educational level, although gender is also a determining variable in a large number of questions. On the other hand, with respect to monthly household income, no significant differences were found in any of the ranges considered. It should also be noted that the daily expenditure variable, although it does appear as a variable that shows significant differences in some questions, is only relevant in the section on the evaluation of the Fiesta de los Patios event and in the section on the experience, specifically in the intention to revisit. As far as gender is concerned, women have shown a greater sensitivity towards artistic preservation and conservation, valuing aspects such as security and attention in the accommodation more highly. Likewise, women show a greater emotional connection with the visit. Regarding the age variable, it is striking that young people between 18 and 35 show a more critical attitude towards aspects such as the friendliness and hospitality of the residents and the appreciation of the historic and monumental center. In addition, they are more appreciative of the opportunity to buy traditional crafts. Likewise, people without a university education are more positive about waiting times and accessibility, while university graduates are more critical. In relation to the sociodemographic profile, it could be concluded that the profile of the tourist who comes to this UNESCO's ICH event corresponds to that of a young woman with a university degree, who is employed and has a medium-high income.

6.2. Practical Implication

The results obtained are very useful for local authorities and organizers of this event, as they are the key to tailoring tourism strategies to the needs and preferences of different demographic groups, thus improving overall tourist satisfaction. Knowing these data, measures can be taken to increase the satisfaction of those sociodemographic groups whose scores are lower.

Knowing the sociodemographic profile of assistants has many uses: Firstly, it allows to design tourism policies and offer services adapted to the requirements of potential clients. Secondly, it serves to implement continuous improvements to increase satisfaction and loyalty towards the destination, and thirdly, it gives the possibility of creating tourism strategies aimed at attracting people whose profile is different from those who usually visit the city, thus guaranteeing the attraction of tourists with different sociodemographic profiles. That is, knowing the sociodemographic profile allows it to be used as a basis for tourism promotion campaigns, contributing to their effectiveness.

Likewise, the tourist position in which this event finds itself after its declaration and which has been reflected in the research makes it necessary to promote responsible and sustainable tourism policies, to avoid the overcrowding of the event. Examples of these could be encouraging visits to lesser-known patios further away from the center, encouraging attendance at times other than peak hours, or establishing entrances to limit mass influx, since currently it is a free event.

This study constitutes a post-event evaluation that provides data on the lived experience, satisfaction and behavioral intention, so that its comparison with previous studies makes continuous improvement possible.

6.3. Limitations

In relation to the limitations offered by the results of this study, it is necessary to consider the following aspects. The study is based on convenience sampling, so the results may not be representative of the entire population of visitors, in addition to offering a bias in relation to the people who were available and willing to participate at the time of the collection. On the other hand, the period to carry out the field work was limited to the days that the event lasted. Furthermore, providing the questionnaire to tourists while they wait to access a patio is a reflection of the long waiting times to begin the visit, something that has been reflected in the results.

6.4. Future Lines of Research

The following are suggested as future lines of research:

(1) Carry out an analysis of the different dimensions of perception and attributes, which will serve as a basis for segmenting tourists through a cluster analysis.
(2) Apply predictive methodologies, such as structural equation systems based on PLS-SEM and multi-layer perceptron-type artificial neural networks (ANN).
(3) Expand the spectrum of the analysis to the perception by the residents of the city of Córdoba at the level of heritage conservation, economic, social, environmental and identity of the city's residents with the Intangible Heritage.
(4) Address, from a broader perspective, the study of the event and its relationship with heritage conservation, through qualitative interviews and focus group methodology, collecting the opinions of local officials, as well as the owners of the patios and businesses of the neighborhoods where the event is held.
(5) Conduct research in future editions to identify possible changes in the behavior of tourists and their perceptions.
(6) Connect the results obtained with other research on similar events.

6.5. Main Contribution of the Study

The main contribution of this study lies in the detailed analysis of the sociodemographic profile of the visitors to the Fiesta de los Patios. Enriching the academic literature in this sense provides a clear line of action to promote public–private collaboration between local authorities responsible for the management and conservation of intangible heritage, hand in hand with companies in charge of attracting tourists to the destination. Having identified this target audience, their perceptions and satisfaction, therefore, facilitates the optimization of resources and contributes to sustainable management.

Author Contributions: Conceptualization, J.E.R.-R. and M.A.-R.; methodology, L.G.-G. and L.C.-P.; software, J.E.R.-R. and L.G.-G.; validation, L.C.-P. and L.G.-G.; formal analysis, M.A.-R. and L.C.-P.; investigation, J.E.R.-R., L.C.-P., M.A.-R. and L.G.-G.; resources, L.G.-G. and J.E.R.-R.; data curation, L.C.-P. and L.G.-G.; writing—original draft preparation, J.E.R.-R. and M.A.-R.; writing—review and editing, J.E.R.-R., M.A.-R., L.G.-G. and L.C.-P.; visualization, L.G.-G.; supervision, L.C.-P. All authors have read and agreed to the published version of the manuscript.

Funding: This research received no external funding.

Data Availability Statement: Data are contained within the article.

Conflicts of Interest: The authors declare no conflicts of interest.

References

1. Leite, F.C.D.L.; Ruiz, T.C.D. O turismo cultural como desenvolvimento da atividade turística: O caso de Ribeirão da Ilha-Florianópolis (SC). In *VII Fórum Internacional de Turismo do Iguassu*; Foz do Iguaçu: Paraná, Brazil, 2013.
2. Prada-Trigo, J.; Pérez Gálvez, J.C.; López-Guzmán, T.; Pesantez, S. Tourism and motivation in cultural destinations: Towards those visitors attracted by intangible heritage. *Almatourism-J. Tour. Cult. Territ. Dev.* **2016**, *7*, 17–37. [CrossRef]
3. Organización de la Naciones Unidas para la Educación, la Ciencia y la Cultura. *Convention for the Safeguarding of the Intangible Cultural Heritage*, 7th ed.; Organización de la Naciones Unidas para la Educación, la Ciencia y la Cultura: Paris, France, 2012.
4. Qiu, Q.; Zuo, Y.; Zhang, M. Intangible cultural heritage in tourism: Research review and investigation of future agenda. *Land* **2022**, *11*, 139. [CrossRef]
5. Tortul, M.; Elías, S.; Leonardi, V. World heritage sites, international tourism demand and tourist specialization in Latin America and the Caribbean 1995–2019. *JTA* **2022**, *29*, 72–100. [CrossRef]
6. López-Guzman, T.; González Santa-Cruz, F. International tourism and the UNESCO category of intangible cultural heritage. *Int. J. Cult. Tour. Hosp. Res.* **2016**, *10*, 310–322. [CrossRef]
7. Chen, J.; Guo, Z.; Xu, S.; Law, R.; Liao, C.; He, W.; Zhang, M. A Bibliometric Analysis of Research on Intangible Cultural Heritage Tourism Using CiteSpace: The Perspective of China. *Land* **2022**, *11*, 2298. [CrossRef]
8. Qiu, Q.; Liang, X.; Zuo, Y. Identifying European and Chinese styles of creating tourist destinations with intangible cultural heritage: A comparative perspective. *Int. J. Tour. Res.* **2023**, *25*, 266–278. [CrossRef]
9. Kim, S.; Whitford, M.; Arcodia, C. Development of intangible cultural heritage as a sustainable tourism resource: The intangible cultural heritage practitioners' perspectives. In *Authenticity and Authentication of Heritage*; Routledge: London, UK, 2021; pp. 34–47. [CrossRef]

10. Albertoni, P. Mercantilización y autenticidad en la frontera uruguayo-brasileña: El portuñol en el siglo XXI. *Trab. Linguística Apl.* **2021**, *60*, 410–424. [CrossRef]
11. UNWTO. *Tourism and Intangible Cultural Heritage*; Servicio de publicaciones de la Organización Mundial del Turismo: Madrid, Spain, 2012.
12. López-Guzmán, T.; González Santa Cruz, F. The Fiesta of the Patios: Intangible cultural heritage and tourism in Cordoba, Spain. *Int. J. Intang. Herit.* **2016**, *11*, 182–192.
13. Organización de las Naciones Unidas para la Educación, la Ciencia y la Cultura. Convention Concerning the Protection of the World Cultural and Natural Heritage. In Proceedings of the World Heritage Committee, Eighth Ordinary Session, Buenos Aires, Argentina, 29 October–2 November 1984; UNESCO. Available online: https://whc.unesco.org/archive/1984/sc-84-conf004-inf3e.pdf (accessed on 15 June 2024).
14. Organización Mundial del Turismo. *Definiciones de Turismo de la OMT*; OMT: Madrid, Spain, 2019. Available online: https://www.e-unwto.org/doi/book/10.18111/9789284420858 (accessed on 15 June 2024).
15. Nguyen, T.H.H.; Cheung, C. The classification of heritage visitors: A case of Hue City, Vietnam. *J. Herit. Tour.* **2014**, *9*, 35–50. [CrossRef]
16. De Santana, J.C.; Maracajá, K.F.B.; de Araújo Machado, P. Turismo cultural y sostenibilidad turística: Mapeo del desempeño científico desde Web of Science. *Tur. Soc.* **2021**, *28*, 95–113.
17. Manjavacas Ruiz, J.M. Patrimonio cultural y actividades turísticas. Aproximación crítica a propósito de la Fiesta de Patios de Cordoba. *Rev. Andal. Antropol.* **2018**, *15*, 127–155. [CrossRef]
18. Adie, B.A. Franchising our heritage: The UNESCO World Heritage brand. *Tour. Manag. Perspect.* **2017**, *24*, 48–53. [CrossRef]
19. Richards, G. Cultural tourism: A review of recent research and trends. *J. Hosp. Tour. Manag.* **2018**, *36*, 12–21. [CrossRef]
20. Lew, A.A. A framework of tourist attraction research. *Ann. Tour. Res.* **1987**, *14*, 553–575. [CrossRef]
21. Chahal, H.; Devi, A. Destination Attributes and Destination Image Relationship in Volatile Tourist Destination: Role of Perceived Risk. *Metamorphosis* **2015**, *14*, 1–19. [CrossRef]
22. Sangpikul, A. The effects of travel experience dimensions on tourist satisfaction and destination loyalty: The case of an island destination. *Int. J. Cult. Tour. Hosp. Res.* **2018**, *12*, 106–123. [CrossRef]
23. McCartney, G.; Butler, R.; Bennett, M. Positive tourism image perceptions attract travellers—Fact or fiction? The case of Beijing visitors to Macao. *J. Vacat. Mark.* **2009**, *15*, 179–193. [CrossRef]
24. Moon, H.; Han, H. Destination attributes influencing Chinese travelers' perceptions of experience quality and intentions for island tourism: A case of Jeju Island. *Tour. Manag. Perspect.* **2018**, *28*, 71–82. [CrossRef]
25. Valduga, M.C.; Breda, Z.; Costa, C.M. Perceptions of blended destination image: The case of Rio de Janeiro and Brazil. *J. Hosp. Tour. Insights* **2019**, *3*, 75–93. [CrossRef]
26. Biswas, C.; Deb, S.K.; Hasan, A.A.T.; Khandakar, M.S.A. Mediating effect of tourists' emotional involvement on the relationship between destination attributes and tourist satisfaction. *J. Hosp. Tour. Insights* **2021**, *4*, 490–510. [CrossRef]
27. Hall, C.M.; Prayag, G.; Amore, A. *Tourism and Resilience: Individual, Organisational and Destination Perspectives*; Channel View Publications: St Nicholas House, UK, 2017.
28. Lai, M.T.H.; Yeung, E.; Leung, R. Understanding tourists' policing attitudes and travel intentions towards a destination during an ongoing social movement. *J. Hosp. Tour. Insights* **2022**, *6*, 874–891. [CrossRef]
29. Kim, J.H. The antecedents of memorable tourism experiences: The development of a scale to measure the destination attributes associated with memorable experiences. *Tour. Manag.* **2014**, *44*, 34–45. [CrossRef]
30. Dabphet, S. Applying importance-performance analysis to identify competitive travel attributes: An application to regional destination image in Thailand. *J. Community Dev. Res. (Humanit. Soc. Sci.)* **2017**, *10*, 7–21.
31. Bowling, A.; Rowe, G.; McKee, M. Patients' experiences of their healthcare in relation to their expectations and satisfaction: A population survey. *J. R. Soc. Med.* **2013**, *106*, 143–149. [CrossRef]
32. Del Bosque, I.R.; San Martín, H. Tourist satisfaction a cognitive-affective model. *Ann. Tour. Res.* **2008**, *35*, 551–573. [CrossRef]
33. Olivera, A. Patrimonio inmaterial, recurso turístico y espíritu de los territorios. *Cuad. Tur.* **2011**, *27*, 663–677.
34. Carvache Franco, W.; Torres Naranjo, M.; Carvache Franco, M. Análisis del perfil y satisfacción del turista que visita montañita–Ecuador. *Cuad. Tur.* **2017**, *39*, 113–129. [CrossRef]
35. Noyola Aguilar, L.Y.; Campón-Cerro, A.M. La percepción de la autenticidad del destino cultural y su relación con la satisfacción y lealtad. *ROTUR. Rev. Ocio Tur.* **2016**, *11*, 65–76. [CrossRef]
36. Shahrivar, R.B. Factors that influence tourist satisfaction. *J. Travel Tour. Res. (Online)* **2012**, *12*, 61–78.
37. Zhao, Y.; Zhan, Q.; Du, G.; Wei, Y. The effects of involvement, authenticity, and destination image on tourist satisfaction in the context of Chinese ancient village tourism. *J. Hosp. Tour. Manag.* **2024**, *60*, 51–62. [CrossRef]
38. Romao, J.; Neuts, B.; Nijkamp, P.; Van Leeuwen, E. Culture, product differentiation and market segmentation: A structural analysis of the motivation and satisfaction of tourists in Amsterdam. *Tour. Econ.* **2015**, *21*, 455–474. [CrossRef]
39. Battour, M.; Battor, M.; Bhatti, M.A. Islamic attributes of destination: Construct development and measurement validation, and their impact on tourist satisfaction. *Int. J. Tour. Res.* **2014**, *16*, 556–564. [CrossRef]

40. Nasir, M.N.M.; Mohamad, M.; Ghani, N.; Afthanorhan, A. Testing mediation roles of place attachment and tourist satisfaction on destination attractiveness and destination loyalty relationship using phantom approach. *Manag. Sci. Lett.* **2020**, *10*, 443–454. [CrossRef]
41. Davis, F.D. Perceived usefulness, perceived ease of use, and user acceptance of information technology. *MIS Q.* **1989**, *13*, 319–340. [CrossRef]
42. Fu, H.; Ye, B.H.; Xiang, J. Reality TV, audience travel intentions, and destination image. *Tour. Manag.* **2016**, *55*, 37–48. [CrossRef]
43. Tajeddini, K.; Gamage, T.C.; Hameed, W.U.; Qumsieh-Mussalam, G.; Chaijani, M.H.; Rasoolimanesh, S.M.; Kallmuenzer, A. How self-gratification and social values shape revisit intention and customer loyalty of Airbnb customers. *Int. J. Hosp. Manag.* **2022**, *100*, 103093. [CrossRef]
44. Arteaga Sánchez, R.; López, F.M.; García Ordaz, M.; Sánchez-Franco, M.J.; Yousafzai, S.Y. Adoption of online social networks to communicate with financial institutions. *J. Promot. Manag.* **2017**, *23*, 228–257. [CrossRef]
45. Seetanah, B.; Teeroovengadum, V.; Nunkoo, R.S. Destination satisfaction and revisit intention of tourists: Does the quality of airport services matter? *J. Hosp. Tour. Res.* **2020**, *44*, 134–148. [CrossRef]
46. Afshardoost, M.; Eshaghi, M.S. Destination image and tourist behavioural intentions: A meta-analysis. *Tour. Manag.* **2020**, *81*, 104154. [CrossRef]
47. Menor-Campos, A.; Fuentes Jiménez, P.A.; Romero-Montoya, M.E.; López-Guzmán, T. Segmentation and sociodemographic profile of heritage tourist. *Tour. Hosp. Manag.* **2020**, *26*, 115–132. [CrossRef]
48. Valverde-Roda, J.; Gómez-Casero, G.; Medina-Viruel, M.J.; López-Guzmán, T. *Categorización del Turista Patrimonial. El Caso de Granada (España)*; Fundación Dialnet: Logroño, Spain, 2021.
49. Remoaldo, P.C.; Vareiro, L.; Ribeiro, J.C.; Santos, J.F. Does gender affect visiting a World Heritage Site? *Visit. Stud.* **2014**, *17*, 89–106. [CrossRef]
50. Antón, C.; Camarero, C.; Laguna-García, M. Towards a new approach of destination royalty drivers: Satisfaction, visit intensity and tourist motivation. *Curr. Issues Tour.* **2017**, *20*, 238–260. [CrossRef]
51. Chen, G.; Huang, S. Understanding Chinese cultural tourists: Typology and profile. *J. Travel Tour. Mark.* **2018**, *35*, 162–177. [CrossRef]
52. Ramires, A.; Brandao, F.; Sousa, A.C. Motivation-based cluster analysis of international tourists visiting a World Heritage City: The case of Porto, Portugal. *J. Destin. Mark. Manag.* **2018**, *8*, 49–60. [CrossRef]
53. Moreira- Gregori, P.E.; Martín, J.C.; Oyarce, F.; Moreno-García, R. Turismo y patrimonio. El caso de Valparaíso (Chile) y el perfil del turista cultural. *PASOS. Rev. Tur. Patrim. Cult.* **2019**, *17*, 901–914. [CrossRef]
54. McKercher, B. Towards a classification of cultural tourists. *Int. J. Tour. Res.* **2002**, *4*, 29–32. [CrossRef]
55. Correia, A.; Kozak, M.; Ferradeira, J. From tourist motivations to tourist satisfaction. *Int. J. Culture. Tour. Hosp. Res.* **2013**, *7*, 411–424. [CrossRef]
56. Hair, H.J.F.; Anderson, R.E.; Tatham, R.L.; Black, W.C. *Análisis Multivariante*; Prentice Hall: Madrid, Spain, 1999.
57. Chen, H.; Rahman, I. Cultural tourism: An analysis of engagement, cultural contact, memorable tourism experience and destination loyalty. *Tour. Manag. Perspect.* **2018**, *26*, 153–163. [CrossRef]
58. Moore, Z.; Harrison, D.E.; Hair, J. Data Quality Assurance Begins Before Data Collection and Never Ends: What Marketing Researchers Absolutely Need to Remember. *Int. J. Mark. Res.* **2021**, *63*, 693–714. [CrossRef]
59. Finn, M.; Elliott-White, M.; Walton, M. *Tourism and Leisure Research Methods: Data Collection, Analysis and Interpretation*; Pearson Education: Harlow, UK, 2000; ISBN 9780582368712.
60. Kolmogorov, A. Determinazione empirica di una lgge di distribuzione. *Inst. Ital. Attuari Giorn.* **1952**, *4*, 83–91.
61. Smirnov, N. Table for estimating the goodness of fit of empirical distributions. *Ann. Math. Stat.* **1948**, *19*, 279–281. [CrossRef]
62. Shapiro, S.S.; Wilk, M.B. An analysis of variance test for normality (complete samples). *Biometrika* **1965**, *52*, 591–611. [CrossRef]
63. Mann, H.B.; Whitney, D.R. On a test of whether one of two random variables is stochastically larger than the other. *Ann. Math. Stat.* **1947**, *18*, 50–60. [CrossRef]
64. Kruskal, W.H.; Wallis, W.A. Use of ranks in one-criterion variance analysis. *J. Am. Stat. Assoc.* **1952**, *47*, 583–621. [CrossRef]
65. Phan, T.T.H.; Tran, H.L.; Dao, M.N.; Trang, D.T.M. Tourist loyalty and intangible cultural heritage: Casestudy in Danang, Vietnam. *Ann. For. Res.* **2023**, *66*, 234–256.
66. Solano-Sánchez, M.Á.; Arteaga-Sánchez, R.; Castaño-Prieto, L.; López-Guzmán, T. Does the Tourist's Profile Matter? Perceptions and Opinions about the Fiesta De Los Patios in Cordoba, Spain. *Enlightening Tour. A Pathmaking J.* **2022**, *12*, 436–469. [CrossRef]
67. Kim, J.; Fesenmaier, D.R. Measuring emotions in real time: Implications for tourism experience design. *J. Travel Res.* **2015**, *54*, 419–429. [CrossRef]
68. Hosany, S.; Prayag, G.; Van Der Veen, R.; Huang, S.; Deesilatham, S. Mediating effects of place attachment and satisfaction on the relationship between tourists' emotions and intention to recommend. *J. Travel Res.* **2017**, *56*, 1079–1093. [CrossRef]
69. Eagly, A.H. *Sex Differences in Social Behavior: A Social-Role Interpretation*; Lawrence Erlbaum: Hillsdale, NJ, USA, 2013.

70. Okazaki, S.; Hirose, M. Does gender affect media choice in travel information search? On the use of mobile Internet. *Tour. Manag.* **2009**, *30*, 794–804. [CrossRef]
71. Karim, K.; Ilyas, G.B.; Umar, Z.A.; Tajibu, M.J.; Junaidi, J. Consumers' awareness and loyalty in Indonesia banking sector: Does emotional bonding effect matters? *J. Islam. Mark.* **2023**, *14*, 2668–2686. [CrossRef]

Disclaimer/Publisher's Note: The statements, opinions and data contained in all publications are solely those of the individual author(s) and contributor(s) and not of MDPI and/or the editor(s). MDPI and/or the editor(s) disclaim responsibility for any injury to people or property resulting from any ideas, methods, instructions or products referred to in the content.

Article

Economic Contribution, Characterization, and Motivations of Tourists: The Raymi Llaqta in Peru

Franklin Omar Zavaleta Chavez Arroyo [1], Alex Javier Sánchez Pantaleón [1], Milena Leticia Weepiu Samekash [1], Jhunniors Puscan Visalot [1] and Rosse Marie Esparza-Huamanchumo [2,*]

[1] Research Institute of Economics and Development, Universidad Nacional Toribio Rodríguez de Mendoza de Amazonas, Chachapoyas 01000, Peru; franklin.zavaleta@untrm.edu.pe (F.O.Z.C.A.); alex.sanchez@untrm.edu.pe (A.J.S.P.); milena.weepiu@untrm.edu.pe (M.L.W.S.); jhunniors.puscan.epg@untrm.edu.pe (J.P.V.)

[2] Faculty of Hotel Administration, Tourism and Gastronomy, Universidad San Ignacio de Loyola, Lima 15024, Peru

* Correspondence: resparza@usil.edu.pe

Abstract: This study assesses the economic contribution and motivations of tourists attending the Raymi Llaqta festival in Chachapoyas, Peru. This study used an econometric analysis based on the application of two types of regression models: non-zero truncated Poisson regression and zero-truncated negative binomial regression. Data were collected through face-to-face structured interviews with domestic and foreign tourists who visited Chachapoyas during the festival. Results indicate significant spending on accommodation, food, transportation, and activities. Tourist satisfaction averaged 3.7, with notable appreciation for the festival's variety, authenticity, local hospitality, and safety. While both foreign and domestic tourists expressed positive views on the cultural representation and organization of the event, they suggested improvements in hygiene and promotional efforts. Despite using a structured questionnaire, response bias could affect the accuracy of self-reported experiences. This research provides valuable insights into the festival's direct economic impact on the local economy and highlights the importance of high tourist satisfaction for effective marketing strategies.

Keywords: cultural tourism; economic contribution; Raymi Llaqta; tourist profile; visitor satisfaction

Citation: Zavaleta Chavez Arroyo, F.O.; Sánchez Pantaleón, A.J.; Weepiu Samekash, M.L.; Puscan Visalot, J.; Esparza-Huamanchumo, R.M. Economic Contribution, Characterization, and Motivations of Tourists: The Raymi Llaqta in Peru. *Heritage* **2024**, *7*, 6243–6256. https://doi.org/10.3390/heritage7110293

Academic Editors: Fátima Matos Silva and Isabel Vaz de Freitas

Received: 25 September 2024
Revised: 29 October 2024
Accepted: 30 October 2024
Published: 5 November 2024

Copyright: © 2024 by the authors. Licensee MDPI, Basel, Switzerland. This article is an open access article distributed under the terms and conditions of the Creative Commons Attribution (CC BY) license (https://creativecommons.org/licenses/by/4.0/).

1. Introduction

Tourism is an activity that emphasizes the social (values, individual behaviors, family relationships, collective lifestyles, and moral rules), cultural (artistic expressions and traditional ceremonies), economic (trade and food and hospitality services), and environmental (commitment to biodiversity) aspects of all the parties [1]. This largely depends on the human element, which allows for expansion, creation, and innovation within its scope [2]. Thus, focusing on a tourism market could increase the number of visitors, tourist information could be provided, and the destination's identity could be promoted. This could be linked to sustainability and the search for more authentic and meaningful tourist experiences [3], directing it to the potential for innovation and allowing active participation in the spaces and resources that the destination has to offer [4].

Tourism has been defined as a multidimensional, dynamic, and contextualized process that offers people, both individually and collectively, opportunities to improve their life quality through their participation in society [5]. However, the "tourist" concept can be used to describe and possibly explain a type of human behavior that is categorically different from that of a traveler or a migrant (or any other type of person on the move). This behavior can be directly measured by recording the purposes, experiences, and meanings that people associate with their travels [6], thereby underlining the need to question the ideology of traditional economic growth, which is often deeply rooted in economic rather than social well-being [7], reflected in both growth and economic development (economic and social well-being) indicators. Tourism tends to perform poorly in terms of innovation, and there is little evidence to suggest that the policy initiatives promoted have achieved significant success [8]. Thus, by moving toward an innovative approach, regenerative tourism emerges as a guide to "reboot tourism", challenging traditional and binary economic models by integrating living systems' ideas with ecological perspectives [9]. The tourism industry is an important part of the global economy, representing the services of a diversified industry that generates a multiplier effect on activities in several economic sectors [10], and its nature is extremely complex due to its close relation with social, political, and cultural factors [11]. Thus, this sector favors boosting regional economic vitality and improving life quality in many neglected areas [12]. It promotes the development of innovation through behavioral perspectives, breadth, and valuable knowledge [13], thereby showing a significant increase in regular production within the tourism sector [14].

Tourism, one of the most important economic sectors in the world, has a positive impact on the demand of related sectors. However, these economic benefits often carry with them negative externalities [15]. This is the case of experiential tourism, which brings a large number of tourists to these destinations, making them an attractive economic booster. However, the negative externalities of the activities, such as water and air pollution, overcrowding, and increased stress on local communities and ecosystems, have cast doubts about the industry's practices in these destinations [16] as the sustainability of the destinations is seriously challenged. The tourism and hospitality industry are affected by several internal and external variables, such as natural disasters and crises [17]. Currently, integrating tourism into recovery initiatives marks a controversial shift in industry priorities that risks perpetuating instrumentalist views of nature that suggest positive change [18] by considering making the design and management of destinations more sustainable, resilient, and inclusive [19].

In the field of social development, the tourism industry also contributes to alleviating socioeconomic problems like unemployment. According to the World Economic Forum report, the tourism industry generated more than 319 million jobs in 2018 [20], and mitigates inequality and poverty by providing opportunities and social value at the local level [21]. To achieve sustainable tourism, it is necessary to strike a balance between three dimensions, namely, environmental, economic, and socio-cultural, to ensure the short- and long-term sustainability of the tourism sector in the light of climate change [22]. However, given the clear growth and development of tourism, the demand for energy, whether electricity or fuel, is increasing due to the rise in the number of visitors, thereby increasing mobility, infrastructure, consumption, and so on [23]. Therefore, management actions toward sustainable tourism must be ensured, focusing on specific impacts [24].

At the beginning of the 21st century, specifically in 2020, the COVID-19 pandemic caused great disruptions to the flow of currency brought by tourism. In 2019, the tourism sector accounted for 10.6% of global employment and 10.4% of global GDP [25], and the imposition of travel bans resulted in swapping foreign destinations for domestic ones, which led to a boom in domestic tourism [26]. In both sending and receiving communities and/or destinations, rural tourism has the potential to not only improve the well-being of local people, but also alter livelihoods and cultural traditions [27], as well as partially or substantially affecting the ecological environment through soil compaction, erosion, and disturbance of animal habitat [28]. Thus, challenges to tourism management have been identified along with traditional government strategies for dealing with such challenges [29]. Chachapoyas, one of the seven Amazonian provinces in Peru, has several characteristics that make it one of the most important tourist destinations [30] due to its architecturally beautiful plazas and cathedrals of the bygone era, the Gilberto Tenorio Ruiz Exhibition Hall, the Pozo de Yanayacu, the Luya Urco Viewpoint, and the Temple of the Señora de la Buena Muerte, and because it is a gateway to other major sites such as the Kuelap archaeological site, the Gocta Waterfall, the Quiocta Caves, and the Mausoleums of Revash, among many other attractions.

In 1996, the Sub-Regional Directorate of Industry, Tourism, Integration, and International Trade Negotiations promoted creating the Chachapoyas Tourism Week, under the direction of Professor Luis Herrera Castro. This event is held annually from 1–7 June, enhancing the commemoration of the anniversary of the Battle of Higos Urco, strengthening cultural identity, and positioning it in the calendar of national festivities [31].

In 1997, an event called "Raymi Llaqta de los Chachapoya" (Great Festival of the People of Chachapoyas) was created for recovering the cultural heritage of the entire department. This folkloric and traditional celebration gathers between 35 and 40 communities from the department of the Amazonas, bringing together approximately 2500 people. Around the main square and along the main streets of Chachapoyas, the capital of the department of Amazonas, these communities present their millenary traditions to the world. Law No. 27,425 of 16 February 2001 made the Raymi Llaqta an official ritual festival of national identity, to be held in the month of June as part of the Chachapoyas Tourism Week. Regional Ordinance No. 172—Amazonas Regional Government/CR of 17 April 2007 approved the Rules of Organization and Functions of the Raymi Llaqta of Chachapoya. Since then, the Regional Administration of Amazonas, through the Regional Department of Foreign Trade and Tourism, has been working on consolidating and promoting this event as a unique folk and traditional festival in the north and east of the country [32].

The main goal of this study was to determine the economic impact and analyze the characteristics of those visiting Raymi Llaqta. This study covers, among others, several aspects such as tourists' origin, average expenditure, reason for the trip, length of stay, and satisfaction. These data will provide a comprehensive view of the visitor profile and the economic impact of the event, which will allow for the implementation of more effective tourism development strategies in the Amazon region.

1.1. Literature Review

1.1.1. Economic Impact and Tourism

The perception of the impact generated by tourism activities on destinations is one of the most studied areas in the field of tourism [33]. Economic benefits are a key element in the exchange process of tourism development [34], and several studies refer to economic benefits as the most important element sought by local residents [35].

This study used the economic impact of tourism to describe its overall impact [36]. The economic impact of tourism is a part of the broader economic benefits of tourism and serves as a measure of social welfare in the tourism destination, which includes both market and non-market values [37]. Three key concepts evaluate the economic performance of tourism: economic growth, economic development, and economic impact. Growth refers to the long-term increase in national output; development encompasses economic, social, and environmental welfare; and impact describes the impact of tourism on regional and national economies in terms of quantity and quality.

It is necessary to promote economic growth models that consider the positive impact on employment and income. This will allow us to better understand how tourism contributes to decent work and the reduction of inequalities, in line with Sustainable Development Goals 8 and 10, respectively, and thus achieve socioeconomic sustainability in the tourism sector [38]. Similarly, global tourism demand and added value would increase by 106.4% in an appropriate international tourism scenario, if there were complete security and safety in all countries. If the level of uncertainty and insecurity in each country were reduced to a minimum, tourism demand and added value would increase by 14.3% [39].

COVID-19 was a major and unprecedented blow to tourism spending, which rippled through the global economy, severely impacting smaller ones. The global disruption of tourism was due to restrictions on international travel, travel bans, and a significant decline in tourism activity [40]. This situation has had a negative impact on destinations in the Amazon region, such as the Kuelap archaeological site, where the sector's demand was affected and, as a result, the provision of services by suppliers and employment decreased in the tourism sector associated with this site [41].

1.1.2. Profile of the Tourists Visiting Chachapoyas

Because nature-based experiences generate recreational income, shape human identity and traditions, and support conservation, it is important to recognize the differences in classifying tourism profiles and nature-based experiences. Ignoring the importance of nature-based experiences can result in the loss of cultural identity and heritage, environmental education, and enjoyment of nature [42]. The profile of tourists includes the main reason for their stay and the type of tourist impact they will have. Thus, it is possible to identify which profiles are more or less accepting of these measures and to determine where they are more feasible from a socio-environmental perspective [43].

Several approaches focused on analyzing tourist profiles have been developed to evaluate human activities and preferences for cultural and natural activities. Tourists are very satisfied with visiting touristic cities, emphasizing the relationship between customs, local gastronomy, and travelers' growing interest in culture [44]. In turn, this study focuses on the interpretation and description of the relationship between motivations to travel, tourist destinations, and tourist engagement [45].

Seeking to better understand demand through market segmentation analysis will be mutually beneficial to both the visitor and the party receiving the service. Ref. [46] identified five segments of potential tourists in rural tourism: four value nature, environment, and tranquility, two are interested in activities (outdoor and cultural activities for one and typical rural life activities for another), and a fifth, the smallest, motivated only by spending time with friends. To achieve this properly, it is essential to have an understanding of tourists' attitudes and knowledge, as well as their behavior, perceptions, and motivations [47].

2. Materials and Methods

To identify the characteristics of the sample, a descriptive analysis of the study variables was conducted first. The econometric analysis performed in this study was mainly based on the application of two types of regression models: zero-truncated Poisson regression [48,49] and zero-truncated negative binomial regression [50–52]. These models were chosen to study the impact of different variables in a tourism context, where Poisson regression was initially used due to its feature to model count data [51], but due to over-dispersion and the absence of zeros in the data, zero-truncated negative binomial regression was chosen to adequately handle the additional variability and specific structure of the data. This approach is preferable to ordinary least squares (OLS) because it assumes normality and homoscedasticity, which are not appropriate for count data that exhibit over-dispersion.

The data were gathered through a survey of tourists who visited the city of Chachapoyas for the Raymi Llaqta from 5–8 June 2024, considering tourists who stayed in the city for the night. A sample was formed of 186 tourists (134 nationals and 52 foreigners).

The instrument was developed by the authors and validated by an expert panel of three professionals in the fields of economics, management, and tourism. Moreover, the main dimensions and indicators were considered and are described below (Table 1).

Table 1. Dimensions and indicators.

Planning	Characteristics	Economy	Satisfaction
Length of stay	Means of transport and route	Number of people	City
Information	Organization	Total expenditure	Food
	Travel companion	Transport	Event
		Accommodation	People
		Food	
		Consumption	
		Others	

Source(s): Authors' own creation.

The reliability of the instrument showed a high Cronbach's alpha (0.9148). Subsequently, the data were tabulated and coded to be processed by applying descriptive statistics to indicators related to the research topic, together with econometric analyses developed in the R programming language version 4.3.2.

3. Results

Table 2 shows the average expenditure of foreign and domestic tourists during their stay for Raymi Llaqta. It was found that the highest expenditure was on food and beverages with an average of PEN 595.83 (USD 156), followed by accommodation and air transportation used to reach the destination from their countries; these items had the greatest impact on total budget.

Table 2. Average spending per tourist in soles (PEN) in the city of Chachapoyas during Raymi Llaqta (four days).

Indicator	Foreign (USD)	Peruvian (PEN)
Air transport	450.00	499.23
Ground transport	200.91	207.71
Fuel	0.00	234.00
Taxi	105.71	50.49
Accommodation	584.50	295.06
Food and drink	595.83	249.74
Visits	326.18	241.82
Nightlife	187.50	188.28
Handicrafts	187.50	92.96
Clothing	170.00	105.60
Typical foods	100.00	78.76
Medicines	75.00	57.20
Others	0.00	346.11
Total	**2983.14**	**2646.94**

Source(s): Authors' own creation.

On the other hand, it represents a contribution to the growth of the tourism sector and the local Gross Domestic Product (GDP), where the income generated by the festival strengthened the regional economy by stimulating the demand for local services, benefiting small and medium-sized businesses, and directly impacting the sectors involved, generating temporary employment and improving economic conditions in Chachapoyas and nearby areas. The presence of national and international tourists increases the visibility of the region, attracting future investment in tourism infrastructure, which ensures long-term sustainable growth.

Sightseeing and ground transportation were other significant expenses for foreign tourists, which highlights the importance of the costs associated with these indicators. Regarding domestic tourists, the highest expenses were for air transportation due to the high cost of airlines in the city, accommodation, food and beverages, tours, and fuel (if applicable), and are the main expenses that a national tourist should consider if they wish to participate in the Raymi Llaqta activities. Nighttime expenses were related to consumption in clubs and bars. In addition, concerning expenses for handicrafts and the use of cabs, foreign tourists spent more frequently and in greater quantities than Peruvian tourists.

Table 3 shows the social characteristics of tourists who participated in the Raymi Llaqta festival. Males (58.93%) represented the largest number of tourists, and recreation and vacation (59.14%) was the main reason for visiting. Of those considered tourists by this study, 64.23% came specifically for the Raymi Llaqta festival. Additionally, awareness of the event among tourists was recorded: 51.69% knew about the event. Most tourists who came to the city stayed mainly in the homes of family or friends, hostels, or 1-star or 2-star hotels (28.57%, 19.64%, and 19.64%, respectively).

Table 3. Social characteristics of tourists who participated in the Raymi Llaqta.

Variables	Indicator	Mean or %	Variables	Indicator	Mean or %
Gender	Male	58.93	Marital status	Married	25.6
	Female	41.07		Widower	17.86
Level of education	Complete Primary	4.17		Divorced	0.6
	Completed secondary school	16.07		Cohabitant	4.76
	Complete technical higher	26.79	Satisfaction	Quality of food and beverages offered at the event	3.8
	Complete university college	48.81		Representation and promotion of local culture	3.9
	Postgraduate/Master's Degree	2.98		Order of community presentations	3.9
	Doctorado/Phd	1.19		Total Event Duration	3.6
Visiting Time	First time	38.69		Variety of customs offered	4
	No first time	61.31		Originality and authenticity	4
Main reason	Vacation/Recreation	59.14		Kindness and attention	4
	Visiting Family/Friends	18.28		Opportunity to interact	3.8
	Bless you	1.08		Clarity and access to information	3.7
	Working Committee	3.23		Comfort when observing the event	3.1
	Business	1.61		Facilities to observe the event	3
	Work	9.68		Quality of the show presented	4
	Other	6.99		Sound during the event	3.7
Accompaniment	Alone (unaccompanied)	43.45		Availability of toilets	2.7
	With my partner	22.02		Event promotion and advertising	3.5
	Group of participants	1.79		Event Security	4
	In direct family group	16.67		General organization of the event	3.9
	With friends, no children	10.71		Accommodation	3.8
	With friends, with children	5.36		Feeding	4
Travel organization	Cuenta propia	89.88		Souvenirs or handicrafts	3.7
	I bought a tour package	4.17		Tours	3.6
	Travel with a group of participants	1.19		Guide service	3.6
	I hired a tourist agency in the place visited	4.17		Hospitality of the local people	4
	Other	0.6		Bachelor	51.19

Source(s): Authors' own creation.

The main motivation for participating was leisure or recreation (59.14%), although most of the attendees organized their trip independently (89.8%), and there was twice as many people who traveled accompanied or with a partner compared to those who travel alone (43.45% vs. 22%, respectively). In terms of satisfaction with the event on a scale of 1 to 5, positive elements such as the quality of the show, local hospitality, and originality were identified; however, points for improvement were focused on the availability of restrooms and comfort for observing the event.

Table 4 present the economic characteristics of tourists who participated in the Raymi Llaqta festival. It shows a varied distribution in the expenses and means of transportation used, where, on average, most of the tourists' budget was spent on food and beverages (22.8%) and lodging (18.36%). As for the means of transportation, 54.76% of the tourists used interprovincial buses, while a smaller percentage used air transportation (8.93%) or private vehicles (6.55%). The accommodation that tourists opt for in the event was largely the homes of relatives or friends (28.57%), followed by hotels and hostels (19.64%). As for spending per person, more than half spent on average for two people (53.19%); however, it was found that spending on purchases of handicrafts and clothing was relatively low (5.02% and 1.87%, respectively). Most of the tourists who came to the event made expenditures mainly related to food and beverages (restaurants and bars, 22.81%), lodging (18.36%), ground transportation (16.69%), and sightseeing (museums, archaeological sites, 13.83%).

Table 5 shows the cultural characteristics of tourists who participated in the Raymi Llaqta festival. The majority of tourists attended the event mainly for the festival (64.23%), and in terms of prior knowledge of the event, 51.69% of tourists knew what the event was before attending; social networks and websites were the main means by which tourists were informed about the event (43.1% and 31.03%, respectively), indicating the importance of digital platforms for the dissemination of information about the event.

Table 4. Economic characteristics of tourists who participated in the Raymi Llaqta festival.

Variables	Indicator	Mean or %	Variables	Indicator	Mean or %
Expenditure	Air transport	4.4	Means of transport	Airplane	8.93
	Ground transportation	16.69		Interprovincial bus	54.76
	Fuels and tolls	1.97		Home car	6.55
	Internal transfers	3.56		Particular mobility	2.38
	Accommodation	18.36		Mobility of a tourism agency	7.14
	Food and beverages	22.81		Other means of transport (specify)	20.24
	Sightseeing	13.83	Accommodation	Three-star hotel (tourist)	7.14
	Nightclubs	3.49		One- and two-star hotel (budget)	19.64
	Purchase of handicrafts	5.02		Hostel	19.64
	Purchase of clothing/footwear	1.87		Lodging	13.1
	Typical food or candy shopping	3.77		Housing of relatives, friends	28.57
	Purchase of medicines	0.73		Airbnb	0.6
	Other expenses	3.5		Rented house or apartment	1.79
Number of people	One	1.72		Own home	1.79
	Two	22.41		Hiker	0.6
	Three	25.86		Other	7.14
	Four	20.69	Spending for people	One	14.89
	My co	15.52		Two	53.19
	Six	6.9		Three	17.02
	Eight	1.72		Four	7.45
	Ten	1.72		Five	4.26
	Twelve	1.72		Six	2.13
	Twenty-seven	1.72		Twenty	1.06

Table 5. Cultural characteristics of tourists who participated in the Raymi Llaqta festival.

Variables	Indicator	Mean or %	Variables	Indicator	Mean or %
Especially for the Raymi Llaqta	Yes	64.23	Knowledge medium	Through family or friends	24.14
	No	35.77		Websites	31.03
Knew what the Raymi Llaqta is	Yes	51.69		Social Media	43.1
	No	48.31		Travel agencies	1.72

Table 6 shows the level of satisfaction of the tourists reflected in different level indicators (1–5), with an average level of satisfaction of 3.7 (between a regular and a satisfied indicator). On a general level, the aspects with the highest average satisfaction were related to the variety, originality, and authenticity of the customs represented, the friendliness and attention of the community members who participated in the event, the quality of the show performed, and the safety of the event (average level = 4). Table 3 shows the average level of satisfaction of foreign and domestic tourists, and their assessment revealed greater satisfaction in relation to the representation and promotion of local culture, order of presentation of the communities, variety of customs, and overall organization of the event. Foreign tourists also thought that improvements were necessary in sanitation (quantity and quality), promotion, and advertising of the event. Domestic tourists were of the opinion that the promotion of the event was adequate, but there was room for improvement in terms of the availability of toilets and facilities to watch the event.

Table 7 shows the results of two regressions analyzing the impact of several variables on total expenditure related to the event. The variables included in the model had an intercept coefficient of 6.624 with a significant p-value (0.000), indicating that in the absence of the other variables, the expected total expenditure in the celebration of the Raymi Llaqta is positive and high. The perceived motivation of the tourists who came specifically for the Raymi Llaqta festival was not significant in the binomial regression model because many of them were in the city when the event took place; however, the Poisson regression model indicates that an increase in motivation mainly for the event was associated with an increase in total expenditure. It was found that the expenditure on the means of transport used to reach Chachapoyas decreased the total expenditure (-5.043) due to an increase in the fares to travel to Chachapoyas. Furthermore, the type of accommodation chosen by the tourist had a strong influence on total expenditure, i.e., cheaper accommodation decreases total expenditure (-2.508; $p = 0.000$); however, a tourist who organizes their trip is slightly more inclined to increase their expenditure on the event.

Table 6. Indicator of tourist satisfaction with Raymi Llaqta.

Indicator	Foreign	Peru
Quality of food and beverages offered at the event	3.7	3.8
Representation and promotion of local culture	4.3	3.9
Order of community presentations	4.3	3.8
Total event duration	3.3	3.6
Variety of customs offered	4.3	4.0
Originality and authenticity of the customs presented	4.3	4.0
Kindness and friendliness of the community members who participated in the event	4.0	4.0
Opportunity to interact with the local community	4.0	3.8
Clarity and access to information of what the community is presenting	3.0	3.8
Comfort when watching the event	3.0	3.1
Facilities to watch the event	3.7	3.0
Quality of the show performed	4.0	3.9
Sound during the event	3.3	3.7
Availability of toilets (quantity and quality)	2.7	2.7
Event promotion and advertising	2.7	3.5
Event security	4.0	4.0
General organization of the event	4.3	3.9
Accommodation	4.0	3.8
Feeding	4.0	3.9
Souvenirs and/or gifts and/or handcrafts	3.7	3.7
Tours	3.7	3.6
Guide service	4.0	3.5
Hospitality of the local people	4.0	4.0

Source(s): Authors' own creation.

Table 7. Results of the determinants of total expenditure at the Raymi Llaqta festival.

Variables	Non-Zero Truncated Poisson Regression					Zero-Truncated Negative Binomial Regression				
	Coefficients	Std. Error	z Value	Pr (>|z|)		Coefficients	Std. Error	z Value	Pr (>|z|)	
Intercept	6.624	0.0311300	212.809	0.000	***	6.496	0.1482000	43.82	0.000	***
Motivation R.	3.540×10^{-2}	0.0054330	6.516	0.000	***	2.948×10^{-2}	0.0265900	1.108	0.268	
Transport. Trip	-5.043×10^{-2}	0.0016450	-30.657	0.000	***	-4.448×10^{-2}	0.0073940	-6.015	0.000	***
Type of accommodation	-2.508×10^{-2}	0.0010430	-24.055	0.000	***	-2.278×10^{-2}	0.0044380	-5.134	0.000	***
Organization	8.782×10^{-3}	0.0027940	3.143	0.002	**	1.508×10^{-3}	0.0149400	0.101	0.920	
Travel companion	1.504×10^{-3}	0.0014740	1.02	0.049	*	2.504×10^{-3}	0.0069540	0.36	0.619	
Air transport	3.600×10^{-4}	0.0000129	28.002	0.000	***	3.850×10^{-4}	0.0000818	4.707	0.000	***
Ground transportation	3.012×10^{-4}	0.0000067	45.149	0.000	***	3.583×10^{-4}	0.0000485	7.391	0.000	***
Fuel	5.789×10^{-4}	0.0000251	23.053	0.000	***	6.439×10^{-4}	0.0001362	4.726	0.000	***
Taxi	5.541×10^{-4}	0.0000510	10.866	0.000	***	4.268×10^{-4}	0.0002818	1.514	0.130	
Accommodation	2.384×10^{-4}	0.0000117	20.302	0.000	***	3.439×10^{-4}	0.0000661	5.201	0.000	***
Food	2.798×10^{-4}	0.0000096	29.114	0.000	***	2.811×10^{-4}	0.0000629	4.47	0.000	***
Visits	7.016×10^{-4}	0.0000149	46.994	0.000	***	7.515×10^{-4}	0.0000786	9.562	0.000	***
Nightlife	8.566×10^{-4}	0.0000332	25.804	0.000	***	9.518×10^{-4}	0.0001817	5.237	0.000	***
Handicrafts	1.109×10^{-4}	0.0000368	3.015	0.003	**	1.490×10^{-4}	0.0002125	0.701	0.483	
Clothing	9.381×10^{-4}	0.0000456	20.559	0.000	***	9.232×10^{-4}	0.0002462	3.75	0.000	***
Typical food	7.445×10^{-4}	0.0000299	24.874	0.000	***	7.429×10^{-4}	0.0001638	4.534	0.000	***
Medicines	8.985×10^{-4}	0.0000866	10.373	0.000	***	1.215×10^{-3}	0.0004825	2.517	0.012	*
Others	5.291×10^{-4}	0.0000081	65.321	0.000	***	5.729×10^{-4}	0.0000512	11.199	0.000	***
Return C.	1.277×10^{-1}	0.0277900	4.596	0.000	***	1.551×10^{-1}	0.1319000	1.176	0.240	

Degrees of freedom for the fit = 122. Akaike Information Criterion = 4847. Number of Fisher scoring iterations = 4.9.

From the results, it can be seen that travel company had a positive impact on the total expenditure; the specific expenses related to air and ground transportation, fuel, taxi, accommodation, food and drink (restaurants and bars), visiting tourist sites (museums and archaeological centers), nightclubs, buying typical food, medicines, and so on are significant and positive in the total expenditure.

Both models coincided in identifying several factors as significant determinants of total spending on air transportation, indicating that an increase in this expenditure is associated with an increase in total tourist spending along with ground transportation and fuel. It was also found that food is highly significant for visitors to Raymi Llaqta. However, lodging expenses had a negative impact on the model, i.e., when the tourist opts for cheaper lodging, their total expenditure decreases.

4. Discussion

The results indicate significant spending on accommodation, food, transportation, and tourism activities, with an average expenditure of USD 780 for foreign tourists and USD 692 for domestic tourists for the four days of the event. The average expenditure per tourist during Raymi Llaqta in Chachapoyas was significant and stood out in categories such as accommodation, food and beverages, and local tourism activities, and was perceived as positive by the population [53]. This finding is in line with previous research that identified cultural events as key generators of tourism revenue in similar destinations and found similar patterns in other contexts, underlining that spending on accommodation and food services represents a significant part of the total tourist budget during cultural events [6], becoming part of the tourism offer that generates interactions between tourists and residents [35,54].

In comparison with other cultural events in the country, such as the pilgrimage of the Virgen de la Puerta in Otuzco, La Libertad, combining the spiritual and cultural [55]; the Inca city of Machu Picchu, Cuzco, motivated by experiential archeological tourism [56]; and the Nepeña Valley, Ancash, characterized by having cultural tourism as predominant, [57] there is a growth pattern that can be used to promote balanced development by promoting responsible tourism practices and improving infrastructure in a sustainable manner without jeopardizing the natural and cultural resources that sustain them.

This level of spending demonstrates the economic importance of Raymi Llaqta in the Chachapoyas region of Peru, as not only a major cultural event, but also an important local economic driver. The economic impact is clear: the event not only attracts visitors but also boosts demand for local services and products, benefiting businesses in different sectors [35,58].

Regarding tourist profiles, it was found that a significant majority were male (58.93%) and that the main motivation for visiting Chachapoyas during Raymi Llaqta was recreation (59.14%). This profile varies from studies on religious festivities, where female participation predominates, as in the case of the pilgrimage of the Virgin of Montserrat in Ecuador [59]. This demographic and motivation distribution is consistent with studies on cultural and event tourism, where the search for authentic cultural experiences and participation in unique events are common motivation factors among visitors [3,44]. Tourists stay primarily with family or friends, in hostels, or in 1- and 2-star hotels.

Their expenditure was mostly on food and beverages, accommodation, ground transportation, and sightseeing, thereby resulting in increased total profits and job creation [60]. The average length of stay was four days, and the main reason for travelling was to participate in the Raymi Llaqta cultural event of the Chachapoyas. Both foreign and domestic tourists had a positive view of the representation of local culture [35,60] and the organization of the event, but mentioned the need for improvements in hygiene services and greater promotion of the event. Promotional strategies play an important role in attracting tourists to local festivals and events as they contribute to the socioeconomic development of the community [61].

The average level of tourist satisfaction was 3.7, highlighting the variety, originality, and authenticity of the customs, friendliness of the villagers, quality of the show, and safety. An analysis of tourist satisfaction indicates that aspects such as the quality of local food, cultural authenticity of Raymi Llaqta activities, and event organization were highly valued by visitors. This positive assessment is consistent with previous studies on Central-Eastern European countries that have identified cultural authenticity [62] and quality of experience as key determinants of tourist satisfaction with cultural events [52].

A high level of tourist satisfaction has a direct impact on the reputation and future attractiveness of Raymi Llaqta as a tourist destination. Positive perceptions not only foster tourist loyalty [63], but also generate word-of-mouth recommendations and positive reviews on digital platforms, which expands the promotional impact of the event both nationally and internationally [64,65]. Tourists mentioned a high level of satisfaction, especially regarding the hospitality, organization of the event, and cultural richness displayed. Although the vast majority of visitors were, indeed, domestic tourists, it has been observed that the greatest expenditure was made by foreign tourists.

According to the analysis of the coefficients of the different factors determining the total expenditure in the development of a tourist activity in the city of Chachapoyas, the intercept (6.6) indicates that, although there were no changes in the other variables, the total expenditure increased positively and significantly. This finding is in line with the translation of visitors' tourism activities into economic expenditures [66,67]. However, ground transportation presented a lower coefficient in comparison to air transportation, and these differences can be attributed to the accessibility, experience, and satisfaction of tourists [68,69]. Overall, tourist accommodation showed a significant impact on the local economy, which is consistent with similar studies [70,71]. Food consumption showed a high coefficient of total expenditure impact, reflecting the city's gastronomic offerings, emphasizing that culinary offerings attract more tourists and increase spending [72].

5. Conclusions and Limitations

According to the economic impact assessment, the results indicate an average expenditure per tourist during the Raymi Llaqta in Chachapoyas of approximately USD 780 for foreign tourists and USD 692 for domestic tourists. During the entire period of the tourism event, the main expenditures were concentrated on food and beverages (22.81%), lodging (18.36%), and ground transportation (16.69%), demonstrating the importance of the event as a local economic engine. This finding underscores the event's ability to boost key sectors such as accommodation, gastronomy, and smaller-scale tourism activities. The consistency of the revenue stream reflects the confidence and stability of the Raymi Llaqta festival as an economic catalyst, which is essential for sustainable business development in the region.

Over 60% of the tourists mentioned an interest in local culture and history as the main reason for attending Raymi Llaqta. This highlights the authenticity of the event as a touristic attraction and consolidates its reputation as an experience deeply rooted in the traditions of the Chachapoyas. The demand for authentic cultural experiences strengthens the strategic importance of preserving and promoting cultural identity in events of this nature.

The high level of tourist satisfaction during Raymi Llaqta is highlighted by the fact that more than 85% of the participants rated local food quality and the organization of the event positively. These results are not only a reflection of the efficiency of the event's management and operation, but also of its ability to meet visitors' expectations. This positive perception is key for maintaining the event's reputation and competitiveness in the tourism market.

Although most visitors were male (58.93%), the significant presence of women and families illustrates the inclusiveness of Raymi Llaqta as a family event. This demographic diversity broadens the potential visitor base and reinforces the importance of inclusive promotional strategies. The ability to attract and satisfy different population segments strengthens the long-term feasibility of the event as an inclusive tourism destination.

Raymi Llaqta has shown promising potential for future expansion and development, with a 10% annual increase in attendance over the past three years. This upward trend indicates not only the growing popularity of the event, but also its ability to adjust and evolve over time. Through continued growth, Raymi Llaqta is positioned as an ever-evolving event that can generate increasingly significant economic and cultural benefits for Chachapoyas and its surrounding communities.

One limitation was that by using a structured questionnaire, the information obtained could be subject to response bias, such as the tendency of respondents to give socially desirable or inaccurate answers about how they experienced and perceived the event.

This study contributes to the analysis of various customary events of the same or similar nature in Latin America, and, therefore, it is recommended that the tourism infrastructure in such destinations should be strengthened in order to manage the growing number of visitors, including improvements in hotels, transportation and basic services, thus ensuring a comfortable and pleasant experience. For all stakeholders in the tourism and hospitality sector, it is recommended that a focus on marketing strategies in national and international markets could significantly increase the influx of tourists, with actions such as social media campaigns, collaborations with travel agencies, and participation in international tourism fairs, thus ensuring the long-term sustainability of the sector in destinations that seek to promote tourism in a sustainable and responsible manner.

Author Contributions: Conceptualization, F.O.Z.C.A. and A.J.S.P.; methodology, M.L.W.S.; software, M.L.W.S. and J.P.V.; validation, F.O.Z.C.A., A.J.S.P., and R.M.E.-H.; formal analysis, F.O.Z.C.A.; investigation, M.L.W.S. and J.P.V.; resources, A.J.S.P.; data curation, F.O.Z.C.A. and A.J.S.P.; writing—original draft preparation, M.L.W.S. and J.P.V.; writing—review and editing, R.M.E.-H.; visualization, F.O.Z.C.A.; supervision, R.M.E.-H.; project administration, F.O.Z.C.A. and R.M.E.-H.; funding acquisition, A.J.S.P. All authors have read and agreed to the published version of the manuscript.

Funding: This research received no external funding.

Informed Consent Statement: Informed consent was obtained from all subjects involved in this study.

Data Availability Statement: Data are available upon request from researchers who meet the eligibility criteria. Kindly contact the first author privately through e-mail.

Conflicts of Interest: The authors declare no conflicts of interest.

References

1. Alamineh, G.A.; Hussein, J.W.; Endaweke, Y.; Taddesse, B. The local communities' perceptions on the social impact of tourism and its implication for sustainable development in Amhara regional state. *Heliyon* **2023**, *9*, e17088. [CrossRef] [PubMed]
2. Trillo Rodríguez, M.J.; Flores Ruiz, D. Tourism and gender: Bibliometric and bibliographic analysis of open articles. *J. Tour. Anal. (JTA)* **2023**, *30*, 28–65. [CrossRef]
3. Pérez-Priego, M.A.; de los Baños García-Moreno, M.; Jara-Alba, C.; Caro-Barrera, J.R. Local gastronomy as a destination tourist attraction: The case of the 'Chiringuitos' on the Costa del Sol (Spain). *Int. J. Gastron. Food Sci.* **2023**, *34*, 100822. [CrossRef]
4. Mansilla, J.; Yanes Torrado, S.; Espinosa Zepeda, H. The impossible tourist experience: The case of the Festes de Gràcia in Maó, Menorca. *J. Tour. Anal. (JTA)* **2023**, *30*, 66–88. [CrossRef]
5. Moreira dos Santos, E.R.; Pereira, L.N.; Pinto, P.; Boley, B.B. Development and validation of the new resident empowerment through Tourism Scale: RETS 2.0. *Tour. Manag.* **2024**, *104*, 104915. [CrossRef]
6. McCabe, S. Theory in tourism. *Ann. Tour. Res.* **2024**, *104*, 103721. [CrossRef]
7. Fuchs, M. A post-Cartesian economic and Buddhist view on tourism. *Ann. Tour. Res.* **2023**, *103*, 103688. [CrossRef]
8. Makkonen, T.; Williams, A.M. Cross-border tourism and innovation system failures. *Ann. Tour. Res.* **2024**, *105*, 103735. [CrossRef]
9. Pung, J.M.; Houge Mackenzie, S.; Lovelock, B. Regenerative tourism: Perceptions and insights from tourism destination planners in Aotearoa New Zealand. *J. Destin. Mark. Manag.* **2024**, *32*, 100874. [CrossRef]
10. Annamalah, S.; Paraman, P.; Ahmed, S.; Dass, R.; Sentosa, I.; Pertheban, T.R.; Shamsudin, F.; Kadir, B.; Aravindan, K.L.; Raman, M.; et al. The role of open innovation and a normalizing mechanism of social capital in the tourism industry. *J. Open Innov. Technol. Mark. Complex.* **2023**, *9*, 100056. [CrossRef]
11. Devi, R.; Agrawal, A.; Dhar, J.; Misra, A.K. Forecasting of Indian tourism industry using modeling approach. *MethodsX* **2024**, *12*, 102723. [CrossRef] [PubMed]
12. Meng, G.; Wang, K.; Wang, F.; Dong, Y. Analysis of the tourism-economy-ecology coupling coordination and high-quality development path in karst Guizhou Province, China. *Ecol. Indic.* **2023**, *154*, 110858. [CrossRef]
13. Blomstervik, I.H.; Olsen, S.O. Progress on novelty in tourism: An integration of personality, attitudinal and emotional theoretical foundations. *Tour. Manag.* **2022**, *93*, 104574. [CrossRef]
14. Page, S.J.; Duignan, M. Progress in tourism management: Is urban tourism a paradoxical research domain? Progress since 2011 and prospects for the future. *Tour. Manag.* **2023**, *98*, 104737. [CrossRef]
15. Baños, J.F.; Boto, D.; Zapico, E.; Mayor, M. Optimal carrying capacity in rural tourism: Crowding, quality deterioration, and productive inefficiency. *Tour. Manag.* **2024**, *105*, 104968. [CrossRef]

16. Hoarau, H.; Wigger, K.; Olsen, J.; James, L. Cruise tourism destinations: Practices, consequences and the road to sustainability. *J. Destin. Mark. Manag.* **2023**, *30*, 100820. [CrossRef]
17. Liu, L.W.; Pahrudin, P.; Tsai, C.Y.; Hao, L. Disaster, risk and crises in tourism and hospitality field: A pathway toward tourism and hospitality management framework for resilience and recovery process. *Nat. Hazards Res.* **2024**. [CrossRef]
18. Joyce, E. Rewilding tourism in the news: Power/knowledge and the Irish and UK news media discourses. *Ann. Tour. Res.* **2024**, *104*, 103718. [CrossRef]
19. Smit, B.; Melissen, F.; Font, X. Co-designing tourism experience systems: A living lab experiment in reflexivity. *J. Destin. Mark. Manag.* **2024**, *31*, 100858. [CrossRef]
20. Álvarez, M.; Aranda, R.; Rodríguez, A.Y.; Fajardo, D.; Sánchez, M.G.; Pérez, H.; Martínez, J.; Guerrero, R.; Bustio, L.; Díaz, Á. Natural language processing applied to tourism research: A systematic review and future research directions. *J. King Saud Univ.-Comput. Inf. Sci.* **2022**, *34*, 10125–10144. [CrossRef]
21. Aguinis, H.; Kraus, S.; Poček, J.; Meyer, N.; Jensen, S.H. The why, how, and what of public policy implications of tourism and hospitality research. *Tour. Manag.* **2023**, *97*, 104720. [CrossRef]
22. Campos, C.; Laso, J.; Cristóbal, J.; Fullana-i-Palmer, P.; Albertí, J.; Fullana, M.; Herrero, Á.; Margallo, M.; Aldaco, R. Tourism under a life cycle thinking approach: A review of perspectives and new challenges for the tourism sector in the last decades. *Sci. Total Environ.* **2022**, *845*, 157261. [CrossRef]
23. Osorio, F.M.; Muñoz Benito, R.; Pérez Neira, D. Empirical evidence, methodologies and perspectives on tourism, energy and sustainability: A systematic review. *Ecol. Indic.* **2023**, *155*, 110929. [CrossRef]
24. Piano, E.; Mammola, S.; Nicolosi, G.; Isaia, M. Advancing tourism sustainability in show caves. *Cell Rep. Sustain.* **2024**, *1*, 100057. [CrossRef]
25. De Bruyn, C.; Ben Said, F.; Meyer, N.; Soliman, M. Research in tourism sustainability: A comprehensive bibliometric analysis from 1990 to 2022. *Heliyon* **2023**, *9*, E18874. [CrossRef]
26. Yepez, C.; Leimgruber, W. The evolving landscape of tourism, travel, and global trade since the COVID-19 pandemic. *Res. Glob.* **2024**, *8*, 100207. [CrossRef]
27. Yanan, L.; Azzam Ismail, M.; Aminuddin, A. How has rural tourism influenced the sustainable development of traditional villages? A systematic literature review. *Heliyon* **2024**, *10*, E25627. [CrossRef]
28. Zhang, Y.; Wang, L.; Zheng, Y.; Tian, F. Cooperation, hotspots and prospects for tourism environmental impact assessments. *Heliyon* **2023**, *9*, E17109. [CrossRef]
29. Liu, J.; Yu, Y.; Chen, P.; Chen, B.Y.; Chen, L.; Chen, R. Facilitating urban tourism governance with crowdsourced big data: A framework based on Shenzhen and Jiangmen, China. *Int. J. Appl. Earth Obs. Geoinf.* **2023**, *124*, 103509. [CrossRef]
30. Zavaleta Chávez Arroyo, F.O.; Sánchez Pantaleón, A.J.; Navarro-Mendoza, Y.P.; Esparza-Huamanchumo, R.M. Community tourism conditions and sustainable management of a community tourism association: The case of Cruz Pata, Peru. *Sustainability* **2023**, *15*, 4401. [CrossRef]
31. Aponte, S. Get to know the Raymi Llacta, the most outstanding cultural event in the north of the Peruvian jungle. La República. 2023. Available online: https://larepublica.pe/sociedad/2023/06/01/raymi-llacta-de-los-chachapoyas-que-es-y-como-se-celebrara-este-2023-la-fiesta-emblematica-de-amazonas-cronograma-significado-fecha-danzas-programa-general-junio-lrsd-44513 (accessed on 30 June 2024).
32. PromPeru. *Chachapoyas Tourism Week*; Ministry of Foreign Trade and Tourism: Lima, Peru, 2023. Available online: https://www.peru.travel/es/eventos/semana-turistica-de-chachapoyas (accessed on 15 July 2024).
33. García, M.O.; Arias, A.V.V.; Barquín, R.; Del, C.S.; Ontiveros, M.M.M.; Ortega, O.A.S. The perception of the impacts of tourism in the community of Ixtapa-Zihuatanejo, Mexico, from the Causal Mapping Model. *Rev. Rosa Dos Ventos Tur. E Hosp.* **2018**, *10*, 441–463. [CrossRef]
34. Ahirwar, V.; Gupta, R.; Kumar, A. Impacts of tourism on development of urban areas in Indian cities: A systematic literature review. *J. Environ. Manag. Tour.* **2023**, *14*, 2993–3005. [CrossRef] [PubMed]
35. Lobo, C.; Costa, R.A.; Chim-Miki, A.F. Events image from the host-city residents' perceptions: Impacts on the overall city image and visit recommend intention. *Int. J. Tour. Cities* **2023**, *9*, 875–893. Available online: https://www.emerald.com/insight/content/doi/10.1108/IJTC-10-2022-0242/full/html (accessed on 15 July 2024). [CrossRef]
36. Liu, A.; Kim, Y.R.; Song, H. Toward an accurate assessment of tourism economic impact: A systematic literature review. *Ann. Tour. Res. Empir. Insights* **2022**, *3*, 100054. [CrossRef]
37. Majewski, L. Economic impact analysis of nature tourism in protected areas: Towards an adaptation to international standards in German protected areas. *J. Outdoor Recreat. Tour.* **2024**, *45*, 100742. [CrossRef]
38. Kronenberg, K.; Fuchs, M. Aligning tourism's socio-economic impact with the United Nations' sustainable development goals. *Tour. Manag. Perspect.* **2021**, *39*, 100831. [CrossRef]
39. Manrique, C.; Santana Gallego, M.; Valle, E. The economic impact of global uncertainty and security threats on international tourism. *Econ. Model.* **2022**, *113*, 105892. [CrossRef]
40. Allan, G.; Connolly, K.; Figus, G.; Maurya, A. Economic impacts of COVID-19 on inbound and domestic tourism. *Ann. Tour. Res. Empir. Insights* **2022**, *3*, 100075. [CrossRef]

41. Zavaleta Chavez Arroyo, F.O.; Sánchez Pantaleón, A.J.; Aldea Roman, C.E.; Esparza-Huamanchumo, R.M.; Álvarez-García, J. A structural analysis of the economic impact of tourism and the perspective of tourism providers in Kuélap, Peru. *Land* **2024**, *13*, 120. [CrossRef]
42. Moreno, R.F.; Méndez, P.; Ros Candeira, A.; Alcaraz Segura, D.; Santamaría, L.; Ramos Ridao, Á.F.; Revilla, E.; Bonet García, F.J.; Vaz, A.S. Evaluating tourist profiles and nature-based experiences in Biosphere Reserves using Flickr: Matches and mismatches between online social surveys and photo content analysis. *Sci. Total Environ.* **2020**, *737*, 140067. [CrossRef]
43. Gabarda, A.; Garcia, X.; Fraguell, R.M.; Ribas, A. How guest profile and tourist segment explain acceptance of economic-based water-saving measures. A Mediterranean destination case study. *J. Hosp. Tour. Manag.* **2022**, *52*, 382–391. [CrossRef]
44. López, T.; Vieira, A.; Rodríguez, J. Profile and motivations of European tourists on the Sherry wine route of Spain. *Tour. Manag. Perspect.* **2014**, *11*, 63–68. [CrossRef]
45. Villamediana, J.D.; Vila López, N.; Küster Boluda, I. Predictors of tourist engagement: Travel motives and tourism destination profiles. *J. Destin. Mark. Manag.* **2020**, *16*, 100412. [CrossRef]
46. Molera, L.; Albaladejo, I.P. Profiling segments of tourists in rural areas of South-Eastern Spain. *Tour. Manag.* **2007**, *28*, 757–767. [CrossRef]
47. Pérez, A.; Caamaño Franco, I.; Lezcano González, M.E. Conceptual approach to experiential tourism: Presence, use and scope of a new model of tourism. *J. Tour. Anal. (JTA)* **2023**, *30*, 86–119. [CrossRef]
48. Thrane, C. Research note: The determinants of tourists' length of stay: Some further modelling issues. *Tour. Econ.* **2015**, *21*, 1087–1093. [CrossRef]
49. Gurmu, S.; Trivedi, P. Excess zeros in count models for recreational trips. *J. Bus. Econ. Stat.* **1996**, *14*, 469–477. [CrossRef]
50. Greene, W.H. *Econometric Analysis*; Prentice Hall: Boston, MA, USA, 2012.
51. Alén, E.; Nicolau, J.; Losada, N.; Domínguez, T. Determinant factors of senior tourists' length of stay. *Ann. Tour. Res.* **2014**, *49*, 19–32. Available online: https://www.sciencedirect.com/science/article/pii/S0160738314000954 (accessed on 8 April 2024). [CrossRef]
52. Cameron, A.C.; Trivedi, P. *Regression Analysis of Count Data*; Cambridge University Press: New York, NY, USA, 1998.
53. Kourkouridis, D.; Frangopoulos, Y.; Kapitsinis, N. Socio-economic effects of trade fairs on host cities from a citizens' perspective: The case of Thessaloniki, Greece. *Int. J. Event Festiv. Manag.* **2023**, *14*, 113–133. [CrossRef]
54. Godov, M.; Ridderstaat, J. Health outcomes of tourism development: A longitudinal study of the impact of tourism arrivals on residents' health. *J. Destin. Mark. Manag.* **2020**, *17*, 100462. [CrossRef]
55. Siles, J. Estudio sobre las motivaciones de los visitantes a la peregrinación de la Virgen de la Puerta en Otuzco, La Libertad, Perú. ESAN. Graduate Thesis, UNIVERSIDAD ESAN, Lima, Peru, 2024. Available online: https://hdl.handle.net/20.500.12640/4024 (accessed on 22 July 2024).
56. López, C. Turismo Experiencial y su impacto positivo en las Comunidades Locales. Claves para un turismo transformador a través de la inversión sostenible. *Cent. Estud. Diseño Comun.* **2024**, *223*, 155–173. [CrossRef]
57. Chávez, M.; Flores, C.; López, C. Cultural Identity and Tourism in the Nepeña Valley (Ancash Region). *Surandino Rev. Humanidades Cult.* **2024**, *5*, 35–53.
58. Boonsiritomachai, W.; Phonthanukitithaworn, C. Residents' support for sports events tourism development in beach city: The role of community's participation and tourism impacts. *Sage Open* **2019**, *9*, 2158244019843417. [CrossRef]
59. Carvache-Franco, M.; Carvache-Franco, W.; Orden-Mejía, M.; Carvache-Franco, O.; Andrade-Alcivar, L.; Cedeno-Zavala, B. Motivations for the Demand for Religious Tourism: The Case of the Pilgrimage of the Virgin of Montserrat in Ecuador. *Heritage* **2024**, *7*, 3719–3733. [CrossRef]
60. Doe, F.; Preko, A.; Akroful, H.; Okai-Anderson, E.K. Festival tourism and socioeconomic development: Case of Kwahu traditional areas of Ghana. *Int. Hosp. Rev.* **2022**, *36*, 174–192. [CrossRef]
61. Whitford, M.; Dunn, A. Papua New Guinea's indigenous cultural festivals: Cultural tragedy or triumph? *Event Manag.* **2014**, *18*, 265–283. [CrossRef]
62. Irimias, A.; Mitev, A.; Michalko, G. Demographic characteristics influencing religious tourism behaviour: Evidence form a Central-Eastern-European country. *IJRTP* **2016**, *4*, 3. [CrossRef]
63. Borges, A.P.; Vieira, E.P.; Romão, J. The evaluation of the perceived value of festival experiences: The case of Serralves em Festa! *Int. J. Event Festiv. Manag.* **2018**, *9*, 279–296. [CrossRef]
64. Esparza-Huamanchumo, R.M.; Quiroz-Celis, A.V.; Camacho-Sanz, A.A. Influence of eWOM on the purchase intention of consumers of Nikkei restaurants in Lima, Peru. *Int. J. Tour. Cities* **2024**, *12*. [CrossRef]
65. Uslu, A. The relationship of service quality dimensions of restaurant enterprises with satisfaction, behavioral intention, e-WOM and the moderator effect of atmosphere. *Tour. Manag. Stud.* **2020**, *16*, 23–35. Available online: https://www.tmstudies.net/index.php/ectms/article/view/1296 (accessed on 8 April 2024). [CrossRef]
66. Ridderstaat, J.; Singh, D.; DeMicco, F. The impact of major tourist markets on health tourism spending in the United States. *J. Destin. Mark. Manag.* **2019**, *11*, 270–280. [CrossRef]
67. Grant, A.; Lecca, P.; Swales, K. The impacts of temporary but anticipated tourism spending: An application to the Glasgow 2014 Commonwealth Games. *Tour. Manag.* **2017**, *59*, 325–337. [CrossRef]
68. Godbey, G.; Graefe, A. Repeat tourism, play, and monetary spending. *Ann. Tour. Res.* **1991**, *18*, 213–225. [CrossRef]

69. Mortazavi, R. The relationship between visitor satisfaction, expectation and spending in a sport event. *Eur. Res. Manag. Bus. Econ.* **2021**, *27*, 100132. [CrossRef]
70. Asgary, N.; De los Santos, G.; Vern, V.; Davila, V. The determinants of expenditures by Mexican visitors to the border cities of Texas. *Tour. Econ.* **1997**, *3*, 319–328. [CrossRef]
71. Brida, J.; Scuderi, R. Determinants of tourist expenditure: A review of micro econometric models. *Tour. Manag. Perspect.* **2013**, *6*, 28–40. [CrossRef]
72. Chiang, C.; Kuo, C. The influence of travel agents on travel expenditures. *Ann. Tour. Res.* **2012**, *39*, 1258–1263. [CrossRef]

Disclaimer/Publisher's Note: The statements, opinions and data contained in all publications are solely those of the individual author(s) and contributor(s) and not of MDPI and/or the editor(s). MDPI and/or the editor(s) disclaim responsibility for any injury to people or property resulting from any ideas, methods, instructions or products referred to in the content.

Article

Tourists' Views on Sustainable Heritage Management in Porto, Portugal: Balancing Heritage Preservation and Tourism

Makhabbat Ramazanova [1,2], Fátima Matos Silva [1,2,3,*] and Isabel Vaz de Freitas [1,4]

1. Department of Tourism, Heritage and Culture, University Portucalense Infante D. Henrique, 4200-072 Porto, Portugal; ramazanova@upt.pt (M.R.); ifc@upt.pt (I.V.d.F.)
2. REMIT—Research on Economics, Management and Information Technologies, University Portucalense Infante D. Henrique, 4200-072 Porto, Portugal
3. CITCEM—Transdisciplinary Culture, Space and Memory Research Centre, Oporto University, 4150-564 Porto, Portugal
4. CIAUD-UPT—Centro de Investigação em Arquitetura, Urbanismo e Design, University Portucalense Infante D. Henrique, 4200-072 Porto, Portugal
* Correspondence: mfms@upt.pt

Abstract: This research explores the perceptions of tourists regarding the state and sustainable management of tangible cultural heritage in the city of Porto (Portugal), the destination recognised for its historical and cultural heritage. Porto attracts a growing number of tourists due to its rich heritage, encompassing cultural, architectural, and scenic elements. This rising demand necessitates sustainable management practices to protect the city's heritage and ensure long-term sustainability. With the increase in tourism and the need to preserve cultural and environmental resources, sustainable management becomes essential to balance economic development and heritage conservation. The study's main objective is to analyse tourists' perspectives on the sustainable management of the cultural heritage of the city of Porto. It focuses on analysing current tourism management practices and calls for new strategies to promote sustainability. Structured questionnaires were distributed to a random sample of tourists in strategic locations between January and April 2023, obtaining quantitative data. The sample consists of 264 respondents. The results revealed four main factors, showing that most respondents show a high level of awareness and appreciation of sustainability and its relevance for heritage conservation. Regarding tourism, the majority agree that it contributes positively to the conservation and restoration of the city's heritage.

Keywords: cultural heritage; tourism; conservation; sustainability; management

Academic Editor: Stella Sofia Kyvelou

Received: 25 November 2024
Revised: 23 December 2024
Accepted: 26 December 2024
Published: 28 December 2024

Citation: Ramazanova, M.; Silva, F.M.; Vaz de Freitas, I. Tourists' Views on Sustainable Heritage Management in Porto, Portugal: Balancing Heritage Preservation and Tourism. *Heritage* **2025**, *8*, 10. https://doi.org/10.3390/heritage8010010

Copyright: © 2024 by the authors. Licensee MDPI, Basel, Switzerland. This article is an open access article distributed under the terms and conditions of the Creative Commons Attribution (CC BY) license (https://creativecommons.org/licenses/by/4.0/).

1. Introduction

Heritage is a term that uses past experiences to represent tangible and intangible socio-cultural expressions in the present and develop a future vision. Strategically, it links rural, urban, or cultural landscapes 'in general,' the main stakeholders, the community, and visitors. In this way, heritage becomes a form of marketing that can help preserve cultural identity and simultaneously be a source of socio-economic development.

The concept of heritage has evolved significantly over time. Initially, it was associated with the preservation of historical monuments and works of art, later extending to intangible aspects. For the subjects and aims of this study, some international orientations resulting from the international reflection of UNESCO and ICOMOS are essential. The Convention Concerning the Protection of the World Cultural and Natural Heritage [1] identifies international protection of the world's cultural and natural heritage in an international cooperation landscape concerning heritage identity conservation and safeguarding at a universal level, involving all countries to adopt measures of heritage conservation and safeguard. UNESCO's 1972 Convention was a very important milestone in heritage history, as it created an effective and permanent system for the collective protection of heritage of universal value. In the Convention, cultural and natural heritage was recognised as having a unique and irreplaceable value, with properties of exceptional interest that need to be protected, thus becoming a collective responsibility of the international community.

Sustainable Development Goal 11 (SDG 11) calls for countries to "make cities and human settlements inclusive, safe, resilient and sustainable". Target 11.4 of this goal aims to "strengthen efforts to protect and safeguard the world's cultural and natural heritage".

The need to preserve heritage is also crucial for tourism activity, which has been growing more and more over the years all over the world, so effective management strategies to balance the demands of tourism with the safeguarding of heritage sites become essential. The conservation of heritage could be an opportunity for local and regional development through tourism, stimulating diverse activities and promoting sustainable development [2]. In this perspective, places enter a development circle where heritage conservation grows tourism, and the increase in tourism numbers could be managed by promoting local investment in heritage conservation and renewal of aesthetic investment in order to regenerate urban landscapes. The findings from Soeiro and co-authors' [3] study emphasise the crucial role of tangible and intangible heritage in crafting a comprehensive urban regeneration strategy.

However, cultural heritage requires careful planning and responsible management strategies and policies to prevent adverse impacts like overcrowding, environmental degradation, and the commodification of cultural practices. Effective and responsible management ensures respect for heritage sites' cultural and historical value, promotes awareness, provides opportunities and resilience, and shapes respect for diversity [4]. In this context, cities are essential for development and intercultural dialogue.

Culture and heritage could be important to collective groups and communities to explore intrinsic place values and place identity. According to Apaydin [5], the diversity of the heritage could promote intercultural cooperation and build trust among communities, creating a national discourse for identity. In this way, heritage management in an international panorama could also be a promoter of cultural significance. The scenic area, the communication of local history, the spiritual values, and the relationship with residents create emotional dimensions that could bring another involvement between tourism and local culture. As Yang [6] argues, managing and stimulating positive emotion and managing the cultural identity and heritage conservation behaviour from the positive tourist experiences are essential.

According to previous studies, as high tourist demand helps to preserve the heritage and contributes to the local economy, it also jeopardises heritage conservation in situations of high tourist pressure [7]. The sustainable management of tourist heritage has become a highly relevant issue today, especially in historic cities such as Porto (Portugal), where tourism plays a crucial role in the local economy [8]. However, this high tourist demand can lead to the degradation of cultural and natural resources, thus jeopardising local identity and quality of life. Some authors [9] state that limiting the number of visitors to a destination is essential and that this flow of tourists should be managed by diverting tourist traffic to areas away from the places most vulnerable to tourism pressure. Some UNESCO World Heritage Sites, such as Venice, are already facing problems of overtourism, especially due to the impact of cruises [10]. Liberatore et al. [11], concerning tourist flows in the UNESCO World Heritage Historical Centre of Florence, developed a system of indicators to measure the carrying capacity of cities aimed at evaluating the risk of over-tourism and its impact on these historic sites.

Thus, one of the main challenges associated with sustainable tourism is the carrying capacity of cities [12], and to counteract it, some authors suggest measures and restrictions such as limiting visits and tourist activities. However, this is a strategy that can be considered drastic and recommended only for sites where tourist pressure directly threatens the heritage, as this limitation can also have severe consequences for the local economy. Other authors highlight another opposite strategy for preserving heritage, which aims to increase the capacity of sites by expanding the physical capacity of attractions, increasing the capacity of facilities, and improving the operational capacity of sites [13]. This option is considered the most viable by many authors since it aims to protect the heritage and, at the same time, is not as drastic as the first option, and it ensures that tourist activity is maintained, contributing to the sustainability of the economy. However, there are also limitations associated with this strategy because sometimes it is not possible to increase the physical capacity of a site without altering its heritage value. However, there are still many possibilities for fostering and developing improvements in the operational capacity of the sites.

In short, a good understanding of the importance of heritage preservation is crucial so that tourist activity does not harm but instead promotes conservation by educating tourists and contributing to sustainable economic development [14]. Implementing new strategies and following existing ones must be carefully planned and executed to avoid negative impacts and maximise the benefits for heritage and local communities.

Thus, sustainable cultural heritage management has become a central theme in promoting tourist destinations that balance economic development with cultural and environmental preservation. In Porto, known for its rich historical and cultural heritage, sustainability has become a priority in formulating tourism policies and practices. In the context of tourism, sustainable heritage management aims to ensure that heritage sites and resources are managed responsibly to maximise their benefits for present and future generations while minimising negative impacts [7].

UNESCO [15] considers urban areas "engines" of growth, innovation, and creativity and centres for social opportunities. In this way, it defends the balance between growth and heritage safeguarding since historic urban areas are the most ample and diverse cultural common heritage manifestation and testimony of humankind. To balance, the recommendation considers international cooperation a vital process of strengthening knowledge and capacity building. In the line of UNESCO, Cerisola and Panzera [16] concluded that the urban structure is the most constructive hub for engagement in the cultural life that promotes regional development, particularly cities with a cultural and creative dimension and dynamism.

Recent studies [17] considered that managing heritage places in cities is a driver for economic, social, and cultural development if the conservation of tangible and intangible heritage occurs to guarantee a sustainable heritage safeguard for future generations. Each country has the autonomy to conduct heritage management and safeguarding; however, international policies create ways and directions for governmental decisions and practices. Working together with all the local and regional stakeholders and sharing practices with other international partners is an important beginning. Above all, this process of experiences in the exchange of culture, heritage, and identity creates several opportunities to share practices to develop cultural destination management and could be seen as a sustainable response to globalisation [18]. Nevertheless, as Hassan and co-authors [19] reinforce, sustainable management of heritage sites faces challenges in balancing preservation with tourism demands.

The main foundations of sustainability include greater community participation in planning and a better understanding of the historic environment, keeping activities at levels that do not deteriorate the historic environment and allowing for adaptive reuse of the heritage [20]. It is possible to see in this concept the importance of planning for the sustainable development of tourist regions and the valorisation and preservation of heritage as a reinforcement of cultural identity, reflected in sustainable heritage management.

However, sustainable heritage management still faces many challenges when it comes to balancing conservation and consumption. The study developed by Buonincontri and co-authors [21] on this topic highlighted a concept of sustainable behaviour for heritage consumers, enhancing how visitors' experiences in heritage regions can influence their attachment to the place visited.

The sustainable management of cultural heritage involves various components aimed at guaranteeing the preservation and valorisation of cultural and economic resources in a balanced and lasting way. One of the main components is heritage conservation and protection, which includes actions to maintain the physical integrity and authenticity of assets, such as restoring and rehabilitating historic buildings [22].

The sustainable use of resources is also fundamental, ensuring that the exploitation of heritage does not deplete the resources available for future generations. Flyen and co-authors [23] found that many visitors were unaware of the protected status of cultural heritage sites. They often did not realise the unintentional damage their behaviour could cause in heritage use. The study highlights a gap in visitor education about the significance and fragility of these sites, which can lead to actions that degrade their physical and cultural value. This lack of awareness is particularly problematic, as visitors may unintentionally contribute to destruction and degradation.

Sustainability in tourism influences the preservation of resources and improves the tourist experience, promoting a more authentic and meaningful involvement with the destination. Tourists' perceptions of these practices can influence their satisfaction, loyalty, and future behaviour, making it crucial for tourism managers to understand these perceptions.

Therefore, the main aim of this research is to analyse tourists' perceptions of sustainable heritage management in Porto. Specific objectives include grouping them into four distinct factors, each capturing a unique dimension of their perspectives, such as level of knowledge about sustainable heritage management practices, measuring tourists' satisfaction with the sustainable management practices implemented in the city, exploring the importance tourists attach to sustainability when choosing Porto as a tourist destination, and analysing the impact of sustainable management practices on tourists' overall experience.

Understanding tourists' perspectives on sustainable heritage management is crucial as it provides valuable information for tourism managers and policymakers regarding the effectiveness of sustainability strategies. It also helps identify areas needing improvement or adjustment, contributing to developing policies promoting sustainable tourism and benefiting visitors and the local community.

Through this research, it is hoped to contribute to a deeper understanding of tourists of sustainable heritage management in the city of Porto, providing a basis for improving management practices and promoting more sustainable and enriching tourism. This research is structured in several main sections. The introduction, as provided above, presents the research objectives, as well as the context and its insertion in the literature to provide a theoretical basis for sustainable heritage management and tourists' perceptions. The methodology describes the methods used to collect and analyse the data. The data analysis details the results obtained, followed by a discussion that compares the findings with the existing literature. Finally, the recommendations and conclusions highlight the practical implications, suggesting directions for future research.

2. Materials and Methods

The target population of the empirical study corresponds to all the visitors who visit the Porto historical city centre. Structured survey questionnaires were distributed to a random sample of tourists in strategic locations face to face between January and April 2023. Fewer tourists during this period allowed for more detailed answers and not in a rush. As a result, 264 survey questionnaires were completed.

The method of distribution and data collection was carried out through two tools: distribution of the online survey using a QR code and paper forms filled out at popular tourist spots indicated, covering a variety of tourists. The questionnaire was made available in English.

The survey questionnaire was designed by researchers and included thirteen questions divided into three sections listed below:

SECTION I: Characteristics of the trip, which aimed to collect data and understand the number of visits to Porto, how the respondents knew about Porto, where they stayed (type of accommodation), and which mode of transportation they used in the city.

SECTION II. Sustainable heritage management. This section included questions aimed at understanding respondents' views on the heritage state, the impact of tourism, and tourists' perceptions and behaviour regarding the site's preservation, as well as emphasising its role in promoting sustainable practices. Visitors were asked to indicate their level of agreement with a series of statements using a 5-point Likert scale, where 1 represented "strongly disagree" and 5 "strongly agree".

SECTION III. Respondent's profile. This section included questions about age, gender, marriage status, educational level, country of residence, occupation, and net monthly income.

Before data collection, a pilot survey was conducted with 15 people visiting the area. Their questions and comments were considered, and minor changes regarding writing were introduced in the questionnaire.

The results were analysed utilising Microsoft Excel and IBM SPSS Statistics 27 and presented in Section 3.

Case Study

Porto city, located in Northern Portugal (Figure 1), was declared a World Heritage Site by UNESCO in 1996; more precisely, the Historic Centre of Porto, the D. Luiz I Bridge, and the Serra do Pilar Monastery, is a renowned tourist destination, attracting thousands of visitors yearly. The application for this classification demanded a high level of quality in the interventions in the city, reinforcing the city's notoriety at the national and international levels [24].

Figure 1. Location of Porto city, Portugal (based on the Official Administrative. Map of Portugal and Google Satellite). Source: [25].

Porto's population was 237,591 according to the 2011 census, showing a decrease of 25,540 from the 2001 census. By 2019, the population was 216,606, a number that has remained stable since 2017.

It's important to note the significant difference between Porto's residential population and the "user" population—those who live in neighbouring areas but come to Porto daily for personal activities. The historical city centre of Porto, located in the heart of the city's historical district, covers an area of 5.43 km^2. This area is home to a population of 40,440 inhabitants, making it a vibrant and densely populated part of Porto. The historical centre is renowned for its rich cultural heritage, architectural landmarks, and significant historical sites, reflecting Porto's long and storied past. It serves as a vital hub for both residents and visitors, offering a unique blend of history, culture, and modern urban life [26]. According to Gusman et al. [27], from 2002 to 2017, the number of hotel guests increased by 70%, rising from 560,777 to 1,876,720 annually, with 74.4% of these guests being international visitors.

Thus, the choice of the city of Porto as a case study in this research project is justified by its rich culture and heritage combined with its significant growth in tourist demand. All these factors place the city of Porto where it is essential to examine its heritage management practices. With the growth of tourism comes significant challenges to preserving the city's heritage. Sustainable heritage management is, therefore, essential to ensure that tourism does not jeopardise the cultural and environmental resources that make Porto unique. Urban regeneration in Porto's Historic Centre (Figure 2) is fundamental to preserving the heritage and preserving the city's cultural and historical wealth, fostering economic and tourist development, improving the quality of life of the local community, and promoting sustainability.

Figure 2. Historic Centre of Porto. Area classified by UNESCO in 1996. Source: [8].

3. Results and Discussion

This section will present the results obtained from the answers collected in the questionnaires to measure the perspectives of tourists visiting the city of Porto in relation to the sustainable management of heritage.

3.1. Sample and Trip Characteristics

Regarding the sample profile, the average age of respondents is 30, predominantly female, single, and highly educated (Table 1). In the case of education, basic education covers a broad general education, followed by intermediate education, which is a more specialised stage that prepares students for higher education or the workforce. Higher education is provided at universities, leading to bachelor's, master's, or doctoral degrees.

Table 1. Sample profile.

Gender	N	%	Education Level	N	%
Female	146	55%	Basic level	11	4%
Male	118	45%	Intermediate level	62	23%
			Higher education	191	72%
Age	**N**	**%**	**Occupation**		
Average	30		Public employee	58	22%
Minimum	17		Private organisation	87	33%
Maximum	65		Entrepreneur	29	11%
Marital status	**N**	**%**	Retired	4	2%
Single	162	61%	Student	78	30%
Married	83	31%	Other	8	3%
Divorced	17	6%			
Widower	2	1%			

Source: authors.

Regarding country of residence, the majority are from Spain (87), Portugal (46), Italy (31), France (21), America (19) and United Kingdom (11). The Figure 3 shows how the sample is diverse, representing a different parts of the world.

Figure 3. Country of residence. Source: Authors.

Regarding occupation (Figure 3), most are entrepreneurs or employees in private and public organisations and students.

Regarding travel characteristics (Table 2), most respondents have visited Porto two to three times, with friends and relatives being the main source of knowledge about the city. Most of the respondents stayed in local accommodations and hotels, and to get around the city, most travelled on foot or used public transportation. The findings indicate that repeat visits to Porto are common, showing a high loyalty level based on friends' and relatives' recommendations. The choice of local accommodations and public transportation highlights a growing trend toward sustainable travel.

Table 2. Trip characteristics.

Number of the Visits to Porto?	N°	%
The first visit	127	48.1%
2–3 times	72	27.3%
more than 7	37	14.0%
4–5 times	24	9.1%
6–7 times	4	1.5%
How did you know about Porto?		
Friends and relatives	113	42.8%
Personal experience	71	26.9%
Social media	43	16.3%
Travel agency	17	6.4%
Advertisement	14	5.3%
Other	6	2.3%
Where do you stay in Porto (accommodation)?		
Local Accommodation (*Alojamento Local*)	105	39.8%
Hotel	72	27.3%
Friends/Family house	38	14.4%
Hostel	32	12.1%
Other	16	6.1%
Couch surfing	1	0.4%
Which transport do you use in Porto?		
Public transport	118	44.7%
Walking	95	36.0%
Car	39	14.8%
Bicycle	6	2.3%
Other	5	1.9%
Scooter	1	0.4%
Total	**264**	**100%**

Source: authors.

3.2. Factor Analysis

Diverse factors can explain tourists' perceptions of heritage management and tourism in Porto. Understanding these factors allows heritage site managers to design strategies to align with tourists' perspectives and expectations. For this research, 23 items related to sustainable heritage management in Porto, tourism impacts, sustainability strategies, and heritage management strategies, among others, were assessed. Means for these items ranged from 4.21 (highest) to 2.55 (lowest).

Further, an exploratory factor analysis using a varimax rotation with principal components analysis (PCA) extraction was used to group them into a smaller number of factors that explained tourists' perceptions.

Bartlett's test of sphericity and Kaiser-Meyer-Oklin Measure of Sampling Adequacy are used to assess the feasibility of applied factor analysis (Table 3). The Kaiser-Meyer-Olkin Measure of Sampling Adequacy is a statistic indicating the proportion of variance in the variables underlying factors might cause. High values indicate a significant correlation between original variables, meaning a factor analysis may be useful. The KMO and Bartlett's Test results in our study (0.500 and $p < 0.001$) revealed that the factor analysis was appropriate.

Table 3. KMO and Bartlett's Test.

Kaiser-Meyer-Olkin Measure of Sampling Adequacy		0.500
Bartlett's Test of Sphericity	Approx. Chi-Square	37.070
	Df	1
	Sig.	0.001

Source: Author's research results.

Principal Component Analysis revealed the presence of 4 factors with eigenvalues exceeding 1.0, explaining 57% of the cumulative variance (Table 4). Table 5 presents these 4 factors related to heritage management, preservation, tourism, and threats to heritage, and for each factor, it provides the CA value, reflecting the internal consistency of a set of items. The first factor related to heritage management has the highest CA value (0.900), suggesting that the items used to assess this factor are highly reliable. The CA values of the second and third factors related to heritage preservation and consumer-driven tourism are also high (0.713 and 0.626). The CA value in the case of the factor 4 threats to heritage (0.533) is low; however, it is still reliable in exploratory analysis, considering a limited number of items. This relationship can be further refined with more variables in future research.

The first factor is labelled as *"Heritage management: authenticity, sustainability, and community involvement"*. This factor includes items related to the importance of sustainable development goals, sustainability and environment in designing heritage management strategies, the importance of considering cultural aspects and the role of the authenticity of Porto in enriching the tourist experience.

The results indicate that tourists express a significant concern for sustainability in tourism, with perceptions of sustainable heritage management in Porto being predominantly positive. Many tourists recognise the city's efforts to balance the preservation of its historical and cultural heritage with developing tourist infrastructure. The rehabilitation of old buildings, the maintenance of traditional façades, and the revitalisation of public spaces are often seen as signs of Porto's commitment to sustainability. Regarding management, respondents consider it essential to develop heritage management strategies that consider sustainable development goals (SDGs). All these are critical and should be considered while designing heritage management strategies.

The second factor is titled *"Heritage preservation through tourism"*, which includes items related to tourism's positive impact on heritage conservation and restoration and its contribution to the preservation of the heritage through increasing awareness of tourists. Growth for touristic experiences will add value to heritage. Moreover, it is stated that globalisation and activities attracting tourists lead to the increase of historical city identity. Respondents agree that well-preserved heritage enriches tourists' perceptions and experiences of Porto City. At the same time, growth for touristic experiences will add value to heritage, leading to a greater appreciation of heritage, resulting in preservation efforts and sustainable management practices. Tourists appreciate heritage preservation and value cultural interactions that promote an authentic experience and add value to the experience of the visit.

Table 4. Tourists' perceptions regarding sustainable heritage management.

Factors	Attributes	L	E	CVE
Heritage management: authenticity, sustainability, and community involvement	Sustainability in heritage management is crucial since it guarantees the health of the economy, culture, and environment	0.822	6.589	32.945
	It is better to take environmentally friendly transport within the city	0.793		
	Heritage management should consider current planning strategies to enhance cultural heritage	0.784		
	Well-preserved heritage enriches tourists' experiences in Porto city	0.774		
	It is important to develop heritage management strategies considering sustainable development goals	0.740		
	It is important to support local businesses while travelling (going to restaurants, local markets, souvenir shops, etc.)	0.734		
	The authenticity of Porto heritage and culture is notable	0.721		
	Interaction and socialization with locals enrich touristic experiences	0.546		
	The local community is engaged in the local cultural heritage	0.448		
Heritage preservation through tourism	Globalization and activities attracting tourists lead to the increase of historical city identity	0.728	2.138	43.633
	Tourism has a positive impact on heritage conservation and restoration	0.698		
	Tourism contributes to the preservation of the heritage by increasing awareness of tourists	0.577		
	Local monuments and historical buildings are well-preserved	0.550		
	Growth for touristic experiences will add value to heritage	0.474		
Consumer-driven tourism	Tourists recycle and reduce the amount of rubbish while travelling	0.701	1.379	50.527
	Tourists value modern attractions more than heritage-based ones	0.662		
	Porto is becoming an artificial city lately due to mass tourism	0.618		
	Tourists' well-being is defined by their consumption behaviour during the visit	0.615		
Threats to Heritage	Sustainability is not a concern while travelling since the tourists are enjoying their experience outside of their home country	0.790	1.320	57.129
	Globalization and activities attracting tourists lead to the loss of historical city identity	0.738		

Source: authors. Extraction Method: Principal Component Analysis. Rotation Method: Varimax with Kaiser Normalization. Note: L-loadings And E-eigenvalue-cumulative variance were explained.

The third factor is called *"Consumer-Driven Tourism"*, linking items such as tourists valuing modern attractions more than heritage-based ones, the perception that Porto has become an artificial city due to mass tourism, and the role of consumption behaviour in defining tourists' well-being during their visits. This factor also considers tourists' efforts to recycle and reduce waste while travelling. This factor stresses the growing role of tourist preferences and practices in shaping tourism experiences and their demand-driven characteristics. On the one hand, tourists value modern attractions more than those based on heritage, and this could lead to a risk to heritage and its significance due to destinations seeking to fulfil tourists' demands. In turn, for those appreciating heritage sites, Porto may be perceived as a mass tourism artificial destination. This factor also considers tourists' efforts to recycle and reduce waste while travelling.

Visitors typically appreciate initiatives related to environmental conservation, such as promoting sustainable transport, including trams and bicycles, and the growing offer of eco-friendly accommodation. In our study, the most used transportation method is public transportation. Regarding tourists' perceptions towards recycling and waste reduction, the responses were neutral, suggesting that no clear measures encourage tourists to recycle and reduce waste while visiting Porto. Thus, educating and sensitising society, in this case, tourists, about the importance of sustainable heritage is crucial. Awareness campaigns and educational programs help to strengthen respect for and appreciation of heritage, encouraging more responsible behaviour. However, it should be considered that nowadays, mass tourism and its effects on the quality of life of residents and heritage are already observed [28]. Mass tourism is beginning to show effects on the quality of life of residents and heritage, pointing to the need for more robust strategies to manage tourist flows. The rapid growth of tourism in some areas of the city can be perceived as a threat to the balance between tourism and sustainability.

The fourth factor, titled *"Threats to heritage sustainability"*, reflects the lack of concern among tourists for heritage sustainability while travelling, as they prioritise enjoying their experience away from home. Additionally, globalisation and activities aimed at attracting tourists are identified as contributing to the loss of a city's historical identity. The final factor should emphasise the importance of a deeper understanding of the potential negative impacts of tourism on heritage sites. These impacts often stem from a lack of sustainability concerns among tourists and the influence of globalisation, which can undermine the authenticity and preservation of cultural heritage.

Table 5. Cronbach alpha.

Factors	Heritage Management: Authenticity, Sustainability, and Community Involvement	Heritage Preservation Through Tourism	Consumer-Driven Tourism	Threats to Heritage
Cronbach alpha	900	713	626	533

Source: authors.

In short, tourists in Porto tend to value sustainable management efforts. Still, some challenges remain, especially regarding managing tourist flows and preserving local heritage in the face of the city's growing popularity.

Supervisory and municipal bodies must consider various elements to carry out integrated planning and management, which must consider environmental, social, cultural and economic aspects, promoting local development and social cohesion. This includes implementing participatory policies, involving local communities in the decision-making process on heritage management, and ensuring that their values and needs are respected.

Finally, continuous monitoring and evaluation of the impacts of sustainable management actions are essential to ensure that objectives are achieved and to adjust strategies, guaranteeing long-term effectiveness.

Despite the significant contributions of this research, some limitations can be identified. Hence, the geographical and temporal scope of the research was limited since this study was tested in the historical city centre during a specific period. More contributions are needed to better understand tourists' perceptions of sustainable herniate management in historical cities vulnerable to uncontrolled tourism growth.

In the future, conducting research with heritage site decision-makers to understand their point of view would be worthwhile. They are the ones who develop the strategies and should consider the theoretical results of the research and its application in practice.

4. Conclusions

The study's conclusions on tourists' perspectives concerning the sustainable management of heritage in the city of Porto highlight crucial points that reflect the opinions and expectations of visitors and the challenges to be faced. The results allow us to draw some main conclusions. Thus, regarding the profile of visitors and the characteristics of the trips, the analysed sample is composed of diverse profile tourists with an age range varying between 17 and 65, an average age of 30 years, primarily female, with high education, and from several countries, including Spain, France, and the USA. Regarding the characteristics of the trip, they reveal that many have visited Porto before, highlighting their loyalty to the destination and continuous interest. The recommendation of friends and family is relevant when choosing a destination. Regarding transport and accommodation, a conscious and sustainable choice by tourists prevails, so we can refer to the primary choice of public transport and the preference for local accommodation.

Tourists express perceptions about sustainability and heritage management, recognising the efforts of the various entities in Porto to preserve its historical and cultural heritage while developing tourist infrastructures. The revitalisation of buildings and the maintenance of historic facades are seen as positive actions that reflect the city's commitment to sustainability, which in turn enriches tourists' perceptions and experiences in Porto City.

Data on the challenges of sustainable management point to the perception that, although efforts are effective, the growth of tourism can threaten the balance between development and preservation.

Four factors related to heritage management were defined, considering authenticity, sustainability, and community involvement in the management process, emphasising its importance and shaping tourists' perceptions. The second factor is called heritage preservation through tourism, stressing the positive impact of tourism. The third factor is related to consumer-driven tourism, showing the importance of balancing tourists' needs and heritage conservation. The fourth factor, entitled 'Threats to heritage,' emphasises the need to address the sustainability and preservation of the cultural significance of heritage sites. It stresses the importance of preserving these sites' authenticity and unique identity in the context of globalisation.

Thus, in our perspective, to ensure the effective, sustainable management of heritage in the long term, strategic planning is necessary that involves all stakeholders, including the local community, promoting participatory policies and making decisions that respect the values and needs of residents. Continuous monitoring, namely through this type of survey of tourists and the local population, as well as the evaluation of management actions, are essential to adjust strategies according to needs and ensure the preservation of Porto's authenticity and historical value.

In conclusion, the city of Porto has demonstrated significant efforts in sustainable heritage management, being recognised by tourists as a destination that balances cultural preservation with tourism. However, the city's growing popularity requires extra attention to maintain the harmony between tourism and sustainability. Authorities such as the Porto City Council and other decision-makers should implement policies and collaborate with the research team of the current project in implementing strategies that consider environmental, social, cultural, and economic aspects, ensuring local community engagement and promoting sustainable practices among visitors. Future success will depend on adapting and responding to the pressures of growing tourism, preserving Porto's unique identity. This becomes critical in contributing to achieving SDG 11, "make cities and human settlements inclusive, safe, resilient and sustainable", by strengthening efforts to protect and safeguard the world's cultural and natural heritage.

Author Contributions: All authors contributed equally to the construction of this article. Conceptualization, M.R., F.M.S. and I.V.d.F.; methodology, M.R., F.M.S. and I.V.d.F.; formal analysis, M.R., F.M.S. and I.V.d.F.; investigation, M.R., F.M.S. and I.V.d.F.; resources, M.R., F.M.S. and I.V.d.F.; data curation, M.R., F.M.S. and I.V.d.F.; writing—original draft preparation, M.R., F.M.S. and I.V.d.F.; writing—review and editing, M.R., F.M.S. and I.V.d.F.; visualization, M.R., F.M.S. and I.V.d.F. All authors have read and agreed to the published version of the manuscript.

Funding: This research received no external funding.

Data Availability Statement: The original contributions presented in the study are included in the article.

Conflicts of Interest: The authors declare no conflict of interest.

References

1. UNESCO. Convention Concerning the Protection of the World Cultural and Natural Heritage. 1972. Available online: https://whc.unesco.org/en/conventiontext/ (accessed on 1 September 2024).
2. Madandola, M.; Boussaa, D. Cultural heritage tourism as a catalyst for sustainable development; the case of old Oyo town in Nigeria. *Int. J. Herit. Stud.* **2023**, *29*, 21–38. [CrossRef]
3. Soeiro, D.; Falanga, R.; Martins, J.; Reis Silva, M.; Pomesano, L. Sustainable urban regeneration: The role of cultural heritage in Cultural Ecosystem Services (CES). *Conserv. Património* **2022**, *40*, 9–28. [CrossRef]
4. ICOMOS. International Charter for Cultural Heritage Tourism. 2023. Available online: https://www.icomos.org/images/DOCUMENTS/Secretariat/2023/CSI/eng-franc_ICHTCharter.pdf (accessed on 1 September 2024).
5. Apaydin, V. Introduction: Approaches to Heritage and Communities. In *Shared Knowledge, Shared Power. SpringerBriefs in Archaeology*; Springer: Cham, Switzerland, 2018; pp. 1–8.
6. Yang, Y.; Wang, Z.; Shen, H.; Jiang, N. The Impact of Emotional Experience on Tourists' Cultural Identity and Behavior in the Cultural Heritage Tourism Context: An Empirical Study on Dunhuang Mogao Grottoes. *Sustainability* **2023**, *15*, 8823. [CrossRef]
7. Ferreira, L.; Aguiar, L.; Pinto, J.R. Turismo Cultural, Itinerários Turísticos e Impactos nos Destinos. *Rev. Cult. Tur.* **2012**, *6*, 109–126.
8. Freitas, I.V.; Sousa, C.; Ramazanova, M.; Albuquerque, H. Feeling a historic city: Porto landscape through the eyes of residents and visitors. *Int. J. Tour. Cities* **2022**, *8*, 529–545. [CrossRef]
9. Bramwell, B.; Higham, J.; Lane, B.; Miller, G. Twenty-five years of sustainable tourism and the Journal of Sustainable Tourism: Looking back and moving forward. *J. Sustain. Tour.* **2016**, *25*, 1–9. [CrossRef]
10. González, A.T. Venice: The problem of overtourism and the impact of cruises. *Investig. Reg. = J. Reg. Res.* **2018**, *42*, 35–51.
11. Liberatore, G.; Biagioni, P.; Ciappei, C.; Francini, C. Dealing with uncertainty, from overtourism to overcapacity: A decision support model for art cities: The case of UNESCO WHCC of Florence. *Curr. Issues Tour.* **2023**, *26*, 1067–1081. [CrossRef]
12. Fernández-Villarán, A.; Espinosa, N.; Abad, M.; Goytia, A. Model for measuring carrying capacity in inhabited tourism destinations. *Port. Econ. J.* **2020**, *19*, 213–224. [CrossRef]
13. García-Hernández, M.; Calle-Vaquero, M.; Chamorro-Martínez, V. Can Overtourism at Heritage Attractions Really Be Sustainably Managed? Lights and Shadows of the Experience at the Site of the Alhambra and Generalife (Spain). *Heritage* **2023**, *6*, 6494–6509. [CrossRef]
14. Nowacki, M. Heritage Interpretation and Sustainable Development: A Systematic Literature Review. *Sustainability* **2021**, *13*, 4383. [CrossRef]
15. UNESCO. Recommendation on the Historic Urban Landscape. 2011. Available online: https://whc.unesco.org/en/hul/ (accessed on 1 September 2024).
16. Cerisola, S.; Panzera, E. Cultural cities, urban economic growth, and regional development: The role of creativity and cosmopolitan identity. *Pap. Reg. Sci.* **2022**, *101*, 285–303. [CrossRef]
17. Hassan, G.F.; Rashed, R.; EL Nagar, S.M. Regenerative urban heritage model: Scoping review of paradigms' progression. *Ain Shams Eng. J.* **2022**, *13*, 101652. [CrossRef]

18. Espinosa, C.R.; Múzquiz, E.E.P. The loss of intangible cultural heritage in the "Pueblos Mágicos" as a result of the implementation of public policies. In *Nuevas Estrategias Para un Turismo Sostenible*; Dehnhardt, M., Cortés, R., Matheu, A., Gutiérrez, B., Eds.; Thomson Reuters Aranzadi: Cizur, Spain, 2022; pp. 97–110.
19. Hassan, T.H.; Almakhayitah, M.Y.; Saleh, M.I. Sustainable Stewardship of Egypt's Iconic Heritage Sites: Balancing Heritage Preservation, Visitors' Well-Being, and Environmental Responsibility. *Heritage* **2024**, *7*, 737–757. [CrossRef]
20. Lerario, A. The Role of Built Heritage for Sustainable Development Goals: From Statement to Action. *Heritage* **2022**, *5*, 2444–2464. [CrossRef]
21. Buonincontri, P.; Marasco, A.; Ramkissoon, H. Visitors' Experience, Place Attachment and Sustainable Behaviour at Cultural Heritage Sites: A Conceptual Framework. *Sustainability* **2017**, *9*, 1112. [CrossRef]
22. Silva, F.M.; Arreiol, M.; Fragata, A. The Impact of Pollution on Cultural Heritage in the Historic Centre of Porto, Portugal. *Urban Sci.* **2024**, *8*, 31. [CrossRef]
23. Flyen, A.C.; Flyen, C.; Hegnes, A.W. Exploring Vulnerability Indicators: Tourist Impact on Cultural Heritage Sites in High Arctic Svalbard. *Heritage* **2023**, *6*, 7706–7726. [CrossRef]
24. Carvalho, M.J.E. O Centro Histórico na Dinamização das Cidades: O Centro Histórico do Porto. Master's Dissertation, Universidade do Porto, Porto, Portugal, 2011.
25. Albuquerque, H.; Quintela, J.A.; Marques, J. The Impact of Short-Term Rental Accommodation in Urban Tourism: A Comparative Analysis of Tourists' and Residents' Perspectives. *Urban Sci.* **2024**, *8*, 83. [CrossRef]
26. Porto City Council. Management Plan for Historic Centre of Porto, Luiz I Bridge and Serra do Pilar Monastery: World Heritage. 2021. Available online: https://smarttourism.cm-porto.pt/wp-content/uploads/2024/05/Historic-Centre-management-and-sustainability-plan.pdf (accessed on 16 December 2024).
27. Gusman, I.; Chamusca, P.; Fernandes, J.; Pinto, J. Culture and tourism in Porto City Centre: Conflicts and (Im) possible solutions. *Sustainability* **2019**, *11*, 5701. [CrossRef]
28. Liberat, P.; Silva, A. The Tourism Impact on The Residents' Life Quality—Case Study. In Proceedings of the 36th International Business Information Management Association (IBIMA), Granada, Spain, 4–5 November 2020; ISBN 978-0-9998551-5-7.

Disclaimer/Publisher's Note: The statements, opinions and data contained in all publications are solely those of the individual author(s) and contributor(s) and not of MDPI and/or the editor(s). MDPI and/or the editor(s) disclaim responsibility for any injury to people or property resulting from any ideas, methods, instructions or products referred to in the content.

Article

Cultural Dimensions of Territorial Development: A Plan to Safeguard the Intangible Cultural Heritage of Guano's Knotted Carpet Weaving Tradition, Chimborazo, Ecuador

Claudia Patricia Maldonado-Erazo [1,2], Susana Monserrat Zurita-Polo [1], María de la Cruz del Río-Rama [3] and José Álvarez-García [4,*]

[1] Facultad de Recursos Naturales, Escuela Superior Politécnica de Chimborazo-ESPOCH, Riobamba 060155, Ecuador; claudia.maldonado@espoch.edu.ec or claudia.maldonado@uvigo.gal (C.P.M.-E.); susana.zurita@espoch.edu.ec (S.M.Z.-P.)
[2] Programa de Doctorado Interuniversitario en Protección del Patrimonio Cultural, Escuela Internacional de Doctorado (Eido), Universidades de Vigo, Edificio Filomena Dato, 36310 Vigo, Spain
[3] Business Management and Marketing Department, Faculty of Business Sciences and Tourism, University of Vigo, 32004 Ourense, Spain; delrio@uvigo.es
[4] Departamento de Economía Financiera y Contabilidad, Instituto Universitario de Investigación para el Desarrollo Territorial Sostenible (INTERRA), Universidad de Extremadura, 10071 Cáceres, Spain
* Correspondence: pepealvarez@unex.es

Abstract: The current research article focuses on safeguarding the knotted carpet weaving tradition in Guano, an endangered intangible cultural heritage (ICH) threatened by globalisation and a lack of intergenerational transmission. The research aims to revitalise this artisanal technique through a comprehensive safeguarding plan, using a participatory action research approach. Activities included in-depth interviews, workshops, and the documentation of seven key cultural practises related to the weaving, such as spinning, natural dyeing, and design. The study found that 86% of these practises are highly vulnerable. To address this, the research developed strategies to promote generational transmission, strengthen local collaboration, and connect the craft to territorial identity and sustainable tourism. Proposed actions include intergenerational education programmes, tourism initiatives, and local fairs to boost carpet marketing. The study contributes to the field of ICH by highlighting the role of cultural tourism in preserving at-risk artisanal techniques and community identity. It emphasizes the need for collaborative approaches to safeguard living heritage in a globalized world. The research findings underscore the importance of integrating traditional practices into modern contexts to ensure their long-term sustainability.

Keywords: intangible cultural heritage; social use; safeguarding; tourist use; carpets; Guano; Ecuador

1. Introduction

The safeguarding of intangible cultural heritage (ICH) has become a highly relevant process within all territories due to an accelerated increase in acculturation and globalisation processes that threaten the continuity of local traditions [1–4]. Despite the growing recognition of the importance of ICH, there is still a significant gap in understanding how to address the challenges associated with the preservation of traditional craft techniques in rapidly changing socio-economic contexts. These challenges include the weakening of intergenerational knowledge transmission, the undervaluation of manual practises in contemporary markets, and the lack of integration of these traditions into sustainable development frameworks. Such limitations hinder the effective safeguarding of cultural practises and highlight the need for targeted interventions that go beyond mere documentation to actively engage communities in revitalisation efforts. In this framework, the artisanal technique of knotted carpet weaving in Guano, located in the Guano canton, Chimborazo province, Ecuador, represents not only a significant cultural legacy, but also a challenge to the cultural erosion of ancestral knowledge. This research focuses on implementing a safeguarding plan that arose from the need to revitalise and enhance the value of this textile practice, which has been gradually relegated to the background in the face of more visible and tangible cultural manifestations.

This research work combines three key aspects in order to emphasise the dynamic nature of the ICH, considering both its conservation and its adaptation to new realities, thus ensuring its long-term sustainability. Firstly, the participatory diagnosis, through workshops and interviews, aims to identify the cultural manifestations associated with carpet weaving and their degree of vulnerability. On the other hand, communicative action integrates the active collaboration of artisans and local actors with the aim of designing safeguarding strategies that reflect the needs and aspirations of the community. And thirdly, interdisciplinary integration, in the sense that technical, cultural and social knowledge is combined, is employed to develop a comprehensive plan that not only preserves craft techniques but also promotes their inclusion in sustainable tourism and territorial development.

The theoretical framework underpinning this research integrates elements of stakeholder theory, cultural ethnography, and historical approaches. The stakeholder theory approach will identify the key actors involved in the preservation and revitalisation of the craft technique, including artisans, local communities, government institutions, and tourists. Interaction and collaboration between these stakeholders is essential to ensure the sustainability of the ICH and its integration into the economic and cultural development of the region. The documentation of traditional practises was also based on ethnographic techniques such as semi-structured interviews, participant observation, and collaborative workshops. This approach allowed for capturing the narratives and knowledge of the knowledge bearers, ensuring that safeguarding strategies reflect local perspectives and needs. Finally, the reconstruction of the history of carpet weaving in Guano was supported by primary and secondary sources, providing a context for understanding the evolution of this practice and its cultural importance. This historical component facilitates the valorisation of traditional techniques as an integral part of the territorial identity. Following this approach will make it possible to approach the safeguarding of intangible cultural heritage from an interdisciplinary perspective.

To achieve the proposed objective of documenting, safeguarding, and revitalising the artisanal technique of knotted carpet weaving in Guano, Ecuador, a participatory action research approach was adopted with the knowledge bearers. This has made it possible to document and strengthen the practises related to carpet weaving, creating a space where traditional techniques are revalued and their intergenerational transmission is promoted. However, the study also raises critical questions about the broader implications of these safeguarding efforts: How can these traditional practises be adapted to modern economic and cultural contexts without compromising their authenticity? What mechanisms can effectively counteract the pressures of cultural homogenisation while fostering innovation and economic viability for artisans? Addressing these questions is essential for developing safeguarding strategies that are not only inclusive but also responsive to the evolving needs of the community. This approach not only seeks to preserve the art of weaving but also to integrate it into the social and economic fabric of the community through cultural tourism and creative industries.

The importance of ICH for building local identities is fundamental, as it allows communities to reaffirm their sense of belonging and resistance to cultural homogenisation [5–8].

The methodology was designed in alignment with the guidelines of the National Institute of Cultural Heritage (INPC). In the first stage, cultural manifestations were recorded through in-depth interviews, participatory workshops, and ethnographic observation to document the processes and knowledge associated with the technique of carpet weaving. For this purpose, tools such as diagnostic matrices and record cards were used, which allowed for the collection of exhaustive information on seven key cultural manifestations, such as spinning, scale design, and knotting technique. In a second stage, a vulnerability diagnosis was carried out. Workshops were held with knowledge bearers, and assessment matrices were applied that considered categories such as memory, identity, and heritage. In the last phase, the safeguarding plan was formulated.

This research aims to contribute to the discussion on ICH safeguarding. Although studies have been carried out on ICH, these have focused on social uses, rituals, and festive events, but not on traditional craft techniques. In this sense, this research differs significantly from existing research by focusing on the documentation and revitalisation of a specific craft technique. It offers a unique perspective that directly addresses the challenges associated with traditional manual practises, such as intergenerational transmission, economic sustainability, and cultural integration in contemporary contexts. By problematising these dimensions, this study aims to provide insights into the broader systemic barriers that undermine the sustainability of traditional crafts, such as the lack of institutional support, limited market access, and the undervaluation of traditional knowledge in modern development paradigms. For all these reasons, this study highlights the need for inclusive strategies that recognise the dynamics of cultural heritage as a living process. In this regard, the article provides a detailed analysis of the historical, social, and cultural context surrounding the knotted carpet weaving technique, as well as the methodologies implemented for its safeguarding and revitalisation. The interrelation between intangible cultural heritage and sustainable development becomes a core axis to promote not only the preservation of traditions but also the economic and social welfare of the communities involved [2,9,10].

2. Literature Review

2.1. Intangible Cultural Heritage and Safeguarding

The study of intangible cultural heritage (ICH) focuses on safeguarding the non-physical aspects of culture. According to Lixinski [11], preservation generally refers to and focuses on physical artefacts, while safeguarding recognises the dynamic and living nature of intangible heritage, emphasising that it requires community participation and respect for cultural contexts. Therefore, the term safeguarding is inherent to the intangible because it provides a holistic approach by considering legal, social, and cultural dimensions, with the objective of empowering communities and ensuring the continuity of their cultural practises.

Safeguarding focuses on numerous actions, which enable identification, protection, conservation, revitalisation, and transmission to future generations [1–3,12,13]. As pointed out by UNESCO, it corresponds to actions applied to traditions, practises, and expressions, which are essential for maintaining the identity and cultural diversity of the social groups that own the territories, and even more for fostering sustainable development.

At this point it should be noted that ICH was linked to heritage management strategies at the turn of the new millennium, initially categorised as "poor heritage". Through this expression, it sought to reflect the challenges and complexities associated with the recognition, appreciation, and safeguarding of cultural expressions due to their reduced permanence and visibility as opposed to tangible assets [14,15].

Despite this incipient beginning, the study of ICH has evolved significantly since 2001, following the first UNESCO Proclamation of Masterpieces of the Oral and Intangible Heritage of Humanity. The proclamation recognises that the cultural manifestations of territories are full of dynamic and living aspects of culture, highlighting the importance of this for local communities, while redefining heritage as an active and participatory process rather than a static collection of artefacts [5]. This has also led to generating a legal and institutional framework that allows for the holistic protection and management of cultural heritage, including both tangible and intangible elements.

The relevance that ICH has taken on at international level has provided opportunities for economic development and social cohesion through the use of the cultural resources of each locality. Unfortunately, as Stefano et al. [15] point out, one of the main challenges to safeguarding ICH is based on its inherent vulnerability and the risk it faces, as it is gradually undervalued and minimised in public management. This statement is corroborated by Aikawa-Faure [3], Deacon et al. [16], and Lenzarini [10]. The concept of cultural erosion is particularly relevant in this context, as it refers to the gradual loss or dilution of ancestral knowledge due to external pressures such as globalisation, acculturation, migration, and the lack of intergenerational transmission [17,18]. Cultural erosion occurs when traditional practises are displaced by more dominant or homogenised cultural expressions, often perceived as modern or economically advantageous [18–20]. This process not only weakens the continuity of intangible heritage but also disconnects communities from their cultural roots, leading to a loss of identity and a diminished sense of belonging [18,19,21,22].

The evolution of the relationship between ICH and safeguarding is shown below in a timeline (Figure 1).

1960s-1970s: First debates on intangible culture.
- 1966 (Bolivia): World Conference on Folklore.
- 1972 (UNESCO): Convention on the Protection of the World Cultural Heritage. (focused on tangible heritage, but the first discussions arose on the need to include intangible heritage). Access link: https://whc.unesco.org/en/conventiontext/

1980s: inclusion of intangible heritage in cultural debates.
- 1982: The Intergovernmental Committee for the Promotion of Popular Arts is established within UNESCO.
- 1989: UNESCO Recommendation on the Safeguarding of Culture and Traditions. First attempt to formally include ICH in safeguarding policies. Access link: https://ich.unesco.org/en/masterpiece-proclamation-00103

1990s: Development of the concept of Intangible Cultural Heritage.
- 1993: UNESCO programme "Proclamation of Masterpieces of the Oral and Intangible Heritage of Humanity. Outstanding intangible cultural practices were selected for preservation and promotion.
- 1999 (Stockholm): UNESCO World Conference on Culture. Milestone: underlined the importance of intangible culture in sustainable development and cultural diversity.

2003: Convention for the Safeguarding of the Intangible Cultural Heritage
It is a fundamental milestone that recognizes the importance of ICH and establishes an international legal framework for its safeguarding. Access link: https://ich.unesco.org/en/convention

2010s: Implementation and expansion of PCI safeguarding
- 2010: Representative List of the Intangible Cultural Heritage of Humanity. Includes cultural practices and expressions from diverse communities around the world.
- 2012: Tenth Anniversary of the 2003 Convention. Organized by UNESCO. Reflection on progress and challenges in safeguarding ICH.
- 2015: 2030 Agenda for Sustainable Development. The UN establishes a direct link between safeguarding ICH and sustainable development, integrating culture into social, economic and environmental development. Access link: https://sdgs.un.org/2030agenda

Current decade: PCI as a central axis of sustainable development
- 2020: National and international policies are consolidated, with greater integration of ICH into sustainable development strategies, education and social cohesion. National plans for safeguarding ICH and sustainable cultural tourism (on the agenda of many countries).
- From 2020 onwards: a period characterized by: (1) international efforts are concentrated on combating the vulnerability of ICH to globalization, conflicts and climate change. (2) local communities play an increasingly active role in its safeguarding, encouraged by UNESCO and other international organizations.

Figure 1. Evolution of the relationship between ICH and safeguarding. Source: own elaboration.

2.2. Importance of ICH for Building Local Identities

ICH, which comprises oral traditions, knowledge, customs, social practises, and crafts, is a vital element for the cultural identity of communities and allows communities to reaffirm their identity in today's globalised world [1]. In this regard, it plays a crucial role in building and understanding how communities perceive themselves, as well as how they are perceived by others, addressing organisational, economic, social, and cultural aspects. Goode [23] further argues that communities have managed to build identities (discourses) that reflect their cultural background by means of ICH. This is an element which enables the identification, promotion, and dissemination of distinctive characteristics, where territories can increase their attractiveness to target markets through tourist products that take advantage of these characteristics [24], with the aim of preventing urban development from producing a loss of distinctiveness [25].

ICH is also strongly linked to community cohesion, social organisation, and the system of representations of the human group, since it generates a sense of belonging, i.e., the cultural representations and manifestations generated by communities are reflected through it [26,27].

However, the erosion of cultural knowledge poses a serious threat to this cohesion as it fragments the transmission chain of traditions and weakens the symbolic meaning behind cultural practises [18]. This erosion is often exacerbated by the prioritisation of globalised cultural patterns, the migration of younger generations to urban areas, and the lack of institutional mechanisms to protect and revitalise traditional knowledge. Consequently, the disconnection between communities and their cultural heritage can lead to a sense of alienation and reduced community resilience [28,29].

Nevertheless, it should be clear that all this development is not exempt from external influences such as acculturation, globalisation, and technological advances, which are elements that can reinforce and challenge traditional notions of locality, giving rise to dynamic and evolving identities.

2.3. Strategies for Safeguarding Intangible Cultural Heritage

The main strategies to achieve the safeguarding of ICH are detailed as follows: (1) identification, documentation, and recording [10,30], (2) intergenerational transmission (educational programmes and learning by doing) [2], (3) promotion, visibility, and enhancement (media, cultural festivals, and public events) [7], (4) legal and political support (notably the UNESCO Convention for the Safeguarding of ICH [1], which provides a framework for international cooperation in the protection of ICH) [31], (5) community participation and social use (empowerment of communities and inclusive approach) [32,33], (6) fostering sustainable cultural tourism [34,35], (7) innovation [36,37], (8) financing, and institutional support (funds and grants, and international cooperation) [38], and (9) research and education on ICH [30,39], among others. These are derived from the principles established by UNESCO and other relevant legal frameworks at the international level.

These actions take into consideration the entire process required to achieve an adequate understanding of cultural manifestations, highlighting that the active participation of local communities is essential for achieving this [13,40]. It begins with identification and documentation, which are not possible to achieve without the community because they require the use of information gathering techniques, such as participant observation, ethnographic records, cultural mapping, or backgrounds that allow for obtaining comprehensive data to sustain cultural manifestations over time [41].

An additional strategy to strengthen the safeguarding of ICH is the application of stakeholder theory, which highlights the roles and interactions of key actors in preserving and revitalising cultural practises. Developed initially in the context of corporate governance, stakeholder theory has evolved into a versatile conceptual tool applicable across disciplines, including cultural heritage management [42]. This theory posits that the sustainability and success of any initiative depend on identifying and addressing the needs and expectations of all stakeholders involved [43]. In the safeguarding of ICH, stakeholders include a wide array of actors: (1) local communities, who are the bearers and practitioners of the cultural traditions; (2) artisans and practitioners, who ensure the continuity and authenticity of these practises; (3) governmental institutions, responsible for formulating policies and allocating resources; (4) cultural associations and NGOs, which often play a bridging role between communities and policymakers; and (5) tourists and the private sector, whose engagement can provide visibility and economic sustainability for ICH [42–46]. Recognising the interdependence of these groups is critical to creating inclusive and effective safeguarding strategies.

In the context of the carpet weaving technique in Guano, the stakeholders include artisans, local communities, governmental institutions, cultural associations, and tourists. Stakeholder collaboration is fundamental for ensuring the sustainability of ICH, as it aligns the interests and actions of these groups toward common goals, such as intergenerational transmission, economic development through cultural tourism, and the preservation of identity.

The stakeholder theory provides a framework for understanding the dynamics of these interactions, emphasising that effective safeguarding requires inclusive and participatory approaches. For instance, artisans contribute their technical expertise and cultural knowledge, while communities and institutions provide the support needed to integrate these practises into broader cultural and economic systems. Tourists, as external stakeholders, can also play a role in valuing and promoting the unique aspects of ICH, fostering its appreciation and financial sustainability.

Research and preservation are the product of the first two actions, which play an important role because they enable the permanence of manifestations over time. Furthermore, currently, technology integration, as pointed out by Deng and Mo [47], has improved access to information on cultural expressions, but above all, innovative ways to preserve and promote ICH are provided, enabling its value to be transmitted to all age groups.

The promotion, appreciation, transmission, and revitalisation of ICH are the product of all the above actions, and several studies have established that they can be achieved through tourism, which contributes to the economic development of territories, particularly in rural areas. In this regard, Liu and Qi [48] point out that tourism in China has succeeded in promoting rural revitalisation, contributing to economic growth and cultural preservation.

In addition to this, there is the development of creative industries from which heritage interpretation is achieved, and these can easily be integrated into tourism by using modern storytelling techniques and involving visitors through participatory experiences, while cultural manifestations can enhance their attractiveness and economic viability [49].

Finally, social use is approached as a way to integrate ICH into everyday life, fostering a sense of identity and belonging, ultimately contributing to the sustainability of cultural practises [40].

While all of the previously outlined strategies provide a comprehensive approach to safeguarding intangible cultural heritage, the dynamic nature of cultural heritage must be considered in any process, i.e., it is not static, it transforms with society, and thus, safeguarding efforts must be adaptive and inclusive to address diverse perspectives.

Finally, safeguarding ICH, as documented in recent studies, requires a multidisciplinary approach that not only addresses the preservation of artisanal techniques, but also considers the education and empowerment of local communities [50]. For example, the implementation of educational initiatives that promote the intergenerational transmission of knowledge is encouraged, which constitute fundamental processes to ensure the continuity of these practises [6]. Furthermore, within the legal framework and cultural policies, it is promoted that these be adapted to protect the rights of indigenous and mestizo communities in the face of globalisation and the exploitation to which they are exposed [38]. This study not only documents the carpet weaving technique but also proposes a revitalization model that integrates the social use of PCI into tourism development. Referring to successful examples in other regions, such as the strategies implemented in Indonesia to preserve songket, how community participation and institutional support are key to facing current challenges has been highlighted [51]. In this sense, it is crucial to go back and forth between the findings of the present study and existing theories, creating a dialogue that not only validates local practises but also offers new perspectives on the safeguarding of intangible cultural heritage.

3. Methodology

This study corresponds to a work based on participatory action research with the holders of knowledge and practises related to ICH linked to the technique of knotted carpet weaving in Guano, Chimborazo, Ecuador, with the purpose of enhancing the value of this intangible cultural heritage. The methodological process follows the path outlined by the National Institute of Cultural Heritage through the Ministerial Agreement NO. DM-2018-126 [52] and the base guidelines established in the Methodological Guide for the safeguarding of ICH of the INPC [53], which state that in order to achieve clear actions in the process of safeguarding ICH, three key stages must be achieved (Figure 2), as follows.

Figure 2. Work methodology proposed by the INPC. Source: [51].

The first stage, which corresponds to the recording of cultural manifestations linked to the Guano knotted carpet weaving technique, requires the completion of 4 work stages, which are detailed in Figure 3. The aim of this stage is to document traditional practises related to carpet weaving, from spinning and natural dyeing to scaled designs and knotting technique. Three tools will be used: (1) semi-structured in-depth interviews with knowledge bearers to capture technical, historical, and narrative details; (2) participatory workshops bringing together artisans, community members, and local experts to share knowledge, identify problems, and propose solutions; (3) ethnographic observation to produce detailed records of the processes in the artisan workshops through field notes, photographs, and audio-visual recordings.

The methodological process was rooted in a collaborative and participatory approach, which allowed for the systematic organisation and analysis of data. Initially, data collection was designed to involve artisans and other stakeholders in multiple stages of the research, ensuring their perspectives shaped both the collection and interpretation of the information. The interviews, workshops, and ethnographic observations were analysed using qualitative methods, where data were coded and organised into categories that reflected key themes such as generational transmission, cultural relevance, and market challenges. During the workshops, preliminary findings were presented back to the community for validation, enabling a feedback loop that enhanced the accuracy and reliability of the results. This iterative process ensured that the voices of the knowledge bearers were integrated throughout the research and provided a foundation for co-creating solutions.

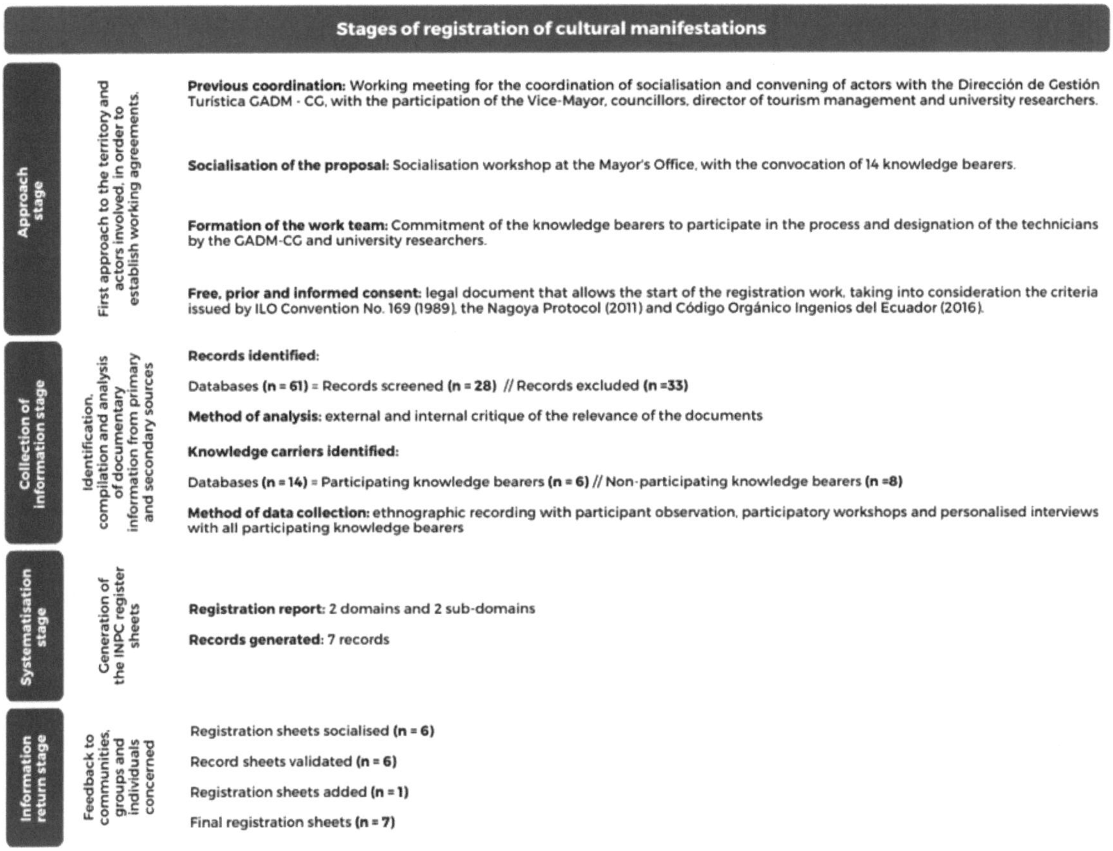

Stages of registration of cultural manifestations		
Approach stage	First approach to the territory and actors involved, in order to establish working agreements	**Previous coordination:** Working meeting for the coordination of socialisation and convening of actors with the Dirección de Gestión Turística GADM - CG, with the participation of the Vice-Mayor, councillors, director of tourism management and university researchers. **Socialisation of the proposal:** Socialisation workshop at the Mayor's Office, with the convocation of 14 knowledge bearers. **Formation of the work team:** Commitment of the knowledge bearers to participate in the process and designation of the technicians by the GADM-CG and university researchers. **Free, prior and informed consent:** legal document that allows the start of the registration work, taking into consideration the criteria issued by ILO Convention No. 169 (1989), the Nagoya Protocol (2011) and Código Orgánico Ingenios del Ecuador (2016).

Collection of information stage	Identification, compilation and analysis of documentary information from primary and secondary sources	**Records identified:**
		Databases (n = 61) = Records screened (n = 28) // Records excluded (n =33)
		Method of analysis: external and internal critique of the relevance of the documents
		Knowledge carriers identified:
		Databases (n = 14) = Participating knowledge bearers (n = 6) // Non-participating knowledge bearers (n =8)
		Method of data collection: ethnographic recording with participant observation, participatory workshops and personalised interviews with all participating knowledge bearers

Systematisation stage	Generation of the INPC register sheets	**Registration report:** 2 domains and 2 sub-domains **Records generated:** 7 records
Information return stage	Feedback to communities, groups and individuals concerned	Registration sheets socialised (n = 6) Record sheets validated (n = 6) Registration sheets added (n = 1) Final registration sheets (n = 7)

Figure 3. Stages performed at the time of recording events.

The systematisation process also involved creating detailed technical files for each cultural manifestation, incorporating narratives, photographic records, and audiovisual documentation. These files were structured using the guidelines provided by the INPC and were validated collaboratively with the participants. The triangulation of data from multiple sources—including interviews, observations, and workshop discussions—helped to cross-check and refine the findings, ensuring consistency and depth in the analysis. This participatory methodology reflects a commitment to empowering the community while producing results that are both rigorous and culturally sensitive.

At this point, it is necessary to clarify that the universe of study was defined from the 14 carriers registered by the Decentralized Autonomous Government of the Municipality of Canton Guano (GADM-CG), taking into consideration that this sectional government is the one that has motivated the process of certification of the cultural manifestation as part of the Cultural Heritage of Ecuador. The work group was made up of both men and women who work as carpet weavers or who carry out some of the steps in the production of carpets, who reside in the urban parish of La Matriz and El Rosario, and who are the last to continue with the practice of this weaving technique. A call was issued to the entire identified universe, but in the end only 6 knowledge bearers participated voluntarily, who were the ones who have a workshop in operation. The rest indicated their refusal to participate in the process because they no longer practice it all the time. The information collection was carried out through personalised interviews and participatory workshops after signing the

Free, Prior and Informed Consent required to safeguard the protection of rights. During this information collection process, the carpet weaving process was exhaustively obtained and the PCI registration forms established by the INPC were completed.

The second stage corresponded to the diagnosis of the recorded ICH, for which a participatory workshop was held with the holders, in which a double-entry assessment matrix was applied to the recorded manifestations. It was structured by taking into consideration the three categories for the identification and management of ICH, proposed in the Methodological Guide of the National Institute of Cultural Heritage [53]: Heritage, Memory, and Identity (Figure 4). The objective at this stage was to assess the current state of cultural manifestations in terms of preservation, transmission, and community relevance.

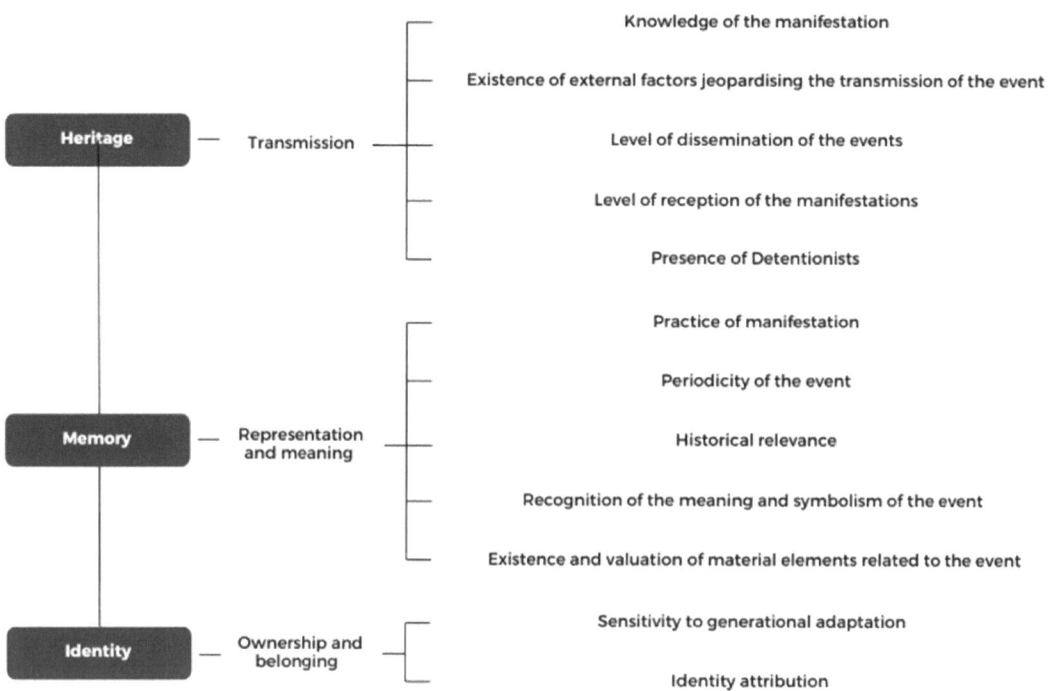

Figure 4. Categories, variables, and criteria considered at the time of diagnosis of the manifestations.

Following the diagnosis of the manifestations, the score achieved in this assessment made it possible to determine the registration category as follows (Table 1).

Table 1. Registration categories according to the INPC.

Registration Category	Description	Score
C1: Existing	Manifestations with low level of sensitivity and vulnerability to change	21–30
C2: Vulnerable	Manifestations with medium level of vulnerability due to threats in the transmission of knowledge and unfavourable conditions	11–20
C3: Highly vulnerable	Manifestations highly vulnerable to change and risk	0–10

Source: [37].

Finally, in phase 3, with the objective of designing comprehensive strategies for the preservation, revitalisation, and sustainability of traditional practises, the strategic approach of the safeguarding plan was formulated through a SWOT analysis, followed by the prioritisation of critical nodes and key success factors. Based on the prioritisation of critical nodes and key success factors, strategic objectives and strategies were formulated to guide the development of programmes and projects for the safeguarding of the community's cultural manifestations.

Thus, the methodological process of safeguarding ICH sought to develop actions aimed at the stimulation, revitalisation, transmission, communication, dissemination, promotion, enhancement, and protection of Intangible Cultural Heritage. The data processing tools applied to achieve the objective were qualitative analysis (coding and analysing the narratives obtained in interviews and workshops), triangulation, which allowed cross-checking the data from the interviews, observations, and workshops to guarantee the validity and reliability of the results, and the technical files through which each cultural manifestation identified was recorded, including technical descriptions, cultural narratives, photographs, and audiovisual records.

4. Results

4.1. Historiography of Guano Carpets

Within the process of recording information, it was consolidated into two components within the bibliographic review. It was possible to establish a historiography in the subsystems proposed by Clarke [54], as shown in Figure 5.

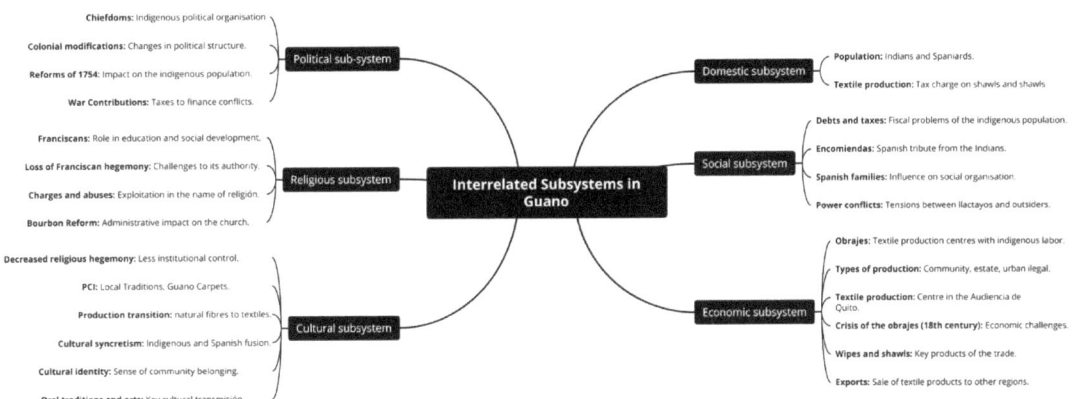

Figure 5. Inter-related sub-systems within Guano according to Clarke [39].

The predominant subsystems within the space were economic and religious due to the representativeness of the obrajes (textile plants) and textile production in the area, with much of this production being generated through the indoctrination process by the Franciscan Order. Guano was characterised for being home to one of the three most important obrajes of Riobamba due to the large number of indigenous labourers [55], a labour force that was the basis of textile production registered by the Audience of Quito. Historical data indicate that about 50% of the labour force in this area came from Riobamba and adjacent areas such as Guano [56].

The level of presence of obrajes was representative, with small obrajes called chorrillos [57], which were considered unfair competition. The most prominent obrajes that withstood the crisis of the obrajes in the mid-18th century were those linked to the haciendas, including Guano.

A unique detail is pointed out by Villavicencio [58], who highlights how the textile quality of the area was recognised by the quality of the fine woven cloth, as well as embroidered blankets, carpets, ponchos, and saddle blankets, which were commercialised in areas such as Nueva Granada. The representativeness of its production meant that during the Independence period, Guano was positioned in the market, even registering specifications in the shipping guides as "rebozos de Guano", which was unusual for that time [59].

Humboldt [60] mentions that by 1802 in "Guano, Riobamba...there is not even a single miserable Indian hut, where you can see a loom or cotton cloth in dye", i.e., production was now linked to "modest mestizos and Indians" with the extinction of large obrajeros [56]. Finally, from 1840 to 1870, Guano was classified as one of the manufacturing centres that most resisted European competition and smuggling.

In the religious sphere, the strong influence of Christianity, ecclesiastical organisation, and conflicts arising from the clergy's secularisation were recognised up until the Bourbon reforms that sought to reduce religious hegemony, especially that of the Franciscans, and the charges and abuses committed by doctrine-makers [55,61].

In the domestic sphere, the population's dedication to textile production is mentioned as a characteristic of the population, which is a condition that provokes changes within the territory, as mentioned by Coronel Feijóo [56], who points out that "obrajes and haciendas, together with a modified indigenous landscape" led to the practice of extensive and intensive grazing in areas populated by forests and native shrubs in high altitude areas, as well as the integration of monoculture practises in haciendas, thus "breaking the ancestral balance of the use of ecological lands".

The social subsystem highlights the condition of bulk labour that the indigenous population represented for the obrajes, which was to a large extent retained due to problems of debts and taxes they owed to the Crown and the Church, in addition to pointing out social disputes during the processes of interaction between indigenous and foreign communities. Within the political sphere, the implementation of colonial modifications and the justice system that modified the dynamics of the territory are highlighted.

Finally, in the cultural subsystem, it is seen that Guano carpets have their origin in the guarlapas, which correspond to rectangular pieces or small rugs, which were used between the horse and the saddle. They served to protect the animal by avoiding damage to its skin. This product was developed in family production obrajeros [62]. In the 20th century, there was a boom in carpet production, especially between 1925 and 1930 in the Espíritu Santo neighbourhood of the Guano canton. The positioning of this carpet resulted in it reaching different latitudes of the national and Latin American territory. Over time, improvements have been evident in the quality of the raw material and in the production processes. However, production was affected by the Ambato earthquake in 1949, which caused migration processes of carpet artisans [63].

From the 1960s to the 1990s, production was taken up again as a family trade, employing between 10 and 15 workers per workshop. In 1969, the first carpet shop was opened in Guano, which contributed to artisanal production becoming one of the pillars of the canton's economy and an important component of its cultural identity. In 2019, the carpet weavers of Guano were certified as intangible cultural heritage of Ecuador [64].

4.2. Recording of the Cultural Events Linked to Guano Knotted Carpets

Over the course of 3 days of personalised interviews, seven record cards were consolidated between the initial recording process and the knowledge return stage. Table 2 shows that 71.42% (5) of the manifestations correspond to the area of traditional craft techniques, while 28.58% (2) correspond to the area of oral traditions and expressions.

Table 2. Recorded cultural manifestations.

No.	Manifestation	Scope	Subscope	Description
1	History of the Guano Carpet	Traditions and oral expressions	Local memory linked to historical events reinterpreted by communities	The first rug called guarlapa is described in detail: "in colonial times, the people of Guano worked with hand-knotted rugs, which were rustic and used as a blanket for the horse's saddle, called guarlapas". Between 1925 and 1930, the production of rugs for rooms began, the main material being cotton. In 1949, production declined due to an earthquake, causing the migration of a large number of artisans. In the years 1960, 1985, and 1990, rug making took off, and most families in the canton engaged in this activity because it was well paid. Rugs were not only sold within the country but were also exported to Colombia. However, nowadays rugs are imported from Colombia and Peru. The decline in the production of Guano carpets is due to factors such as the crisis experienced in Ecuador in 1999 and the ease of entry of Chinese carpets into Ecuador. Currently, Guano carpets are made of 50% wool and 50% cotton and are hand-spun.
2	Narration of the traditional square El Rosario	Traditions and oral expressions	Local memory linked to historical events reinterpreted by communities	This public space was where farmers came to sell only wool or lamb's wool yarn, which at that time were spun by hand using sigse, the raw material used for making carpets, and at that time, wool was also dyed with natural products. Sales began to decline due to the emergence of spinning mills, which left behind those people who spun by hand, and the people who sold wool went to the spinning mills and no longer to the square; thus, the sales and the name of the place disappeared. Today, it is a place where crafts, sweaters, carpets, and shoes are sold. They are sold by the "20 de diciembre" artisans' association.
3	Traditional spinning process	Traditional craft techniques	Traditional craft techniques	Hand spinning is an activity that has been part of the artisan culture of Guano for over 60 years and has been passed down from parents to children. The yarns produced were given to carpet makers, or these in turn could be purchased at the Plaza Roja del Rosario. "People (…) came from Llapo, San Isidro, San Andrés or from the Igualata moor", where they could buy wool or sisal yarns. In some cases, sheep wool was acquired together with leather. Then, they were separated and the wool was used for the traditional spinning process, while the leather was taken to the tannery in Ambato. In the past, wool was washed in streams and sisal was used as a natural detergent, but with the passage of time, it started being washed in hot water and with Alex soap to remove grease. After washing it, it is left to dry for a day or two depending on the climate. Then, the wool is placed in a jute sack, and with the help of a sigse and the artisan's hand, the wool is pulled and twisted little by little to obtain three types of yarn: thick, which required less time; fine, which was more desirable because it lasted longer; and finer, used for the production of blankets and ponchos. To deliver the yarns, they must be skeined with a wooden tool with the same name (skein) and which is composed of a vertical stick in the centre and two sticks placed horizontally at each end, in which the threads are passed in a 'V' shape. This process is performed in order to have already weighed what is going to be delivered, and 18 balls of thread are needed to obtain a 3-pound weft. After this process, the artisan carpet maker can leave the threads in white or can dye them. It is important to mention that nowadays not only woollen yarn is used but it also mixed with acrylic.

Table 2. *Cont.*

No.	Manifestation	Scope	Subscope	Description
4	Technique to scale of the Guano carpet design	Traditional craft techniques	Traditional craft techniques	The design technique on the carpet is made by means of a photo that the customer provides. Previously, the designs were made on the basis of dots, but with time, they changed to squared paper, taking into account the scale and dimension that the carpet will have. An important fact is that each square on the paper represents a knot, i.e., in 10 cm, 50 dots or knots were made. Mañay, Juan Pozo, and Gerardo Llambay were in charge of making the grid drawings. In addition, the designs were previously made by another artisan. However, the process used to have some complications and depending on the complication, it increased the price and time, so deliveries were delayed. This prompted carpet makers to learn this technique and, currently, the same family members of the artisans do it. Some designs have a higher level of difficulty, usually shading or face profiles. As for flowers, figures, or other types of designs, they are not usually so complex, as they have learned the technique of making them over time. In addition, lately, the most common designs are landscapes and animals. They also use the skin of Amazonian animals such as snakes, zebra, and military camouflage, among others. Also, they usually enhance designs with images of political figures.
5	Natural dyeing technique on lamb's wool	Traditional craft techniques	Traditional craft techniques	Obtaining natural dyes is based on the direct observation of plants or insects that produce strong and permanent colours in lamb's wool, which is a technique that has been used for 80 years. Segundo Colcha says that in order to obtain the natural dye, "the plants are placed with boiling water in a bronze pot, after a few minutes the spun wool is placed in the pot and left to cook for two hours until it achieves a colour, it is left to dry and then sheared". Among the species used for the extraction of natural dyes are the following: - Walnut (Juglans neotropica), where a range of brown colours is obtained. - Rumi beard (Tillandsia landbeckii subsp.), used as a natural medicine and as a dye from which a cream colour is obtained. - Moss (Thamniopsis sp.), from which a grey colour is obtained. - Capuli: (Prunus serotin), from which a stronger shade of brown is obtained. In the 1950s and 1960s, artificial dyes were used, but they did not last very long, so they opted to use German and Swiss dyes, the latter being the best quality. Nowadays, this technique is no longer used, as artisans prefer to go and buy the yarn of the colour they need. The thread they buy is acrylic thread, which allows for better colour fixation.
6	Guano carpet knotting technique	Traditional craft techniques	Traditional craft techniques	The process that is applied for this technique begins by placing the warp on the loom. Depending on the size of the carpet, it will be divided into quarters to distribute the fabric and so that the carpet is even. Then, the thread is passed through the middle of the rows that form the warp and the first base knots are started. To make the knots, two rounds must be made. The first one is made at the top and the second one at the bottom, giving the shape of a round knot, and the excess is cut off, thus making the whole row. It is worth mentioning that if this step is not performed carefully, it could cause a serious injury, as the tool used is very sharp. Once the row of knots is finished, a strip of mine is passed over the recently finished row. To ensure that this mine is adjusted with the knots, a striker, which exerts pressure, is used. This process requires practice, as the carpet depends on this to have the same level. The time it takes to tie the knots will depend on the technique and speed with which the official in charge works.

Table 2. Cont.

No.	Manifestation	Scope	Subscope	Description
7	Guano Carpet Shearing Technique	Traditional craft techniques	Traditional craft techniques	The process of finishing a carpet requires extra knowledge "la trasquilada" (shearing). Formerly, this process was very tiring because it was performed only with heavy tin scissors, which tired the craftsman's hands and, therefore, carpet makers chose to pay someone to do the job, people who were called shearers. However, the person in charge had to have some skills to outline, align, and determine the final details so that the drawing was shown as it is. Over the years, this process became much easier with the acquisition of industrial tailor's scissors, and finally machines arrived, which were the same ones used for sheep shearing and were better. Nevertheless, scissors are still used on the edges, i.e., both materials complement each other.

4.3. Vulnerability of Recorded Intangible Cultural Heritage

During the information return workshop, the analysis of the state of vulnerability of each of the previously recorded cultural manifestations was presented and validated. The scores assigned to each criterion were adjusted according to the opinions provided by the knowledge holders, and it was determined that 86% of the manifestations are in a highly vulnerable category and only 14% are in a current state (Table 3).

Table 3. Community intangible cultural heritage vulnerability matrix.

No.	Manifestations	Heritage					Memory			Identity			Total	State of Preservation		
		Knowledge	External Factors that Put Transmission at Risk	Level of Diffusion	Reception Level	Presence of Holders	Practice of Manifestation	Periodicity of the Manifestation	Historical Relevance	Recognition of Meaning and Symbolism	Existence and Valuation of Related Material Elements	Sensitivity to Generational Adaptation	Identity Attribution		Highly Vulnerable	Vulnerable	Current
1	History of the Guano Carpet	2	1	0	0	1	1	1	0	1	1	1	1	10	X		
2	Narration of the traditional square El Rosario	2	1	1	1	1	0	0	0	1	1	1	1	10	X		
3	Traditional spinning process	2	2	0	0	1	0	0	0	1	1	1	1	9	X		
4	Technique to scale of the Guano carpet design	1	3	0	0	1	0	0	1	1	1	1	1	10	X		
5	Natural dyeing technique on lamb's wool	2	2	0	0	1	0	0	0	1	1	1	1	9	X		
6	Guano carpet knotting technique	2	4	1	1	1	1	1	2	2	3	3		22			X
7	Guano carpet shearing technique	1	2	0	0	1	1	1	0	1	1	1	1	10	X		
Total															6	0	1
Total (%)															86%	0%	14%

Highly vulnerable	0–10
Vulnerable	11–20
Current	21–30

4.4. Strategic Approach

A SWOT analysis with knowledge holders and stakeholders of the territory was used as the starting point for building this approach, followed by the identification of internal factors (strengths and weaknesses) and external factors (opportunities and threats) present in the territory, which were combined for an assessment process that determined seven critical nodes based on weaknesses and threats, and five key success factors based on strengths and opportunities.

To develop a strategic planning adjusted to the reality of the territory, it was necessary to establish the stakeholders involved within the territory. During this process, three types of organisational power figures were identified: (a) social base is made up of the population of the Guano canton and is represented by a yellow circle; (b) associative fabric is made up of the Association of Carpet Makers, ESPOCH, and UNACH, which are represented by a purple rectangle; and (c) power images, which possess high power, can make decisions and invest resources within the territory, which are represented by an orange octagon and are formed by the Ministry of Culture and Heritage of Ecuador, National Institute of Cultural Heritage of Ecuador (INPC), Ministry of Tourism of Ecuador, GADM-CG, Directorate of Tourism Management of GADM-CG, Permanent Committee of Guano Festivities, and Citizen Committee of Guano Magical Town.

Figure 6 shows the relationships that the stakeholders involved have and the relationships that exist between them. There is a strong presence of strong relationships of dependence and collaboration between the images of power, established to a large extent by the national legal framework that regulates cultural development and management at the national level through the public policy designed for this purpose. With regard to the associative fabric, ESPOCH maintains collaborative relationships with the Directorate of Tourism Management of GADM-CG due to an agreement between the two institutions, which seeks the development of tourism in the canton, as well as safeguarding its ICH and conserving its tangible cultural heritage. In addition to this, ESPOCH collaborates with the INPC and the Ministry of Culture and Heritage due to the support for the consolidation of the safeguarding plan and other collaboration processes that are maintained between the institutions.

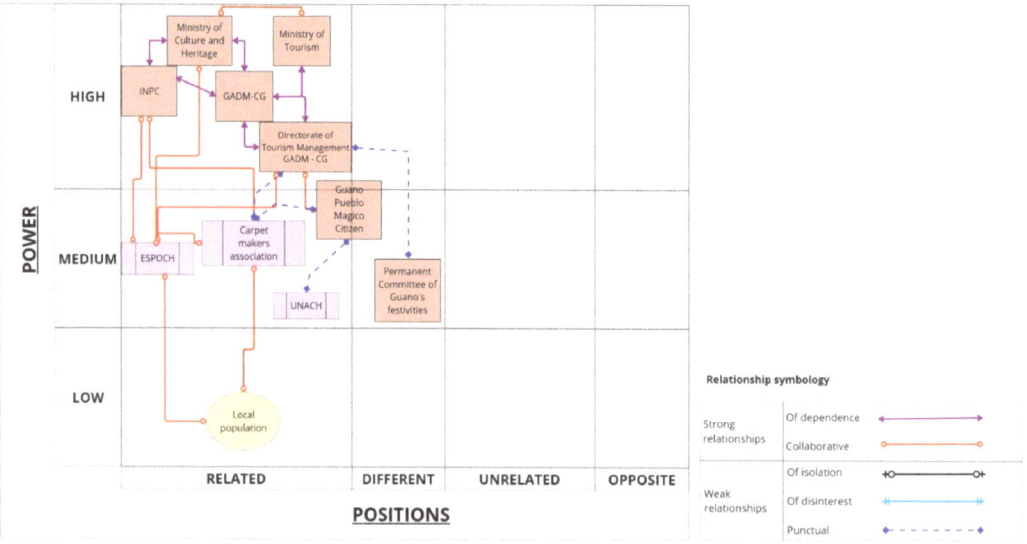

Figure 6. Mapping of stakeholder networks.

Among the weak or specific relationships are the images of power and the associative fabric, mainly between the Association of Carpet Makers and the Directorate of Tourist Management of GADM-CG, which have worked together in specific processes to keep the practice of Guano Carpet making alive. On the one hand, they have achieved certification as intangible cultural heritage (ICH) and on the other hand, coordination of a cultural tourism experience in which the craft technique is exhibited, complemented by the Citizen Committee of Guano Pueblo Mágico because the latter uses this tourist product to reinforce the reaccreditation process of the territory as a magical town.

After identifying the stakeholders that can contribute to strategic planning, as well as the 12 prioritised elements, the definition of programmes and projects began with all the information gathered and the ideas provided by the GADM-CG, knowledge holders, and stakeholders linked to the Association of Carpet Makers.

It is necessary to specify that the safeguarding process implies a multifaceted approach involving a set of actions such as the application of legal frameworks, identification, documentation, research, preservation, protection, promotion, enhancement, transmission, revitalisation, community participation, and the integration of cultural resource exploitation for their social use; taking into consideration the spaces for dialogue on knowledge, it was possible to establish three strategic objectives focused on the following points:

- Strengthening generational links through the transmission of knowledge by holders, through formal and non-formal education.
- Developing alliances between stakeholders interested in artisanal production and the dissemination of carpets and cultural issues.
- Establishing integrative and participatory actions for the strengthening and revitalisation of the craft activity.

Based on these three objectives, three programmes and six inter-related work projects were derived, focused on promoting the preservation of craft techniques while at the same time revitalising and strengthening the cultural identity of the territory (Table 4).

Table 4. Formulation of programmes and projects.

Programmes	Projects	Scope
P1: Transmission and revitalisation programme on the historical and cultural value of the Guano carpet	Generational transmission of handmade carpet weaving	To ensure that the knowledge of the knotted carpet weaving technique is passed on to new generations. It is highlighted that generational transmission is a fundamental pillar in the conservation of ICH, as it allows ancestral skills and knowledge to not only be maintained but also to be adapted to modern times. The implementation of intergenerational workshops, where experienced artisans teach young people, will foster a sense of belonging and cultural pride. Moreover, this approach will contribute to the creation of an active community committed to its heritage.
P2: Programme for the dissemination of the cultural identity of the textile art of carpet making called "The best carpets are in Guano"	Tourist dissemination project "The best carpets are in Guano"	To work on the positioning of Guano as an attractive tourist destination supported by the Magical Towns of Ecuador Programme, highlighting the uniqueness and quality of its knotted carpets. Developing an interpretative space that can exhibit not only carpets but also tools, techniques, and stories related to their making so that visitors can understand the historical and cultural context of carpet weaving better, as well as the implementation of effective communication strategies, is an essential element to increase the visibility of Guano's textile art. This includes the use of social media, digital platforms, and local media to tell stories about artisans and their creations. By generating appealing content that highlights not only the aesthetic beauty but also the cultural importance of weaving, a powerful narrative can be created that connects with the public emotionally, from which both national and international tourists interested in authentic cultural experiences can be attracted.
	Project for the creation of an interpretative space on knotted carpets in the Museum of Guano	
	Project for the implementation of communication strategies for the enhancement of the value of handmade weaving of Guano carpets	
P3: Marketing programme for artisanal participation and stakeholder involvement	Project for the institutionalisation of the local fair called "Marketing what is ours"	It provides an excellent opportunity for artisans to show and sell their products directly to the public, as well as creating a space in which they can diversify their sources of income and attract tourists interested in acquiring more personalised and tangible souvenirs of their visit.
	Project for the production and marketing of souvenirs related to Guano carpets	

This proposal has sought to achieve a comprehensive and multidimensional strategic planning, addressing mainly actions of the transmission and preservation of traditional knowledge, and promotion and revitalisation of the manifestation within the local social and economic fabric. This strategy seeks to ensure that knotted carpets remain a vibrant symbol of Guano's cultural heritage, contributing to sustainable development and strengthening the identity of its community.

The information gathered during the participatory action research process shows that these cultural practises are still relevant and valued by the local community. Through in-depth interviews and participatory workshops with the knowledge bearers, a notable increase in the exchange of knowledge and a revitalization of interest in traditional weaving techniques has been observed. This not only reflects a commitment to cultural preservation but also highlights how this manifestation is fundamental to local identity and the sense of belonging to the Artisanal Capital of Ecuador, as Guano is called.

Finally, it is established that the link between tourism and the safeguarding of ICH is fundamental for the preservation of both artisanal techniques and local cultural identities. The literature has highlighted that tourism can act as a catalyst for cultural revitalization, providing economic resources and visibility to traditions that might otherwise be relegated in the collective memory. In this sense, it was crucial to develop operational strategies that integrate sustainable tourism into the safeguarding plan for Guano knotted carpet weaving. For example, actions such as the creation of tourist workshops where visitors participate in the weaving process have been implemented, which not only provides information about the technique but also generates income for the artisans. In addition, it has been suggested to organise cultural events that celebrate this tradition, attracting both tourists and locals and fostering a sense of community and belonging. These initiatives would not only contribute to the economic sustainability of artisans but would also help to reinforce the importance of ICH in the construction of local identities and territorial development. By strengthening this link between tourism and heritage, an active and dynamic preservation is ensured that benefits the local community and promotes an enriching cultural exchange and encounter.

5. Conclusions

The processes of safeguarding the living heritage linked to the knotted Guano carpet have been shown to involve several actions, but these have not achieved clear results for its sustainability over time. The situation of ICH is increasingly critical because a large percentage of the related manifestations are in a state of high vulnerability of sensitivity to change. Although carpet makers have worked for a process of vindication and positioning of their technique in the local and national market, the transmission processes are becoming less effective, aggravated by the globalisation of spaces and by the lack of respect for the value of traditional craft techniques.

The research process made it possible to record the entire Guano knotted carpet making process, generating a total of seven cultural manifestations associated with the production technique. Based on the records, the following were identified as the main limitations of the manifestation under study:

- Knowledge holders belong to a high age group.
- The vulnerability ranking, as presented, is based on the opinions of a limited number of six knowledge bearers, which may introduce significant biases in the interpretation of the data, but unfortunately there are no more knowledge bearers who continue to practice carpet weaving. To strengthen this section, it is proposed to broaden the scope of the study by including family members of each of the knowledge bearers, thereby strengthening the sample, while at the same time making it possible to consider a more holistic and less biassed view of the vulnerabilities faced by the knotted carpet weaving technique. Furthermore, it is crucial that the results be interpreted in a critical

framework that considers not only individual opinions but also the broader social and cultural context in which this cultural manifestation takes place.
- There is a low generational transmission of knowledge, as the holders' children know the process, but younger generations no longer perform the technique nor continue with re-enacting the technique. This result is corroborated by Sun [65], who found that active transmission mechanisms such as storytelling and parental modelling are essential for preserving cultural values across generations. In the absence of such mechanisms, cultural knowledge becomes fragmented and risks extinction.
- The commercialization has strong limitations for foreigners, in relation to the weight and size of the carpet, making it difficult to be acquired during their visits to the territory.
- The dissemination of the carpets has been concentrated in Guano, limiting their visibility in other cities of the country.
- The local market does not have the purchasing power to cover the costs that knotted Guano carpets represent.
- Guano carpets are not interpreted within the territorial space. Residents, visitors, and tourists do not have information that allows them to know the identity value that this traditional artisan technique has for the social group of Guano.
- The link established between this cultural manifestation and cultural tourism has allowed the creation of tourism products that not only promote the weaving technique but also contribute to its economic sustainability. The implementation of strategies such as workshops where visitors can learn about the weaving process has proven to be effective in validating and revaluing living heritage in practice. These actions not only ensure the continuity of ancestral knowledge but also strengthen social and cultural cohesion within the local community. This finding is consistent with research by Tao [66], who highlights how community-driven tourism initiatives can mitigate cultural erosion by fostering economic incentives and a sense of identity among local participants.

Regarding the term "local communities", it refers to groups of individuals who live and participate actively in the sociocultural dynamics of a specific geographic region, sharing common traditions, values, and practises. In this study, the term is applied to describe the residents of Guano who are directly or indirectly linked to the production, preservation, and promotion of the knotted carpet weaving technique. This includes not only the artisans themselves but also other members of the community who play a role in supporting and sustaining this intangible cultural heritage. By emphasising the role of local communities, the study aligns with the UNESCO [1] definition of community participation as a critical element in safeguarding intangible cultural heritage.

The broader implications of these findings suggest that the challenges faced by the Guano knotted carpet weaving technique are emblematic of global patterns in the loss of intangible cultural heritage. Zhang and Mace [67] warn that the erosion of cultural diversity, including the potential extinction of 90% of the world's languages by the end of the century, reflects an urgent need for proactive preservation strategies. Similarly, the case of the Guano carpet underscores the necessity of addressing both generational disconnection and socioeconomic barriers to sustainability.

Therefore, it is necessary to link the traditional hand-weaving technique of the knotted carpets of Guano to the identity processes of the magical town of Guano. This is because the manifestation requires a process of social use, where tourism is conceived as a privileged area to carry out measures to safeguard intangible cultural heritage and therefore helps local and regional development. The aim of all this is to preserve traditional knowledge in a more lasting way than its simple oral dissemination. By connecting these findings with existing empirical evidence, it is clear that successful safeguarding strategies require a combination of community-based initiatives, market accessibility, and innovative approaches to intergenerational knowledge transfer. This study contributes to the growing body of literature emphasising the need for adaptive and inclusive approaches to ICH preservation.

Author Contributions: Conceptualization, C.P.M.-E., S.M.Z.-P., M.d.l.C.d.R.-R. and J.Á.-G.; formal analysis, C.P.M.-E., S.M.Z.-P., M.d.l.C.d.R.-R. and J.Á.-G.; investigation, C.P.M.-E., S.M.Z.-P., M.d.l.C.d.R.-R. and J.Á.-G.; methodology, C.P.M.-E., S.M.Z.-P., M.d.l.C.d.R.-R. and J.Á.-G.; writing—original draft, C.P.M.-E., S.M.Z.-P., M.d.l.C.d.R.-R. and J.Á.-G.; writing—review and editing, C.P.M.-E., S.M.Z.-P., M.d.l.C.d.R.-R. and J.Á.-G. All authors have read and agreed to the published version of the manuscript.

Funding: This publication has been funded by the Consejería de Economía, Ciencia y Agenda Digital de la Junta de Extremadura, and by the European Regional Development Fund of the European Union through the reference grant GR21161.

Institutional Review Board Statement: Not applicable.

Informed Consent Statement: Not applicable.

Data Availability Statement: Data are contained within the article.

Acknowledgments: To the National Institute of Cultural Heritage of Ecuador for the financial funds provided through the public tender of the Line of Promotion of Social Memory and Cultural Heritage of the year in the sub-line: (a) Research, Modality—Support for research processes and studies on cultural heritage.

Conflicts of Interest: The authors declare no conflicts of interest.

References

1. UNESCO. *Convention for the Safeguarding of Intangible Cultural Heritage*; UNESCO: London, UK, 2003. Available online: https://ich.unesco.org/es/convenci%C3%B3n (accessed on 21 September 2024).
2. Kurin, R. Safeguarding Intangible Cultural Heritage in the 2003 UNESCO Convention: A critical appraisal. *Mus. Int.* **2004**, *56*, 66–77. [CrossRef]
3. Aikawa-Faure, N. From the proclamation of masterpieces to the convention for the safeguarding of intangible cultural heritage. In *Intangible Heritage*; Routledge: London, UK, 2008; pp. 27–58.
4. Szabó, Á.; Ward, C. Acculturation. In *Acculturation*; Routledge: London, UK, 2022. [CrossRef]
5. Bortolotto, C. From Objects to Processes: UNESCO'S 'Intangible Cultural Heritage'. *J. Mus. Ethnogr.* **2007**, *19*, 21–33.
6. Kurin, R. Safeguarding intangible cultural heritage: Key factors in implementing the 2003 Convention. *Int. J. Intang. Herit.* **2007**, *2*, 9–20.
7. Smith, L.; Akagawa, N. *Intangible Heritage*; Routledge: London, UK, 2009.
8. UNESCO. *What Is Intangible Cultural Heritage?* UNESCO: London, UK, 2017. Available online: https://ich.unesco.org/en/what-is-intangible-heritage-00003 (accessed on 21 September 2024).
9. Loulanski, T. Revising the concept for cultural heritage: The argument for a functional approach. *Int. J. Cult. Prop.* **2006**, *13*, 207–233. [CrossRef]
10. Lenzerini, F. Intangible cultural heritage: The living culture of peoples. *Eur. J. Int. Law* **2011**, *22*, 101–120. [CrossRef]
11. Lixinski, L. Selecting Heritage: The Interplay of Art, Politics and Identity. *Eur. J. Int. Law* **2011**, *22*, 81–100. [CrossRef]
12. Blake, J. UNESCO's 2003 Convention on Intangible Cultural Heritage: The implications of community involvement in 'safeguarding'. In *Intangible Heritage*; Routledge: London, UK, 2008; pp. 59–87.
13. Monsalve, L. Gestión del Patrimonio Cultural y Cooperación Internacional. Pregón Ltd.a. 2011. Available online: https://www.guao.org/sites/default/files/biblioteca/Gesti%C3%B3n%20del%20patrimonio%20cultural%20y%20cooperaci%C3%B3n%20internacional.pdf (accessed on 9 September 2024).
14. Olivera, A. Patrimonio inmaterial, recurso turístico y espíritu de los territorios. *Cuad. Tur.* **2011**, *27*, 663–677.
15. Stefano, M.L.; Davis, P.; Corsane, G. *Safeguarding Intangible Cultural Heritage*; The Boydell Press: Martlesham, UK, 2014; Volume 8. Available online: https://boydellandbrewer.com/9781843839743/safeguarding-intangible-cultural-heritage/ (accessed on 9 September 2024).
16. Deacon, H.; Dondolo, L.; Mrubata, M.; Prosalendis, S. *The Subtle Power of Intangible Heritage: Legal and Financial Instruments for Safeguarding Intangible Heritage*; HSRC Press: Pretoria, South Africa, 2004.

17. Deb, D. The erosion of biodiversity and culture: Bankura district of West Bengal as an illustrative locale. *Ecol. Econ. Soc.-INSEE J.* **2022**, *5*, 139–176. [CrossRef]
18. Chapagaee, R.P. Cultural Erosion: Post-Colonial Discourse in Achebe's Things Fall Apart. *Int. J. Humanit. Educ. Soc. Sci.* **2024**, *2*, 98–110. [CrossRef]
19. Chui, J. The Effects of Cultural Assimilation on the Loss of the Tuvan and Seri Languages. *J. Stud. Res.* **2022**, *11*, 1–4. [CrossRef]
20. Lama, K. Transition of indigenous culture: Reading Mann Gurung's "Lost in Transition# 6". *J. Nepal. Stud.* **2024**, *16*, 64–73.
21. Wang, X.; Aoki, N. Cultural Identity at the Community Level and Sustainability of Intangible Heritage: The Case Study of Traditional Handicrafts in Yangliuqing Town, Tianjin. In *International Conference on East Asian Architectural Culture*; Springer International Publishing: Cham, Switzerland, 2017; pp. 334–347.
22. Urbaite, G. The Impact of Globalization on Cultural Identity: Preservation or Erosion? *Glob. Spectr. Res. Humanit.* **2024**, *1*, 3–13. [CrossRef]
23. Goode, J. The contingent construction of local identities: Koreans and Puerto Ricans in Philadelphia. *Identities* **1998**, *5*, 33–64. [CrossRef]
24. Deffner, A.; Metaxas, T. Shaping the Vision, the Identity and the Cultural Image of European Places. In ERSA Conference Papers. European Regional Science Association. 2005. Available online: https://hdl.handle.net/10419/117800 (accessed on 9 September 2024).
25. Erickson, B.; Roberts, M. Marketing local identity. *J. Urban Des.* **1997**, *2*, 35–59. [CrossRef]
26. Guerrero, P. *La Cultura, Estrategias Conceptuales para Comprender la Identidad, la Diversidad, la Diferencia y la Alteridad*; Abya Yala: La Paz, Bolivia, 2002.
27. Tak, H. Longing for Local Identity: Intervillage Relations in an Italian Mountain Area. *Anthropol. Q.* **1990**, *63*, 90. [CrossRef]
28. El Shandidy, M.Z. The power of intangible heritage in sustainable development. *Power* **2023**, *6*, 92–97. [CrossRef]
29. Krasniqi, L. Cultural Resilience in the Face of Globalization: An Anthropological Exploration of Traditional Societies' Adaptation Strategies and Identity Dynamics. *Interdiscip. J. Pap. Hum. Rev.* **2023**, *4*, 31–38. [CrossRef]
30. Blake, J. Safeguarding Intangible Cultural Heritage: Challenges and Approaches. In *Intangible Heritage*; UNESCO Publishing: Bruselas, Belgium, 2009. Available online: https://unesdoc.unesco.org/ark:/48223/pf0000161244 (accessed on 9 September 2024).
31. Aikawa, N. An historical overview of the preparation of the UNESCO International Convention for the Safeguarding of the Intangible Cultural Heritage. *Mus. Int.* **2004**, *56*, 137–149. [CrossRef]
32. Bortolotto, C. Globalising intangible cultural heritage? Between international arenas and local appropriations. In *Heritage and Globalisation*; Routledge: London, UK, 2010; pp. 111–128.
33. Stefano, M.L.; Davis, P.; Corsane, G.; Denes, A.; Cummins, A.; Dixey, A.; Mazel, A.; Hottin, C.; Kreps, C.; Bowers, D.J.; et al. (Eds.) *Safeguarding Intangible Cultural Heritage*; Boydell and Brewer: Woodbridge, UK, 2012.
34. Bouchenaki, M. The interdependency of the tangible and intangible cultural heritage. In Proceedings of the 14th ICOMOS General Assembly and International Symposium, Victoria Falls, Zimbabwe, 27–31 October 2003.
35. Hafstein, V.T. Intangible heritage as a list: From masterpieces to representation. In *Intangible Heritage*; Routledge: London, UK, 2008; pp. 107–125.
36. Smith, L. *Uses of Heritage*; Routledge: London, UK, 2006.
37. Vecco, M. A definition of cultural heritage: From the tangible to the intangible. *J. Cult. Herit.* **2010**, *11*, 321–324. [CrossRef]
38. Logan, W.; Labadi, S. *Urban Heritage, Development and Sustainability: International Frameworks, National and Local Governance*; Routledge: London, UK, 2016.
39. Alivizatou, M. *Intangible Heritage and the Museum: New Perspectives on Cultural Preservation*; Routledge: London, UK, 2016.
40. Jack, M.E. Intangible Cultural Heritage and Participation: Encounters with Safeguarding Practices. *Mus. Manag. Curatorship* **2022**, *37*, 561–563. [CrossRef]
41. Besmonte, E. Identification and Safeguarding of Intangible Cultural Heritage (ICH) of Tabaco City, Philippines, through Cultural Mapping. *J. Educ. Manag. Develop. Stud.* **2022**, *2*, 1–10. [CrossRef]
42. Plichta, J. The co-management and stakeholders theory as a useful approach to manage the problem of overtourism in historical cities–illustrated with an example of Krakow. *Int. J. Tour. Cities* **2019**, *5*, 685–699. [CrossRef]
43. Gningue, M.; Bedoui, W.; Venkatesh, V.G. A port performance measurement approach using a sustainability balanced scorecard based on stakeholders' expectations. *Marit. Policy Manag.* **2024**, *51*, 1861–1883. [CrossRef]
44. Barbour, R.H. Stakeholder safety in information systems research. *Australas. J. Inf. Syst.* **2006**, *14*. [CrossRef]
45. Becker, J.; Niehaves, B.; Plattfaut, R. Stakeholder involvement in business process management agenda-setting and implementation. In Proceedings of the 16th Americas Conference on Information Systems (AMCIS), Lima, Peru, 12–15 August 2010.
46. Kuchler, M. Stakeholding as sorting of actors into categories: Implications for civil society participation in the CDM. *Int. Environ. Agreem. Politics Law Econ.* **2017**, *17*, 191–208. [CrossRef]
47. Deng, Y.; Mo, Z. The Application of New Media Technology in the Safeguarding of Intangible Cultural Heritage. *Appl. Math. Nonlinear Sci.* **2024**, *9*, 1–20. [CrossRef]
48. Liu, J.; Qi, Y. Intangible Cultural Heritage Protection Promotes Rural Revitalization. *J. Manag. Soc. Develop.* **2024**, *1*, 140–144. [CrossRef]

49. Jelinčić, D.A.; Mansfeld, Y. Applying Cultural Tourism in the Revitalisation and Enhancement of Cultural Heritage: An Integrative Approach. In *Cultural Urban Heritage: Development, Learning and Landscape Strategies*; Springer: Cham, Switzerland, 2019; pp. 35–43. [CrossRef]
50. Pemayun, I.; Ridwan, H.; Sholihah, K.; Nasrullah, N.; Pebriana, P. Strategy in Conserving and Marketing Songket Woven Fabrics at Kota Daro I Village. *Indones. Berdaya* **2023**, *4*, 831–836. [CrossRef]
51. Ghose, A.; Ali, S.M.A. Protection and preservation of traditional cultural expressions & traditional knowledge in handicraft trade: Advocating the need for a global cultural policy framework. *Rev. Direito Int.* **2023**, *20*, 473–499. [CrossRef]
52. Ministerial Agreement, NO. DM-2018-126. 2018. Available online: https://contenidos.culturaypatrimonio.gob.ec/wp-content/uploads/ACUERDO-126-2018-Expedir-la-norma-tecnica-salvaguardia-del-patrimonio-cultural.pdf (accessed on 9 September 2024).
53. INPC. Guía Metodológica para la Salvaguardia del Patrimonio Cultural Inmaterial by INPC Ecuador–Issuu. SobocGrafic. 2013. Available online: https://issuu.com/inpc/docs/salvaguardiainmaterial (accessed on 9 September 2024).
54. Clarke, D. *Spatial Information in Archaeology. Spatial Archaeology*; Academic Press: London, UK, 1977.
55. Guerra, S. La Disputa por el Control de las Doctrinas en la Real Audiencia de Quito: Un Estudio Microhistórico Sobre la Tensión Entre y Dentro del Estado, la Iglesia y las Redes de Poder Local, Guano, Siglo XVIII [Universitat Jaume I]. 2008. Available online: https://www.tdx.cat/bitstream/handle/10803/84085/sabrinaguerra.pdf?sequence=1&isAllowed=y (accessed on 15 August 2024).
56. Coronel Feijóo, R. *Poder Local en la Transición de la Colonia a la República: Riobamba 1750–1820*; Universidad Andina Simón Bolivar: Quito, Ecuador, 2009.
57. Büschges, C. Crisis y Reestructuración. La industria textil de la Real Audiencia de Quito al final del período colonial. *Anu. De Estud. Am.* **1995**, *52*, 75–98. [CrossRef]
58. Villavicencio, M. Carta Corográfica de la República del Ecuador: Delineada en Vista de las Cartas de Don Pedro Maldonado, el Baron de Humboldt Mr. Wisse, la de las Sontas de las Costas por M.M. Filzroy i H. Kellet i las Particulares del Autor. F. Mayer y Co. 1858. Available online: http://repositorio.casadelacultura.gob.ec/handle/34000/17611 (accessed on 11 August 2024).
59. Rueda Novoa, R. *El obraje de San Joseph de Peguchi*; Abya-Yala: La Paz, Bolivia, 1988.
60. Humboldt, A.V. *Diarios de Viaje en la Audiencia de Quito*; Cámara Ecuatoriana del Libro-Núcleo de Pichincha: Quito, Ecuador, 2005.
61. Botero, L. *Encomiendas, Guardianías y Doctrinas. Discursos y Representaciones*; El Caso de Chimborazo: Quito, Ecuador, 2020.
62. Decentralised Autonomous Municipal Government of Guano Canton. Alfombras Artesanales de Guano. 2022. Available online: https://visitaguano.com/es-ec/chimborazo/guano/artesanales/alfombras-artesanales-guano-aojxzrsad (accessed on 11 August 2024).
63. Ortiz, C. *Guano, Pasado, Presente*; Edicentro: Quito, Ecuador, 1996.
64. National Institute of Cultural Heritage. Artesanos Tejedores de Alfombras de Guano Recibirán Certificación de Patrimonio Cultural Inmaterial—Instituto Nacional de Patrimonio Cultural. 2019. Available online: https://www.patrimoniocultural.gob.ec/artesanos-tejedores-de-alfombras-de-guano-recibiran-certificacion-de-patrimonio-cultural-inmaterial/ (accessed on 11 August 2024).
65. Sun, L. A narrative inquiry of the intergenerational transmission of cultural family values in mainland China. *J. Beliefs Values* **2023**, *13*, 1–17. [CrossRef]
66. Tao, Y. Research on the Intergenerational Transmission Based on Grounded Theory??"??" A Case Study of Lingqing in Shanxi. In Proceedings of the 2018 4th International Conference on Humanities and Social Science Research (ICHSSR 2018), Wuxi, China, 25–27 April 2018; Atlantis Press: Amsterdam, The Netherlands, 2018; pp. 744–749.
67. Zhang, H.; Mace, R. Cultural extinction in evolutionary perspective. *Evol. Hum. Sci.* **2021**, *3*, e30. [CrossRef]

Disclaimer/Publisher's Note: The statements, opinions and data contained in all publications are solely those of the individual author(s) and contributor(s) and not of MDPI and/or the editor(s). MDPI and/or the editor(s) disclaim responsibility for any injury to people or property resulting from any ideas, methods, instructions or products referred to in the content.

MDPI AG
Grosspeteranlage 5
4052 Basel
Switzerland
Tel.: +41 61 683 77 34

Heritage Editorial Office
E-mail: heritage@mdpi.com
www.mdpi.com/journal/heritage

Disclaimer/Publisher's Note: The title and front matter of this reprint are at the discretion of the Guest Editors. The publisher is not responsible for their content or any associated concerns. The statements, opinions and data contained in all individual articles are solely those of the individual Editors and contributors and not of MDPI. MDPI disclaims responsibility for any injury to people or property resulting from any ideas, methods, instructions or products referred to in the content.

www.ingramcontent.com/pod-product-compliance
Lightning Source LLC
LaVergne TN
LVHW072336090526
838202LV00019B/2428